ALLE ZEIT WACH 1842

35. Colloquium der Gesellschaft für Biologische Chemie
12.–14. April 1984 in Mosbach/Baden

The Impact of Gene Transfer Techniques in Eukaryotic Cell Biology

Edited by
J. S. Schell and P. Starlinger

With 76 Figures

Springer-Verlag
Berlin Heidelberg New York Tokyo 1984

Professor Dr. JOZEF STEPHAAN SCHELL, Max-Planck-Institut für Züchtungsforschung, Egelspfad, 5000 Köln 30

Professor Dr. PETER STARLINGER, Institut für Genetik, Universität Köln, Weyertal 121, 5000 Köln 41

ISBN-13: 978-3-642-70067-5 e-ISBN-13: 978-3-642-70065-1
DOI: 10.1007/978-3-642-70065-1

Library of Congress Cataloging in Publication Data. Gesellschaft für Biologische Chemie. Colloquium (35th: 1984: Mosbach, Baden-Württemberg, Germany) The impact of gene transfer techniques in eukaryotic cell biology. Includes index. 1. Recombinant DNA – Congresses. 2. Genetic engineering – Congresses. 3. Eukaryotic cells – Congresses. I. Starlinger, P. (Peter), 1931– . II. Schell, Josef S. III. Title. [DNLM: 1. Cells. 2. Cytology – congresses. 3. Genetic Intervention – congresses. 4. Transcription, Genetic – congresses. W3 GE382R 35th 1984 / QH 450.2 G389 1984i] QH442.G48 1984 574.87'3282 84-23582

Softcover reprint of the hardcover 1st edition 1984

2131/3130-543210

Contents

Introduction of DNA into the Germ Line of Animals

Applications of Genetic Engineering

Preface

The 35th Mosbach Colloquium *"The Impact of Gene Transfer Techniques in Eukaryotic Cell Biology"* brought together a number of speakers interested in various aspects of cellular and developmental biology and over 600 other scientists, who listened to the lectures and participated in the lively discussions. The questions and experiments described were very varied, but all of them illustrated the importance of recombinant DNA technology. The powerful techniques of identifying and isolating DNA sequences, followed by their introduction into living cells and even into the germ cells of multicellular organisms, have pervaded nearly every branch of molecular biology.

The presentations and discussions that followed showed that recombinant DNA has tremendously increased our potential for fundamental research. Now, and for some time to come, these contributions and the resulting increase in our understanding of life will be the main result of gene manipulation.

There will, however, also be applications that will lead to new industrial processes. One section was devoted to novel ways of vaccine production and another to herbicide resistance. These applications are a matter of intense debate in the public domain today. Although they reach beyond the scope of the research laboratory at a university or research institution, scientists have the knowledge necessary to judge these developments and are sometimes directly involved. Therefore the development of industrial gene technology requires the attention of the whole scientific community. We hope that this Symposium has also served this purpose.

November 1984

J.S. SCHELL
P. STARLINGER

Contributors

You will find the addresses at the beginning of the respective contribution

Introduction of DNA into Animal Cells and Its Use to Study Gene Function

Chemical DNA Synthesis and Its Applications in Eukaryotic Gene Transfer Techniques

E. L. WINNACKER[1]

1 Introduction

The recent and dramatic advances in the molecular biology of eukaryotic cells are based on technical developments commonly described as genetic engineering. These include, for example, restriction enzyme technology, vector design, transformation techniques, cDNA synthesis, and last, but not least, chemical synthesis of oligonucleotides. The latter field has long been ignored by organic chemists and molecular biologists alike, although oligonucleotide synthesis was instrumental, e.g., in deciphering the genetic code. Today, of course, many applications have been recognized.

1. Gene synthesis.
2. Screening of libraries (cDNA or genomic).
3. Functional analysis of control elements.
4. Structural analysis of DNA.
5. Diagnostics of genetic disorders.
6. Synthesis of linkers and adapters.
7. In vitro mutagenesis, enzyme engineering.

This chapter will describe in its first part the present state of the art in chemical oligonucleotide synthesis and subsequently turn into the discussion of several recent applications.

2 The Chemical Synthesis of DNA

From a chemical point of view, oligonucleotides are poly-dialkyl phosphate esters in which the 3'-hydroxy group of one nucleoside unit is linked to the 5'-hydroxy group of another unit through an internucleotide phosphodiester linkage. The general strategy leading to such a reaction product involves:

a) the preparation of the fully protected monomeric building blocks,
b) the phosphorylation reactions, and
c) the linkage of the building blocks to the derived oligonucleotide.

The preparation of the building blocks cannot be the subject of this general presentation. Suffice it to say that the choice of appropriate protecting groups is indeed a critical prerequisite for the success of the subsequent reaction scheme. Protecting groups guard the funcationally multivalent nucleoside bases from undue attacks of solvent and reagents during oligonucleotide synthesis; they have to be stable enough to survive the repertoire of reactions leading to an oligonucleotide but have to be labil enough to be removed under conditions which leave the newly formed oligonucleotide intact.

[1]Institut für Biochemie, Universität München, Karlstrasse 23, 8000 München 2, FRG

35. Colloquium - Mosbach 1984
The Impact of Gene Transfer Techniques in Eukaryotic Cell Biology

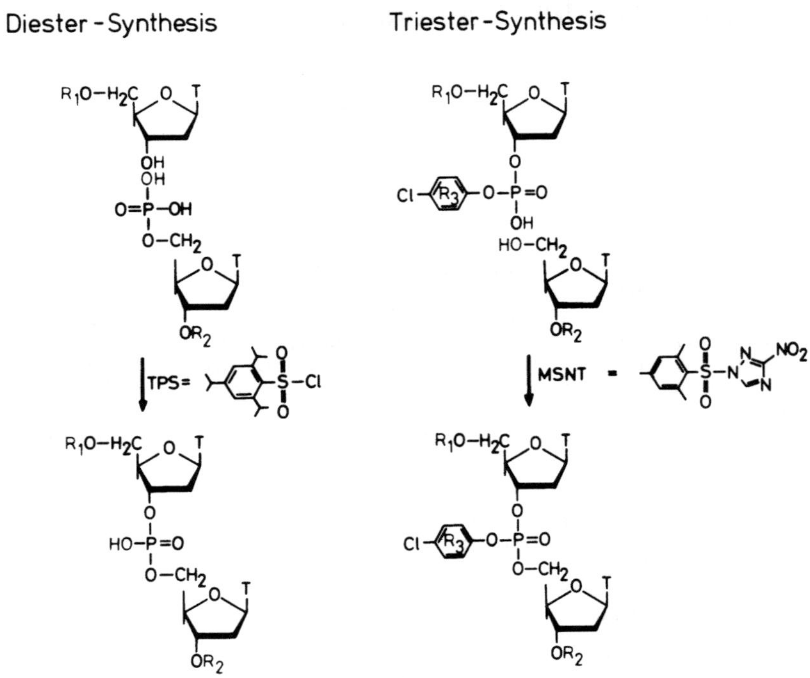

Fig. 1. Strategies for the chemical synthesis of DNA. Diesters and triesters are shown as dinucleotides. The modern coupling reagents TPS and MNST have replaced the more classical dicyclohexylcarbodiimid

The phosphorylation reactions also cannot be a subject of this presentation. The eventual choice will depend on the general strategy of oligonucleotide synthesis employed (see also Reese 1978).

There are two strategies for the chemical synthesis of DNA, the phosphodiester and the phosphotriester approach. Phosphodiesters and phosphotriesters are both protected at their respective 5'- and 3'-hydroxy functions while triesters contain, in addition, a protecting group at the remaining hydroxy function of the internucleotide phosphate group (Fig. 1). The charged phosphodiesters have been used successfully by Khorana and his collaborators in his classical synthesis of the genus coding for alanine- and tyrosine-tRNA's. Nevertheless, phosphotriesters are considered today as preferable reaction procucts, since they demonstrate increased solubility in organic solvents.

Numerous methodical variations have been described for the various chemical reaction steps involved in the construction of phosphotriester. Two basic and recent developments deserve particular attention

a) the solid-phase synthesis approach and
b) the use of reaction intermediates containing derivates of phosphorus III instead of phosphorus V.

2.1 Polymer-Supported Oligonucleotide Synthesis

Solid-phase methodology has been introduced into peptide synthesis by Merrifield (1963) and into oligonucleotide synthesis by Letsinger

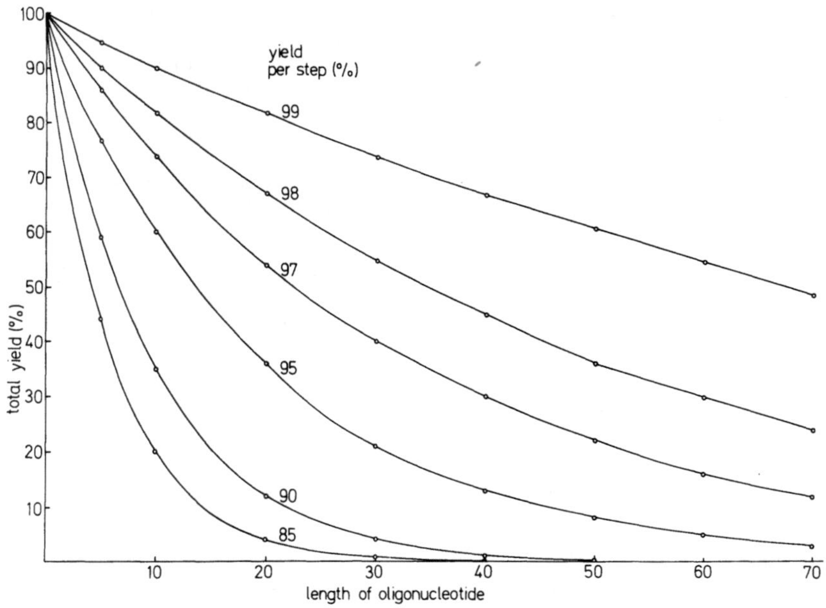

Fig. 2. Dependency of the overall reaction yield of an oligonucleotide on the yield of the single reaction steps. The overall reaction yield N (in %) was calculated from the relation $N = X^n \cdot 100$, where X represents the yield in the single reaction steps and n in the length of the oligonucleotide

(Letsinger and Kornet 1963). As compared to solution chemistry, the solid-phase approach has the advantage that all the various operations involved in chemical synthesis are reduced to mere washing and filtrations steps. However, it is also true that reaction intermediates cannot be purified in the course of a multistep synthesis. It is thus of paramount importance that each reaction step proceeds with the highest possible yield. The strong dependency of overall reaction yield on the yield of a single condensation reaction is shown in Figure 2, which demonstrates that even at a 90% yield of each single synthesis step, the overall yield drops after 44 cycles to only 1%, while an increase to 99% of the single-step yield could permit the synthesis of a 450-mer in an overall yield of 1%. At present, an average yield of 95% is achieved in practice, permitting the synthesis of 30-mers in approximately 20% yield.

A typical size-distribution of 5'-^{32}P-end-labeled oligonucleotides in crude reaction mixtures is shown in Figure 3 for the synthesis of different oligonucleotides 32 bases in length. The separation according to size is achieved by electrophoresis on denaturing polyacrylamide/urea gels. It is quite obvious that the reaction yields are rather low for the initial 2 to 5 steps, increasing to almost quantitative yields in the subsequent reaction steps. Whether the introduction of different solid-phase carriers, e.g., controlled pore glass beads as shown by Köster et al. (1984), could alleviate this problem remains to be seen.

2.2 The Phosphite Triester Approach

The second innovation to be discussed represents the phosphoramidite chemistry (Fig. 4). Intermediates containing III-valent phosphorus

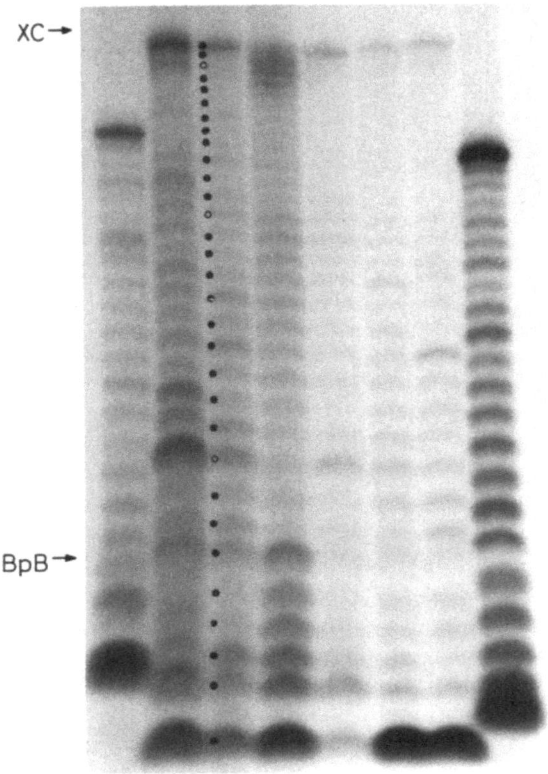

Fig. 3. Size distribution of chemically synthesized oligonucleotides. The cruce reaction mixtures from automated synthesis were 5'-end-labeled and subjected to polyacrylamide/urea gel electrophoresis. The reaction mixtures from a 32-mer and a 24-mer are shown in the *second* and *eighth* slot from the left. *Open dots* indicate positions 10, 20, and 30, respectively. In the *fourth* slot from the left, the capping step has been omitted

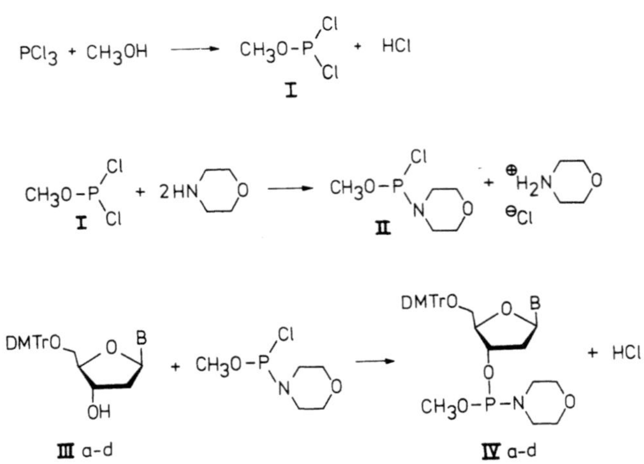

Fig. 4. Chemical reactions in the synthesis of nucleotide-phosphoramidites

appeared attractive as phorylation reagents for a long time, since these compounds are rather reactive and thus could be suspected of reducing the long reaction times necessary in classic phosphotriester chemistry. An initial attempt with methoxy-dichloro-phosphin, however,

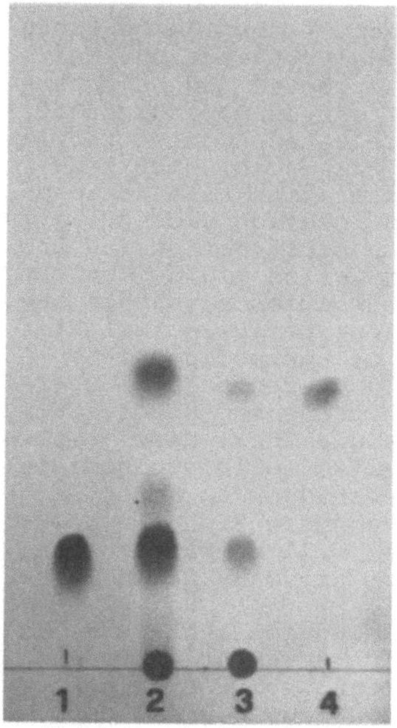

Fig. 5. Silicagel thin-layer chromatography of nucleotide-phosphoamidites. *Slot 1* fully protected 5'-dimethoxytrityl-thymidine; *slots 2* and *3* crude reaction mixtures from the reaction of dimethoxytrityl-T with chloro-methoxy-morpholinophosphin; *slot 4* purified phosphoamidite (T)

proved of only limited and questionable usefulness due to its extremely high reactivity and the corresponding side-reactions. The introduction, however, of the phosphoramidites by Caruthers (Beaucage and Caruthers 1981) led to immediate success and a concomitant revolution in oligonucleotide synthesis. Improvements from our own laboratory (Dörper and Winnacker 1983) now permit routine applications of this method. Starting material is methoxy-chloro-morpholinophosphin (II) which reacts spontaneously with fully protected nucleosides (III) to form the corresponding phosphorylated nucleotides (IV). These compounds are stable under laboratory conditions and can be purified by standard procedure, e.g., silicagel chromatography (Fig. 5).

Oligonucleotide synthesis is performed in a reaction cycle which lasts approximately 12 - 15 min and involves, in addition to the condensation step, an oxidation and a capping reaction. The latter reaction, an esterification with aceticanhydrid, blocks unreacted 5-hydroxy groups and protects corresponding growing chains from entering into further reaction cycles. The capping reaction can be deliberately omitted in order to obtain oligonucleotides with random one or two nucleotide deletions.

The reaction cycle described yields only a fully protected oligonucleotide which is still linked to the solid-phase carrier. Its transformation into a biologically active compound requires at least three additional reaction steps, the removal of the methoxy group at the internucleotide linkage with thiophenol as a strong nucleophilic reagent, the removal of the base-protecting groups and of the oligonucleotide from the carrier with aqueous ammonia and finally, the removal of the 5'-terminal, acid-labile protecting group. These reactions may take longer than the total synthesis of the fully protected oligonucleotide. Although they can be satisfactory handled, improvements in the methodology of the protecting group appear highly desirable.

It becomes apparent that the reaction cycle described for the chain elongation reaction is amenable to automatization. In fact, the corresponding automats are now available and permit the preprogrammed synthesis of oligonucleotides of any desired sequence. As mentioned above (Fig. 2), the length of the oligonucleotide strongly depends on the yields of the individual reaction steps and the expected and necessary overall yield. The yield at each individual step can be monitored and estimated from the amount of dimethoxytritanol released upon detritylation of the terminal 5'-protecting group toward the end of each reaction cycle. Present synthesizers of the first generation have no feedback systems to react, e.g., to a sudden drop in reaction yields. In the course of synthesis of several hundred oligonucleotides during the past year we have never seen dramatic drops in reaction yields, unless for trivial reasons. However, occasional discontinuities have been observed which are unrelated to any known and obvious deficiencies. It appears that certain combinations and sequences of nucleotides lead reproducibly to lower reaction yields which could be raised by longer condensation times, etc. It is to be expected that the second generation automates have sufficient flexibility to handle these difficulties.

In summary, it can be stated that the synthesis of oligonucleotides has become a routine procedure, as many other techniques in molecular biology. In fact I would like to predict that the synthesis of genes from known amino acid sequence data, for example, will be faster than cDNA cloning for proteins of molecular weights of up to 30,000 molecular weight.

This summary of novel developments in oligonucleotide synthesis would certainly not be complete without mentioning a recent improvement of the classical phosphotriester synthesis introduced by Blöcker et al. These authors developed a strategy whereby large numbers of oligonucleotides can be synthesized on a solid phase matrix. In this procedure, marked filter papers, one each for each oligonucleotide are moved between four reaction vessels specific for either A, T, G, or C-specific condensations. It is possible that this ingenious approach could be adapted to phosphoramidite chemistry with the concomitant increase in yields and speed.

3 Applications of Oligonucleotide Synthesis

Among the numerous examples of oligonucleotide synthesis, this presentation will concentrate on the following recent developments in

1. gene synthesis
2. screening of libraries
3. analysis of genetic disorders
4. structural analysis of DNA
5. in vitro mutagenesis

3.1 Gene Synthesis

Present strategies for gene synthesis are still based on the classical work by Khorana and his coworkers. His approach involved the synthesis of overlapping oligonucleotides which are synthesized separately and are subsequently permitted to reassociate in mixtures to the desired sequences (Fig. 6). The overlap positions have to be carefully chosen in order to permit only one single arrangement in which the many DNA fragments can associate to form an entire gene. The present world record appears to be the synthesis of the β-interferon gene, although longer stretches of DNA could easily be envisaged to be synthesized in the near future.

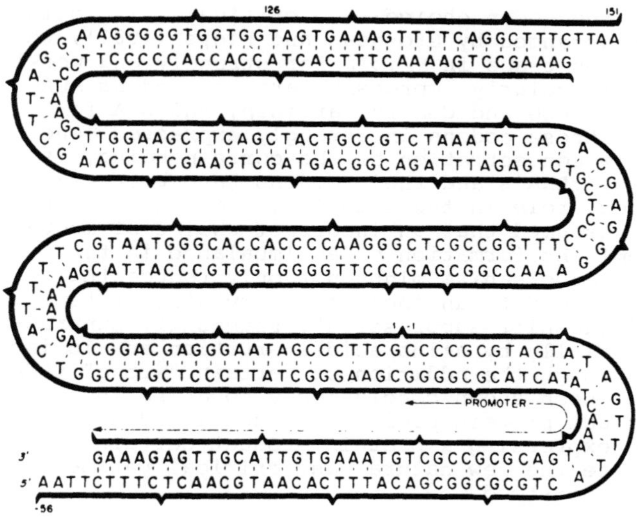

Fig. 6. Structure of the gene for a tyrosine tRNA-suppressor. *Small triangles* indicate the size of individual oligonucleotides which reassociate to the total gene. (Khorana et al. 1976)

Comparison of codon usage in strongly and moderately to weakly expressed genes in E.coli

	U			C			A			G			
		STRONG	WEAK		STRONG	WEAK		STRONG	WEAK		STRONG	WEAK	
U	Phe	39	**151**	Ser	**93**	36	Tyr	34	**96**	Cys	13	34	U
		113	102		87	**49**		**98**	65		**23**	**39**	C
	Leu	12	71		6	37	ochre			opal			A
		16	64		12	62	amber			Trp	25	66	G
C	Leu	26	**73**	Pro	**21**	29	His	19	**95**	Arg	**223**	99	U
		33	69		2	**46**		**75**	59		101	**133**	C
		→ 3	22		26	45	Gln	38	90		→ 3	27	A
		345	294		162	101		169	166		→ 1	42	G
A	Ile	67	**156**	Thr	103	48	Asn	13	**101**	Ser	10	56	U
		262	118		**137**	**119**		**159**	98		**49**	**61**	C
		→ 2	27		15	32	Lys	259	163	Arg	→ 3	28	A
	Met	140	130		28	76		106	44		→ 1	17	G
G	Val	**192**	**108**	Ala	**173**	87	Asp	116	**183**	Gly	**226**	124	U
		41	66		48	**178**		**204**	106		174	**140**	C
		119	48		119	107	Glu	333	210		→ 4	42	A
		83	123		129	149		106	98		→ 14	66	G

Grosjean and Fiers, Gene 18, 199 (1982)

Fig. 7. Codon usage in strongly and moderately expressed genes of *E. coli* (Grosjeans and Fiers 1982)

A unanswered question is the codon choice. How critical this could be can be seen from a table arranged by Grosjean and Fiers (1982), which shows a comparison of the codon usage in *E. coli* for strongly and weakly expressed genes (Fig. 7). Particularly impressive are the codons AUA for isoleucin, CUA for leucin, CGG and CGA for arginin, and GGA for glycin. The limited usage of these codons in strongly expressed genes clearly correlates with a low concentration of the corresponding, iso-accepting tRNA-species. These tRNA's and their codons have been assigned a modulating regulatory role in the translation mechanisms. For practical purpose it can be assumed that an incorrect choice of modulating codons and its concomitant formation of an imbalance in the tRNA population of a particular organism might not only reduce the rate of translation but might lead to an increase in mutation rate through the phenomenon of "framshift suppression". Frameshift suppression has been induced artificially by amino acid starvation and in in vitro translation systems through the addition of certain "wrong" tRNA's. To my knowledge it has never been shown conclusively to arise in *E. coli* overproducing a recombinant DNA protein product. Enough incidences, however, of difficulties in expressing a protein product certainly make it advisable not to manipulate the few natural and critical parameters which appear sensitive and understood.

3.2 In Vitro Mutagenesis and Enzyme Engineering

Oligonucleotides have long been used in a technique known as in vitro or site-directed mutagenesis. The methodology involves the use of single-stranded DNA. It is only recently that this methodology has been applied to the study of structure/function relationships of protein molecules and enzymes.

As shown initially by Hutchinson et al. (1978) in an example with OX174 DNA, and subsequently outlined by Itakura and Riggs for other DNA molecules, it is now possible to introduce specific mutations at any given site in any piece of DNA which can be isolated and cloned. The methodology involves the use of single-stranded DNA which is hybridized to an oligonucleotide carrying the desired mutation. The oligonucleotide serves as a primer for an in vitro DNA synthesis reaction, leading to a heteroduplex which is used to transform competent cells. The mutant yield is usually lower than the theoretical yield of 50%, but can be raised to this value using certain purification procedures, e.g., the isolation of form I DNA (Zoller and Smith 1982).

The potential for site-directed mutagenesis has only recently been recognized for the study of structure/function relationships in protein molecules and enzymes. Here, in particular, the solution of numerous three-dimensional structures by X-rax crystallography has posed more questions than could ever be answered by classical methods. Two examples have recently come to our attention, the dihydrofolate reductase from *E. coli* (Villafranca et al. 1983) and a tyrosyl tRNA synthetase from *Bacillus stearothermophilus* (Wilkinson et al. 1984). In the latter case, the exchange of a single amino acid (Thr 51 - Pro 51) in this 60,000 M.W. protein increased the K_m for ATP by a factor of 100. The former case, dihydrofolate reductase from *E. coli*, has been studied in even more detail. The three mutations in this work targeted the residues of aspartic acid (to asparagin) at position 27, proline (to cysteine) at position 39 and glycine (to alanine) at position 95 (Fig. 8). The rationale for the choice of these residues is derived from the detailed X-ray structure known from the work of Mathews, Kraut and coworkers. In the case of the asp-27 (to arginine), for example, it was assumed that this residue forms a hydrogen bond with the pteridine ring of the substrate, thereby stabilizing the activated transition

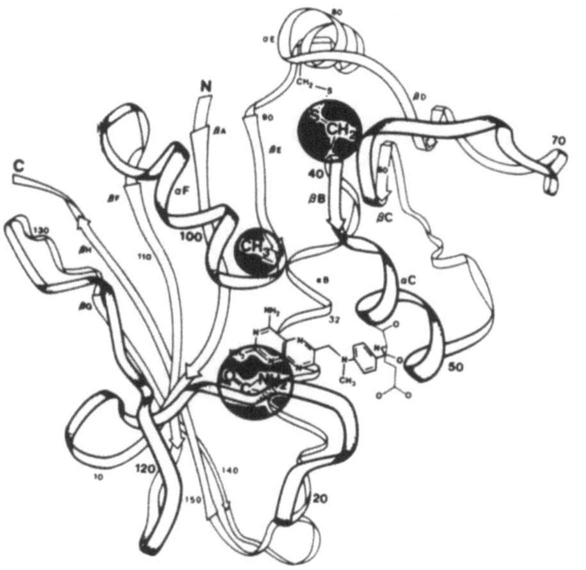

Fig. 8. Three-dimensional structure of dihydrofolate reductase from *E. coli* with the positions of three mutations. (Villafranca et al. 1983)

Residue exchanged	Activity(wt=100%)
Asp-27	0.1%
Cys-39	100%
Gly-95	<1%

state of the DHFR-catalyzed reaction. An exchange to asparagine would remove the source of the negatively charged counterion (carboxyl group), but would probably leave the general geometry of the catalytic site intact. This prediction was verified although some remaining activity (0.1%) seems to indicate that some other, minor contribution to the stability of the transition state appears to exist in this protein molecule.

The possibility of introducing specific mutations will permit an answer to these and related questions, and will thus permit mapping and cataloging the sites of enzyme/substrate interaction. It will add a dynamic aspect to the static solutions of X-ray structural analysis and the assumptions derived therefrom. The question arises, of course, whether we will succeed not only in mapping catalytic sites but whether it will be possible to change specificities and adapt, e.g., a hydrolytic enzyme, to other substrates.

3.3 Functional Analysis of Control Elements

Genetic control elements like promotors, enhancer, ribosomal binding sites, attenuators, operators, terminators, etc. have long been the subject of mutational analysis. The aims of these studies are a size analysis of these elements and a definition of the points of interaction between proteins and nucleic acids. The specific example of this

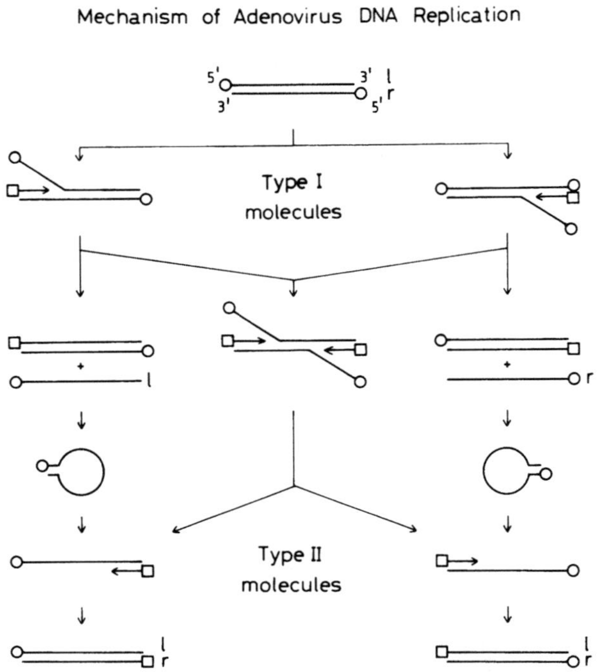

Fig. 9. The mechanism of adenovirus DNA replication

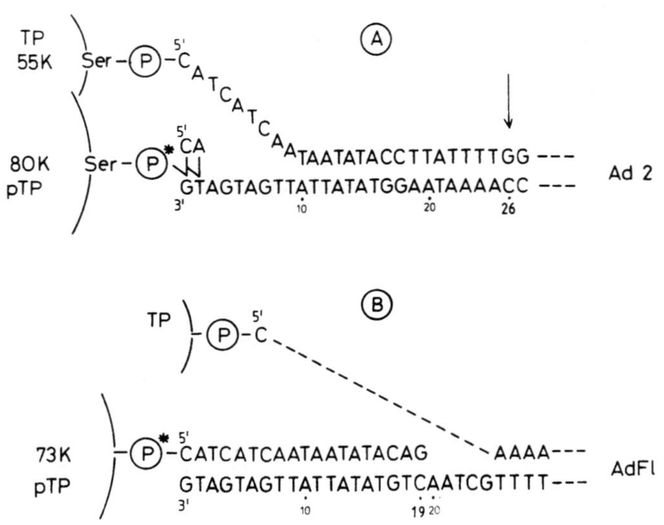

Fig. 10A,B. Assay for the initiation and partial elongation in vitro of adenovirus DNA replication. *A* shows the principle of the initiation assay on adenovirus type 2 as template with dideoxyadenosintriphosphate as inhibitor; *B* shows the partial elongation reaction of mouse adenovirus FL in a complete reaction mixture lacking dGTP. (Lally et al. 1984)

discussion concerns the adenovirus replicon. It is a linear, double-stranded DNA molecule, the genome, of the adenoviruses, which replicates in the nucleus of virus-infected human cells (Fig. 9). Its replication is initiated at either molecular end and leads to the obligatory formation of full length single-strands. Two structural prerequisites guarantee this asymetric, but semi-conservative mode of replication; a terminal repetition and a protein covalently linked to the respective 5'-termini of the viral DNA. The terminal repetitions make

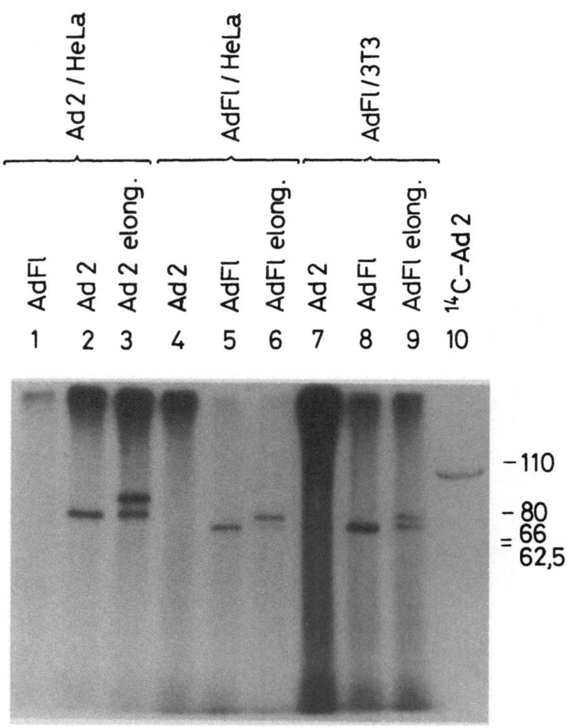

Fig. 11. Initiation and partial elongation assay for adenovirus DNA replication. Assays were performed with full reaction mixtures containing ^{32}P-dCTP, the indicated templates and different nuclear extracts (from HeLa or 3T3 cells). Panels *1*, *4*, *7* AdFL and Ad2 DNA/protein complexes in heterologous reaction systems. Panels *2*, *5*, *8* initiation reactions homologous extracts; panels *3*, *6*, *9* partial elongation reactions. Panel *10* ^{14}C labeled marker proteins

both molecular ends structurally and functionally identical. The terminal protein has been proposed and shown to be involved in the mechanism of initiation of DNA replication. According to this mechanism a complex of deoxycytidine triphosphate and the 88,000 M.W. precursor which is virus-coded binds to the molecular termini and provides the 3'-OH primer necessary for the priming of the initiation reaction (Fig. 10). The elongation reaction leads to the formation of double-stranded daughter molecules and the displaced full-length single strands. These, in turn, are replicated through the formation of double-stranded panhandles identical in structure to the genuine ends of the double-stranded genomes. The newly formed daughter molecules are subsequently packaged, while the terminal protein precursor is processed by a virus-coded proteases.

The initiation reaction can be studied in vitro using a nuclear extract derived from adenovirus-infected human cells (Challberg and Kelly 1979). These extracts contain three virus-coded proteins, a 72,000 M.W. ssDNA binding protein, a DNA polymerase and the terminal protein, and a number of nuclear factors which were shown to be required for the replication of adenovirus DNA. In the presence of ^{32}P-labeled dCTP and with adenovirus DNA as template (Fig. 10), the label from dCTP is transferred to the pTP (Fig. 10) which can be analyzed by PAA-gel electrophoresis and identified by autoradiography (Fig. 11). In the presence of dideoxy-ATP (ddATP), the elongation reaction stops immediately after the addition of the first two bases. If the reaction is performed in the absence of dGTP but in the presence of the other three triphosphates, then the chain is extended to the first dG residue, e.g., residue 26 in adenovirus type 2 and residue 19 in mouse adenovirus FL (Fig. 11). This oligonucleotide addition raises the molecular weight of the terminal protein by approximately 8000 to over 90 K. Thus, both

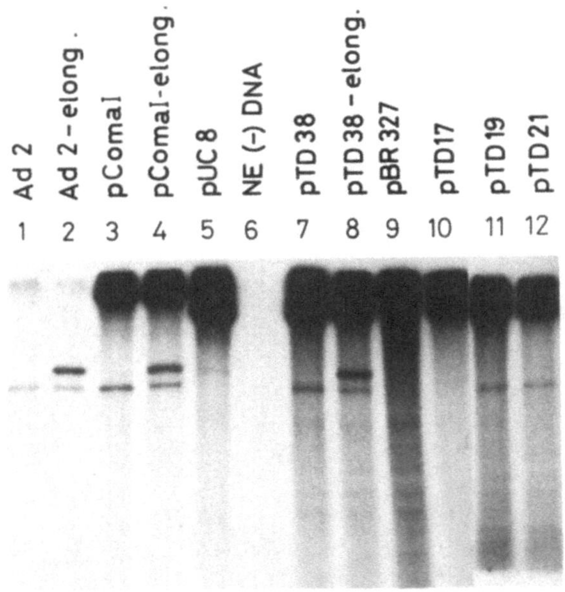

Fig. 12. Initiation and partial elongation reactions of linearized plasmid DNA molecules containing adenovirus termini of different length. (Lally et al. 1984)

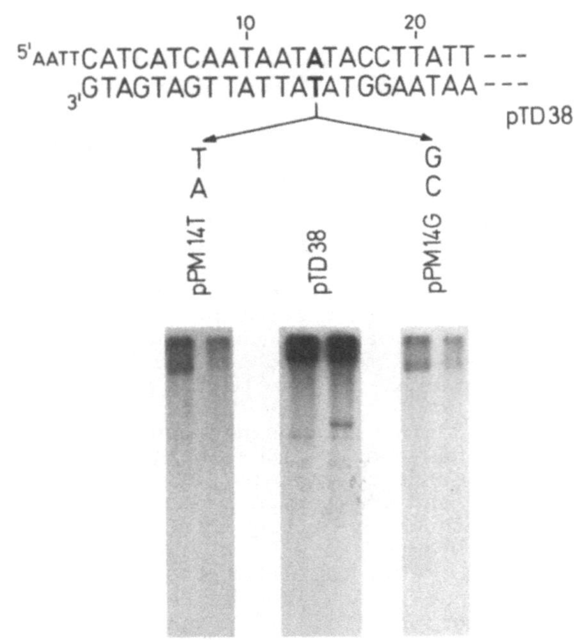

Fig. 13. Transversions and transitions in the adenovirus terminus. While the plasmid pTD38 is fully active, the two point mutations at position 14 lead to inactivity of the adenovirus origin of replication

the initiation and the elongation reaction can easily be assayed by the identification of the two bands, either of a pTP with just a 2-nucleotide addition (80 K) or of a pTP with 28 additional nucleotides.

A critical prerequisite for the subsequent experiments was the finding by Stillman (Tamanoi and Stillman 1982) that linearized plasmid-derived DNA molecules containing adenovirus DNA specific for the molecular ends can serve as substrates in the in vitro initiation and elongation reactions. It has thus been possible to construct sets of plasmids which, after linearization, contain adenovirus specific termini of different length and which can be analyzed for their template activity in the in vitro assay. As shown in Figure 12, a clone (pTD17) with only 17 terminal adenovirus-specific basepairs is essentially inactive (Fig. 12, slot 10), while termini with 19 and 21 base pairs (plasmids pTD19 and pTD21) are active at least for the initiation reaction. Clone pTD38 with 38 terminal base pairs is active in both the initiation and the elongation reaction (Fig. 12, slots 7 and 8). The initiation reaction of adenovirus DNA replication thus requires at least 19 terminal base pairs in vitro. This region includes an AT-rich stretch of DNA between position 9 and 17 which is conserved in all mammalian adenoviruses. In order to establish the significance of this sequence, both transversions and transitions in this area were constructed as described above by synthesizing the appropriate oligonucleotides. As shown in Figure 13, a simple AT-TA transition lead to a reduction of both initiation and elongation activities to less than 5%, while the AT-GC transversion is totally inactive. While these and similar studies have already established the significance of the conserved region between positions 9 and 17, we hope to analyze other regions, e.g., the regions proposed to interact with nuclear factor within the inverted repeats in a similar way. In addition, it will be necessary to study the size requirements of the adenovirus replicon in vivo and as whether the snythesized deletions serve as replicons in the infected cell nucleus.

4 Discussion

All the examples discussed can be summarized under the general term "synthetic biology". As the field of organic chemistry came of age and developed into a synthetic phase with the synthesis of urea by F. Wöhler in 1825, biology is entering its own synthetic phase with the advent of molecular biology. This will permit the answer to many fundamental questions, e.g., in developmental biology, mechanisms of gene expression and structure and function relationships of enzymes. In addition, the "new biology" has practical consequences for the recognition of human genetic disorders and the production of hitherto unavailable, biologically active proteins. According to W. Churchill, the "empire of the future is the empire of the mind". It remains to be seen whether we can live to and also contribute to this challenge in the field of chemical DNA synthesis.

Acknowledgements. The work of the author was supported by the Deutsche Forschungsgemeinschaft through the research group Genomorganisation.

References

Beaucage SL, Caruthers MH (1981) Tetrahedron Lett 22:1859-1862

Challberg MD, Kelly TJ Jr (1979) Proc Natl Acad Sci USA 76:655-659

Dörper T, Winnacker EL (1983) Nucleic Acids Res 11:2575-2584

Frank R, Heikens W, Heisterberg-Montsis G, Blöcker H (1983) Nucleic Acids Res 11:4365-4377

Grosjean H, Fiers W (1982) Gene (Amst) 18:199-209

Hutchison CA, Phillips S, Edgell MH, Gillman S, Jahnke P, Smith M (1978) J Biol Chem 253:6551-6560
Khorana HG, Agarwal KL, Besmer P et al. (1976) J Biol Chem 251:565-570
Köster H, Biernat J, McMannis J, Wolter A, Stumpe A, Narang CK, Sinha ND (1984) Tetrahedron 40:103-112
Lally C, Dörper T, Gröger W, Antoine G, Winnacker EL (1984) EMBO J 3:333-337
Letsinger RL, Kornet MJ (19863) J Am Chem Soc 85, 3045-3046
Merrifield RB (1963) J Am Chem Soc 85:2149-2154
Reese CB (1978) Tetrahedron 34:3143-3179
Tamanoi F, Stillman BW (1982) Proc Natl Acad Sci USA 79:2221-2225
Villafranca JE, Howell EE, Voet DH, Strobel MS, Ogden RC, Abelson JN, Kraut J (1983) Science (Wash DC) 222:782-788
Wilkinson AJ, Fersht AR, Blow DM, Carter P, Winter G (1984) Nature (Lond) 307:187-188
Zoller MJ, Smith M (1982) Nucleic Acids Res 10:6487-6550

An Alternative Gene Cloning Method for the Isolation of Human Genes by Expression in Mouse Cell Clones

W. LINDENMAIER, H. HAUSER, and J. COLLINS[1]

1 Introduction

The present methods for isolating a genomic clone for a given human gene are based primarily on the isolation of DNA clones starting with mRNA from an expressing tissue or cell line or a sequence information from the protein that allows synthesis of a nucleotide probe specific enough for screening of DNA libraries. Isolation of genes from genomic DNA by direct expression in a heterologous animal cell line has been achieved for genes that confer a directly selectable phenotype to the heterologous cell line, e.g., genes for thymidine kinase (Perucho et al. 1980a), adenosine phosphotransferase (Lowy et al. 1980), hypoxanthine guanosine phosphotransferase, or transforming genes (Goldfarb et al. 1982, Shih and Weinberg 1982). The reisolation depends on the linkage of marker sequences that can be screened or selected for in *E. coli*, e.g., human repetitive sequences (Shih and Weinberg 1982) or antibiotic resistance markers (Lund et al. 1982). Normally several rounds (at least two) or transfection as well as preparation and screening of a complete genomic DNA library from the expressing clone(s) is necessary for gene isolation.

The methods presented here describe an alternative approach which incorporates firstly an improved method for the production and maintenance of cosmid gene banks and secondly the "shuttling" of cosmid hybrids into and out of the genome of a eukaryotic cell line, thus allowing retrieval of long contiguous regions containing the gene in question. The gene is identified merely by its expression in the heterologous cell. The principles of the method of repackaging a cosmid gene bank in vivo so as to produce a cosmid bank as a lysate containing cosmid hybrid particles which can be kept in vitro in a stable form has been described, as has the principle of the cosmid gene shuttle (Lindenmaier et al. 1982). We communicate here the formation of a human cosmid gene bank which has been used to establish via DNA mediated gene transfer 8000 mouse cell clones containing on average 30 hybrid cosmids per clone. This bank will allow isolation of human genomic clones via an assay for expression in the mouse cell bank (e.g., via immune fluorescence, bioassay, enzyme assay, radio-binding assay). That the method is efficient enough for isolating single-copy genes from a total genomic cosmid gene bank was shown by the isolation of a human thymidine kinase gene.

[1]GBF - Gesellschaft für Biotechnologische Forschung mbH., Mascheroder Weg 1, 3300 Braunschweig, FRG

35. Colloquium - Mosbach 1984
The Impact of Gene Transfer Techniques in Eukaryotic Cell Biology

Fig. 1a

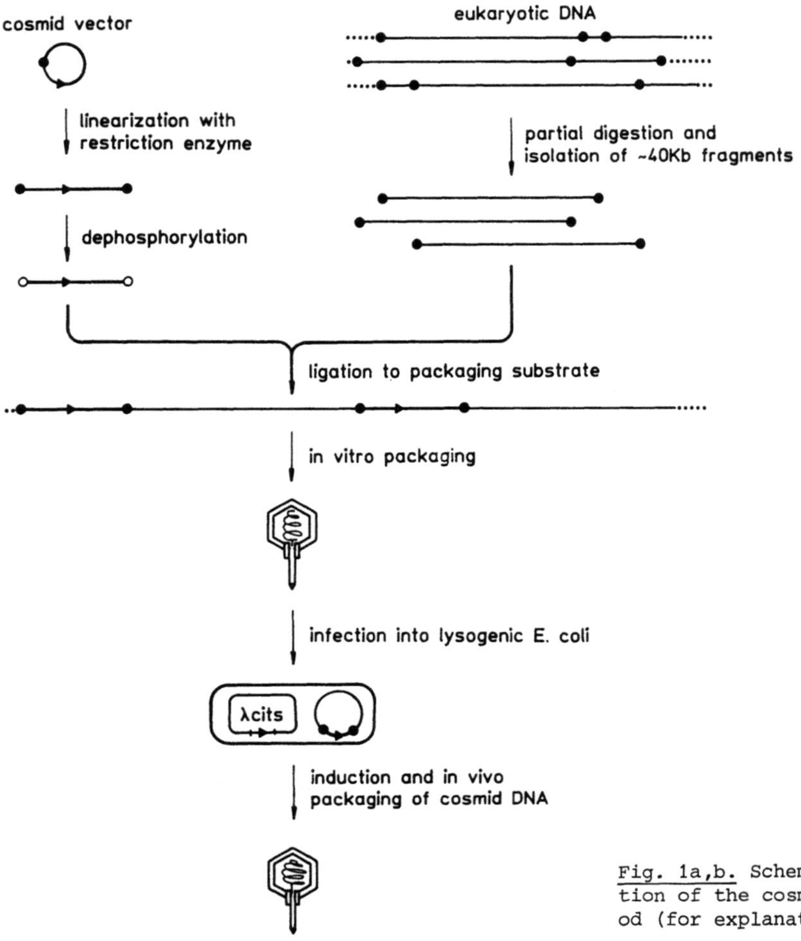

Fig. 1a,b. Schematic presentation of the cosmid shuttle method (for explanation see text)

2 Results

The general design of the method to be described is shown schematically in Figure 1. It involves six major steps, namely:

1. Construction of a cosmid gene bank representing the total genome of the organism under investigation (human) in a cosmid vector that allows selection in *E. coli* and Ltk^- mouse cells in a lysogenic *E. coli* host.
2. In vivo packaging of the cosmid gene bank in λ phage particles after induction of the lysogen. Isolation of packaged cosmids and extraction of linear cosmid DNA. Oligomerization of the linear cosmid-DNA by ligation to restore and protect functional cos-sites.
3. Introduction of oligomerized cosmid-DNA in Ltk^--cells by DNA mediated gene transfer and selection of transformed cell clones.
4. Screening of the animal cell clones for expression of desired gene products.
5. Isolation of high molecular weight DNA from the clone that expresses the desired gene product. Rescue of integrated hybrid cosmids into *E. coli* by direct in vitro packaging from total genomic DNA.

Fig. 1b

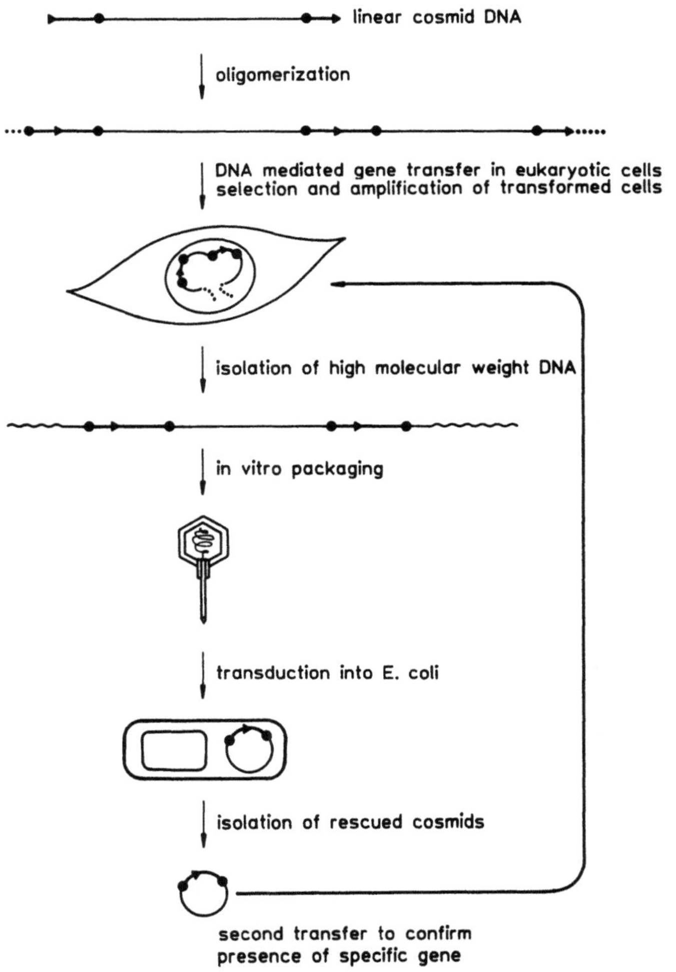

6. Identification of individual cosmids carrying the gene of interest by a second round of DNA mediated gene transfer.

In this contribution we report in the first part on the construction of a human cosmid gene bank in mouse Ltk⁻-cells that should represent most genes of the total human genome in a form that allows screening of individual mouse cell clones for expression of human genes. In the second part we show that a single-copy gene can be isolated from the human genome using the general strategy described.

3 Establishment of a Human Cosmid Gene Bank in Mouse Ltk⁻-Cells

3.1 Construction of a Human Cosmid Gene Bank

High molecular weight DNA was isolated from a human placenta according to the method of Blin and Stafford (1976). DNA was partially digested with MspI using serial dilutions of the enzmye. By salt density gradient

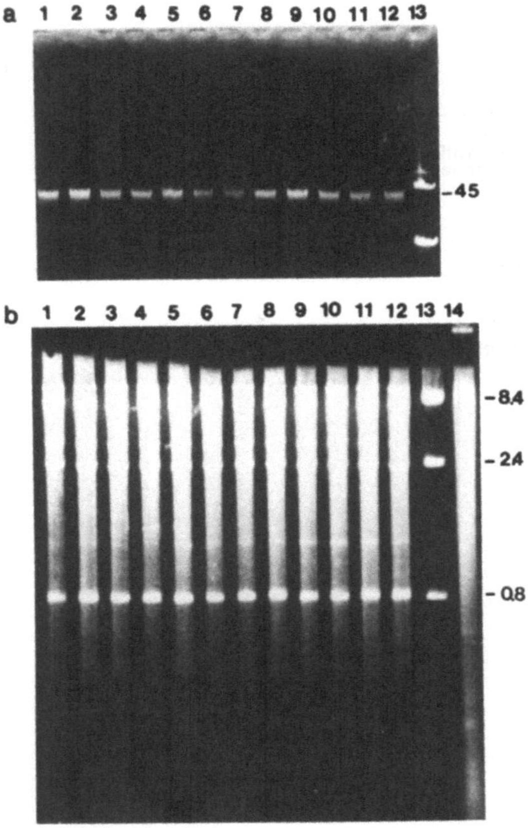

Fig. 2a,b. Agarose gel electrophoresis of linear cosmid gene bank DNA. a) *1–12* undigested linear DNA from gene bank pools, *13* Marker DNA ($\lambda + \lambda$ Hind III); b) DNA's digested with EcoRI and Hind III. *1–12* gene bank pools, *13* vector pHC79-2cos/tk, *14* human placenta DNA

centrifugation (Lin et al. 1980) the DNA was fractionated according to size and fractions containing fragments ≧ 25 kb were pooled. Cosmid vector pHC79 2 cos/tk (Lindenmaier et al. 1982) was linearized by digestion with ClaI and 5'phosphates removed by treatment with calf intestine phosphatase. The DNAs were mixed at a molar ratio of 4:1 vector: insert DNA and ligated at a concentration of about 700 $\mu g\ ml^{-1}$ to produce vector insert concatemers suitable for in vitro packaging. In vitro packaging was performed according to Hohn (1979) and packaged cosmids were transduced into the λ-lysogenic strain *E. coli* 1400 (Cami and Kourilsky 1978). This strain is not very efficient for preparative isolation of in vivo packaged particles but gave a higher number of clones and more reproducible results for primary infections of in vitro packaged cosmids. About 300 000 clones were isolated and subdivided into twelve pools representing about 25 000 colonies each for further amplification and screening.

3.2 In Vivo Packaging of Hybrid Cosmids

The colonies of each pool were resuspended and diluted in L-broth to an OD_{600} of 0.1 and in vivo packaging was performed as described (Lindenmaier et al. 1982). The titers of these primary lysates were about

$1 - 2 \times 10^7$ cfu ml^{-1}. For preparative isolation of linear cosmid DNA about 10^6 cfu were used to infect BHB 3064. In vivo-packaged cosmids were isolated and purified by CsCl step and bouyant density gradient centrifugation (Maniatis et al. 1982). DNA was extracted from each of the 12 cosmid pools. It migrates as linear DNA of about 45 kb in agarose gels. When this DNA is cut with EcoR I and Hind III, enzymes that nearly precisely cut out the integrated fragment, a smear of cloned human DNA fragments is produced superimposed by the fragments expected from the vector (Fig. 2). Southern blots of these digests can be used to identify the pool(s) that contain fragments of the desired gene and thus simplify the screening if a specific probe is available.

3.3 DNA-Mediated Gene Transfer

The number of mouse cell clones necessary to represent the human genome after transfer of the human cosmid bank can roughly be estimated from experiments selecting single copy genes after transfer of total genomic DNA (Wigler et al. 1978). Taking into account the integration of 1500 - 2000 kbp of DNA (ca. 30 - cosmids) per mouse cell clone (Perucho et al. 1980b), about 10^4 clones would be sufficient to represent the total human genome. We decided to establish the transformed mouse cells as single-cell clones, because, for ease of screening, clones must be testable individually. This also has the advantage that clones that often have very different growth properties can be "synchronized". In this way comparable cell numbers are achieved for different clones. 100 µg of DNA from each pool was ligase-treated and purified again by phenol extraction and ethanol precipitation. Using the calcium phosphate precipitation method (Wigler et al. 1977) the ligated DNA was transfected into mouse Ltk$^-$-cells without added carrier-DNA and HAT-resistant clones were selected. In test experiments the transfection efficiency for each DNA preparation was determined. To establish the library in Ltk$^-$-cells, transfections were done with the predetermined amounts of DNA necessary to produce about 1000 clones per pool. Transformed Ltk$^-$-cells were cloned directly into microtiter plates by limited dilution. The dilution necessary to produce mainly single clones per microtiter well was calculated from the test experiments. Using a relatively simple 96-piece pipetting device (Tecnomara), feeding, harvesting, and reduplication of large numbers of microtiter cultures is possible. Upon selection and synchronization of Ltk$^-$-cell clones in microtiter wells, one copy of replica plates from each gene bank pool was trypsinized, pooled, grown for two cell cycles and stored. These cell mixtures can be used for isolation of genes using fluorescence activated cell sorting (FACS) or direct selection. For conservation of individual clones a method has been developed to allow freezing and storage of frozen clones directly on microtiter plates. Thus, the master copy of the mouse-human cosmid-gene bank can be kept frozen while replicas are available for different tests. Up to now about 8000 clones representing 10 cosmid pools have been produced. Currently we are completing and improving the mouse human cosmid gene bank and setting up immunological tests for specific gene products.

4 Isolation of a Human Thymidine Kinase Gene

To show that the cosmid shuttle system allows direct isolation of eukaryotic single-copy genes, we decided to isolate the human thymidine kinase gene. It had been shown that DNA-mediated gene transfer is effective enough to select for single-copy genes after transfer of total

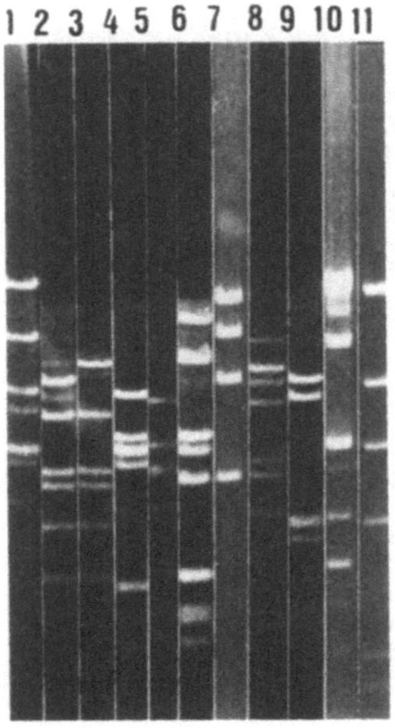

Fig. 3. EcoRI-digests of different cosmids rescued from high molecular weight DNA of HAT resistant mouse L-cell clone M 1 by in vitro packaging and infection into *E. coli*. *1-10* cleared lysates of Ap^R-colonies; *11* λ Hind III-Marker. Lane *7* corresponds to cos HTK *11* and transfers HAT resistance to Ltk cells with high efficiencies

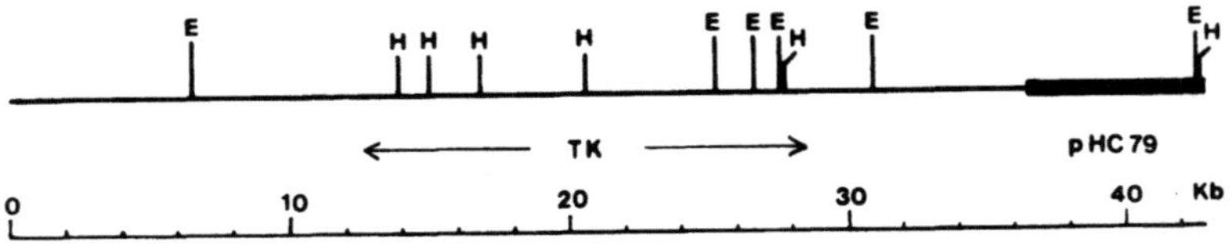

Fig. 4. Restriction map of cos HTK 11, sites for EcoRI *(E)* and Hind III *(H)* are indicated, as is the region of the thymidine kinase genes *(TK)*

genomic DNA (Wigler et al. 1977). Using the human placental DNA, we found 2 - 20 colonies per 10 μg of total genomic DNA. To isolate the human thymidine kinase gene by cosmid shuttle, another human cosmid bank was prepared in pHC 79-2cos. In this case Bam H I-cleaved, phosphatase-treated vector DNA was ligated to partially Sau 3A-digested placental DNA, packaged in vitro and transduced into *E. coli* 1400. About 360,000 primary cosmid clones have been packaged in vivo and amplified as described for the first cosmid library. Linear cosmid DNA was isolated, treated with ligase, and used to transform mouse Ltk^--cells. Three HAT-resistant clones were found after HAT selection. From mass cultures of these clones, high molecular weight DNA was isolated. In vitro packaging of total genomic DNA of these HAT-resistant mouse cell clones was performed. Packaged cosmids were used to infect *E. coli* HB 101 and Ap^R-colonies were selected. By restriction analysis of small cosmid preparations (Birnboim and Doly 1979) at least 10 different cosmids could be found (Fig. 3). By repeated testing of cosmid mixtures and single cosmids, one clone could be identified that gave rise to high numbers of HAT-resistant clones in further rounds of gene transfer. Restriction mapping and southern blots show that this clone (cos HTK 11, Fig. 4) consists of human DNA and contains the region of the human thymidine kinase gene that recently had been reported by Bradshaw

(1983). From these results there is strong evidence that we actually were able to isolate the single copy gene for human thymidine kinase by cosmid shuttle.

5 Discussion

A human cosmid gene bank has been established in a cosmid vector that contains the Herpes Simplex Virus thymidine kinase gene thus allowing selection in Ltk^- mouse cells. With an average insert size of 35 kbp it should contain any human gene with a probability of 0,95. From this gene bank a number of human genes have been isolated by conventional techniques (Lund et al. 1984, Jaenichen et al. 1984, Lindenmaier, unpublished). In the case of human V_K gene sequences some clones contained more than one cosmid vector (Jaenichen et al. 1984). This might be due to incomplete separation of small genomic DNA fragments during size fractionation, but in the case of other genes isolated this phenomenon did not occur. The gene bank was amplified under conditions where competitive growth was minimized as far as possible.

For transfer in Ltk^- mouse cells linearized hybrid cosmid DNA was ligated to protect functional cos-sites. This ligation may not be absolutely required, since after transfer with circular cosmid DNA repackaging was possible as well (Hauser et al. 1982). It has been shown that by using the calcium phosphate-precipitation method large amounts of exogenously added DNA are integrated into the L-cell genome (Perucho et al. 1980). The amount of hybrid cosmids in individual clones is rather variable (2 - 100, data not shown). Assuming that about 30 cosmids per clone are stably integrated, the 8000 mouse L-cell clones should contain any human sequence with a probability of 0,82, presupposed that the representation of individual clones has not been altered during amplification of the library and that stable integration occurs at random.

For the general application of the method described a number of requirements have to be met. (1) The gene in question has to be cloned in its entirety in the cosmid library, since the cosmid cloning system allows efficient cloning of large (up to 45 kb) contiguous regions of DNA. Thus complete functional genes are cloned with high probability, but in some special cases the length of the functional gene will reach or exceed even the cloning capacity of the cosmid system (De Saint Vincent et al. 1981). (2) The gene representation must not be drastically altered after gene transfer into mouse Ltk^--cells. As far as we know, there is no preferential transfection of special clones; but we cannot exclude that some clones, e.g., those containing replicators, are preferred. (3) The gene in question has to be expressed in the heterologous cells after gene transfer in amounts sufficient to be detected and distinguished from L-cell products by immunological or functional tests. Using isolated genes from heterologous donors, e.g., chicken (Lai et al. 1980), rabbit (Wold et al. 1979), rat (Kurtz 1981), and man (Hauser et al. 1982) or mouse genes that are normally expressed only in specialized tissues (Goodenow et al. 1982, Mayo et al. 1982) functional expression in mouse Ltk^--cells could be detected. In some cases (e.g., β-microglobulin (Kurtz 1981), human IFNβ, (Hauser et al. 1982) metallothionein (Mayo et al. 1982)), the regulatory regions of the genes seem to be correctly recognized in mouse cells. Expression of cell-surface proteins (Kavathas and Herzenberg 1983) could be detected in a population of cells after transfer of total genomic DNA by fluorescence-activated cell sorting. On the other hand, our experience using the human IFN β_1 gene showed that expression was rather variable in individual clones and some genes, e.g., immunoglobulins, require rearrangement to be activated and may therefore be expressed only in specialized cells.

At the moment it is difficult to predict how far the requirements discussed above will be met. Nevertheless, we believe that the method of cosmid shuttle will provide a possibility of isolating genes that are not easily accessible by conventional techniques.

References

Birnboim HC, Doly J (1979) A rapid alkaline extraction procedure for screening recombinant plasmid DNA. Nucleic Acids Res 7:1513-1522

Blin N, Stafford DW (1976) Isolation of high-molecular-weigth DNA. Nucleic Acids Res 3:2303-2308

Bradshaw HD (1983) Molecular cloning and cell cycle-specific regulation of a functional human thymidine kinase gene. Proc Natl Acad Sci USSA 80:5588-5591

Cami B, Kourilsky P (1978) Screening of cloned recombinant DNA in bacteria by in situ colony hybridization. Nucleic Acids Res 5:2381-2390

De Saint Vincent BR, Delbrück S, Eckhart W, Meinkoth J, Vitto L, Wahl G (1981) The cloning and reintroduction into animal cells of a functional CAD gene, a dominant amplifiable genetic marker. Cell 27:267-277

Goldfarb M, Shimizu K, Perucho M, Wigler M (1982) Isolation and preliminary characterization of a human transforming gene from T24 bladder carcima cells. Nature (Lond) 296:404-409

Goodenow RS, McMillan M, Örn A, Nicolson M, Davidson N, Frelinger JA, Hood L (1982) Identification of a Balb/c H-2L gene by DNA mediated gene transfer. Science (Wash DC) 215:677-679

Graham F, Van der Eb L (1973) A new technique for the assay of infectivity of human adenovirus 5 DNA. Virology 52:456-467

Hauser H, Gross G, Bruns W, Hochkeppel HK, Mayr U, Collins J (1982) Inducibility of human β-interferon gene in mouse L-cell clones. Nature (Lond) 297:650-655

Hohn B (1979) In vitro packaging of λ and cosmid DNA. In: Wu R (ed) Methods in Enzymology, vol 68. Academic Press, New York, p 299

Jaenichen HR, Pech M, Lindenmaier W, Wildgruber N, Zachau HG (to be published 1984) Human V_K gene sequences and their implication for V gene evolution. Nucleic Acids Res

Kavathas P, Herzenberg LA (1983) Stable transformation of mouse L cells for human membrane T-cell differentiation antigens and β_2-µicro-globulin: Selection by fluorescence-activated cell sorting. Proc Natl Acad Sci USA 80:524-528

Kurtz DT (1981) Hormonal inducibility of rat β2µ globulin genes in transfected mouse cells. Nature (Lond) 291:629-631

Lai EC, Woo SLC, Bordelon-Riser ME, Fraser TH, O'Malley BA (1980) Ovalbumin is synthesized in mouse cells transformed with the natural chicken ovalbumin gene. Proc. Natl Acad Sci USA 77:244-248

Lindenmaier W, Hauser H, Greiser de Wilke I, Schütz G (1982) Gene shuttling: moving of cloned DNA into and out of eukaryotic cells. Nucleic Acids Res 10:1243-1256

Liu C-P, Tucker PW, Mushinski JF, Blattner FR (1980) Mapping of heavy chain genes for mouse immunoglobulins M and D. Science (Wash DC) 209:1348-1353

Lowy I, Pellicer A, Jackson JF, Sim GK, Silverstein S, Axel R (1980) Isolation of transforming DNA: cloning of the hamster apart gene. Cell 22:817-823

Lund B, Edlund T, Lindenmaier W, Ny T, Collins J, Lundgren E, von Gabain A (1984) Novel cluster of α-interferon gene sequences in a placental cosmid DNA library. Proc Natl Acad Sci USA 81:2435-2439

Lund T, Grosveld FG, Flavell RA (1982) Isolation of transforming DNA by cosmid rescue. Proc Natl Acad Sci USA 79:520-524

Maniatis T, Fritsch EF, Sambrook J (1982) Molecular cloning. Cold Spring Harbor New York, pp 80-84

Mayo KE, Warren R, Palmiter RD (1982) The mouse metallothionein-I gene is transcriptionally regulated by cadmium following transfection into human or mouse cells. Cell 29:99-108

Perucho M, Hanahan D, Lipsich L, Wigler M (1980a) Isolation of the chicken thymidine kinase gene by plasmid rescue. Nature (Lond) 285:207-211
Perucho M, Hanahan D, Wigler M (1980b) Genetic and physical linkage of exogenous sequences in transformed cells. Cell 22:309-317
Shih C, Weinberg RA (1982) Isolation of a transforming sequence from a human bladder carcinoma cell line. Cell 29:161-169
Wigler M, Silverstein S, Lee L-S, Pellicer A, Cheng Y, Axel R (1977) Transfer of purified herpes simplex virus thymidine kinase gene to cultured mouse cells. Cell 11:223-232
Wigler M, Pellicer A, Silverstein S, Axel R (1978) Biochemical transfer of single-copy eukaryotic genes using total cellular DNA as donor. Cell 14:725-731
Wold B, Wigler M, Lacy E, Maniatis T, Silverstein S, Axel R (1979) Introduction and expression of a rabbit-β-globin gene in mouse fibroblasts. Proc Natl Acad Sci USA 76:5684-5688

Enhancers as Transcriptional Control Elements

P. GRUSS[1]

Understanding transcriptional control mechanisms is a major problem in molecular biology. Recent studies of the structure and function of a number of eukaryotic transcriptional units have brought to light several different controlling elements that play a part in the initiation of transcription by RNA polymerase II (Shenk 1981). One element with a number of surprising characteristics will be the focus of attention in this chapter. It is now commonly referred to as enhancer or activator.

This element was first discovered within a 72 base-pair (bp) repeat in the DNA tumor virus simian virus 40 (SV40, Fig. 1). Although deletion of one of the two 72-bp units did not abolish its activity, when both repeats were deleted there was a loss of gene activity in vivo - the amount of T antigen was greatly reduced and it failed to complement early defective temperature-sensitive helper virus (Benoist and Chambon 1981, Gruss et al. 1981). This led research workers to define the 72-bp unit as a cis-essential element required in the activation of the SV40 early genes in vivo. However, the activity of heterologous genes could also be increased if there was at least one copy of the 72-bp repeat on the same molecule (Banerji et al. 1981) and so this activation was not restricted to SV40 genes. These authors demonstrated that the SV40 72-bp repeat increased the transcriptional activity of the heterospecific rabbit β-globin promotor approximately 200-fold. Moreover with chicken conalbumin or the adenovirus-2 major late promotor sequences has heterologous promotor regions similar potentiation

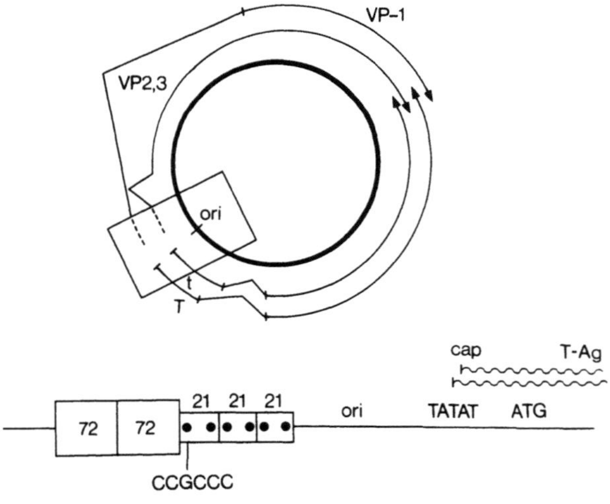

Fig. 1. Schematic representation of the SV40 genome organisation. *Boxed area* is enlarged at the bottom of the figure and represents the control elements for transcriptional initiation. *TATAT* Goldberg-Hogness box; *21* the 21 bp repeats; *72* the 72 bp repeats. *Black dots* within each 21 bp repeat represent a CCGCCC sequence repeated twice in each 21 bp repeat. Details of the SV40 genome organisation can also be found in Tooze (1980)

[1]Institute for Microbiology and ZMBH, Im Neuenheimer Feld 364, 6900 Heidelberg, FRG

35. Colloquium - Mosbach 1984
The Impact of Gene Transfer Techniques in Eukaryotic Cell Biology

effects could be noted in Chambon's laboratory (Moreau et al. 1981, Wasylyk et al. 1983). On the basis of these data, earlier observations that a fragment that included the origin of replication and the SV40 72 bp repeats was able to increase the transformation efficiency of the herpes simplex virus thymidine kinase gene (HSV-TK) 100-fold, could be understood more clearly (Cappecchi 1980).

The fact that enhancement was observed even when the SV40 72-bp repeats were located in inverted orientation or at different positions relative to the gene assayed was surprising, but made it possible to differentiate operationally between enhancer elements and promotor elements: Promoters are control elements that require a position at the immediate 5' end of genes, whereas activators or enhancers act relatively independently of distance and orientation with respect to the coding region of the gene. The question of whether enhancer elements can also exert their activity if located downstream (3') of the respective promoter was also of interest to investigators and, although no firm conclusion has been reached, there are indications that enhancers can indeed activate promoters at this location. Long terminal repeats (LTR) of retroviruses also contain enhancer elements, which increase the transformation frequency of tumor genes with the same efficiency regardless of whether they are located at the 5' end or the 3' end of the gene (Blair et al. 1980, Payne et al. 1982). Transient expression studies with chimeric plasmids gave further indications of 3' activation. Furthermore, the enhancer activity of SV40 is not detectable in these constructions when a 3.7-kb copy of plasmid pJC-1 DNA separates viral enhancer sequences on both sides from the globin gene (De Villiers et al. 1983), but if an enhancer element is located at the 3' end of the β-globin gene, which is now separated in 5' direction by 5.6 kb of plasmid sequences, enhancer activity can be detected. Thus, if the enhancer effect in this construction is not readily transmitted through plasmid DNA, one must conclude that the activation is brought about by the enhancer, which is located at the 3' position downstream of the promoter (De Villiers et al. 1983).

Although the observation that plasmid sequences inhibit enhancer function at first seemed puzzling, when enhancer-mediated transcriptional activities were examined more closely, it was seen that proximal promoter sequences are activated in preference to distal promoter sequences (Wasylyk et al. 1983, De Villiers et al. 1983). If proximal natural promoters are mutated, "substitute" elements can take over their function as promoters and the SV40 72-bp repeats were even observed to have an activator effect on plasmid pBR322 substitute start sites (Moreau et al. 1981, Wasylyk et al. 1983). This might explain the inhibitory effect of plasmid sequences. It still remains uncertain, however, why the activity of genes such as human α-globin cannot be potentiated by enhancers (Mellon et al. 1981, Humphries et al. 1982).

One problem that attracted wide interest was a more detailed characterization of the essential sequences within the SV40 72-bp repeats responsible for the activity. Similar conclusions were reached in several laboratories where deletion mutants were produced (Moreau et al. 1981, Fromm and Berg 1982, Gheysen et al. 1983). The deletion of portions of a single copy of the 72-bp sequence from the 3' end to about the *Eco* RII site (see Tooze 1980) resulted in decreased activity and, in spite of this reduction, the deletion mutants which kept only 20 - 30 bp from the 5' end of the 72-bp repeat retained their viability (Fromm and Berg 1982, Gheysen et al. 1983). The fact that deletions from the 5' end caused abolition of enhancer activity, however, led to the assumption that some essential nucleotides are situated in the immediate 5' region of the SV40 enhancer.

In view of the deletion mapping data, we attempted to define the essential nucleotides more clearly and were able to produce a series of point mutations which affect the SV40 72-bp repeat. When analysing

Table 1. Sequence homologies within DNA sequences mediating enhancer activity. Possible core sequence: $GXTGTGG^{AAA}_{TTT}G$

SV40	AGTTAGGGTGTGGAAAGTCCCC
MSV	AGGATATCTGTGGTAAGCAGTT
Polyoma	AGAGGGCGTGTGGTTTTGCAAG
BK	CCTGAGGTCATGGTTTGGCTGC
BPV1	AGCGATGATGTGGACCCTGCTG
BPV2	CGGAGTGGTGTGTACCTTGGTG
LPV	GAGGAAGCTGTGGTTAGACTCG
Ad2	GCGCCGGATGTGGTAAAAGTGA
Ad12	GCGCCGGATGTGACGTTTTAGA
IgCμ1	CTTGTAGCTGTGGTTTGAAGAA
IgCμ2	GCAGGTCATGTGGCAAGGCTAT
IgCK	TTGTTTTTCTTGGTAAGAACTC
IgCλ	AATCAAGAGCTGGAAAGAGAGG

Sequence comparison of characterized or potential enhancer elements. References are detailed in the text

these mutants we found one critical nucleotide (underlined) at the 5' end of the SV40 72-bp repeat in the sequence 5'-GGTGTGGAAAG (Weiher et al. 1983). We further found that transition of the first three G residues to A residues only resulted in decreased activity, while G/A transition of the underlined nucleotide led to abolition of the enhancer activity. We then attempted to find sequence homologies in other potential enhancer elements with the aid of this nucleotide stretch containing the critical G resdiue and this led to the discovery of short stretches of related sequences in the control region of several other potential enhancer elements (Table 1). The most homologous sequence we found was in a 72,73-bp repeat in the LTR of MSV. In an earlier functional analysis of a 72-bp fragment containing this MSV 72,73-bp repeat there were indications that this fragment could replace the 72-bp repeat of SV40 and, further, that, on infection of CV-1 cells, DNA containing this substitution produced viable virus (Levinson et al. 1982). This suggests that there is an enhancer-like element present in the MSV repeat. In these experiments the SV-rMSV recombinant was seen to have less gene activity in cells from monkey kidney than in mouse fibroblasts, which are the natural host for MSV. In detailed studies it appeared that the SV40 tandem repeats activate gene expression in monkey kidney cells to a greater extent and MSV repeats are more active in a number of mouse cell lines (Laimins et al. 1982).

The host cell preference of the enhancer elements indicated by these findings could be one factor determining the host range of the viruses studied. Kriegler and Botchan (1983) reached a similar conclusion from studies of an SV40-Harvey murine sarcoma virus (HaSV) recombinant which carries the LTR of HaSV in the late region of SV40. Like the SV-rMSV (Levinson et al. 1982), this recombinant transformed mouse cells with greater efficiency than the SV40 wild-type virus (Kriegler and Botchan 1983, Chen and Pollack, personal communication). In the same way sequences responsible for thymotropism were found in the U3 region of murine leukaemia virus (MuLV) (Des Groseillers et al. 1983) and feline sarcoma virus (FeSV) recombinants (MSV-FeSV) with substitutions in this region transformed mouse cells more effectively than the wild-type FeSV

Gardner-Arnstein strain (Even et al. 1983). These data together show that the 73,72-bp repeat of MuLV, MSV, and HaSV-LTR is an element bearing host-specific enhancer activity, which was further supported both by experiments using the HSV-TK gene to examine the increased frequency of stable transformants by MSV LTR and by deletion mapping data (L. Laimins et al., personal communication).

Enhancers were also shown to be present in Rous sarcoma virus (RSV) LTR's (Luciw et al. 1983), polyoma virus (De Villiers and Schaffner 1981, Tyndall et al. 1981), BK virus (Rosenthal et al. 1983), adenovirus EIa (Hearing and Shenk 1983, Weeks and Jones 1983, Sassone-Corsi et al. 1983) and bovine papilloma virus (Campo et al. 1983, Lusky et al. 1983). Interestingly, host specificity was also seen with the polyoma virus enhancer element (De Villiers et al. 1982). Polyoma virus does not grow in embryonal carcinoma cells, but polyoma mutants selected for growth in these cells carry a sequence alteration in the enhancer/ promoter region (Fujimura et al. 1981, Katinka et al. 1981, Sekikawa and Levine 1981). However, firm evidence that enhancer-like elements are involved in this altered function is not yet available.

The natural location of most viral enhancers thus far described is upstream at approximately position -100 from the cap site. Bovine papilloma virus is exceptional, however, in that it contains an activator of gene expression which is contained within 60 bp of a Bam HI site (Weiher and Botchan 1984) and is located at the distal end of the early transcription unit (Lusky et al. 1983).

Because they only have a small coding capacity, viruses depend on the cellular machinery for their vital processes. As the processes of transcriptional initiation are the same in both viruses and cells, it seemed most likely that cellular genomes would also contain elements with enhancer-like function and, in fact, Grosschedl and Birnstiel (1980) demonstrated the importance of upstream elements for the transcription of the sea urchin histone H2a gene in their original studies. This element appears to be relatively independent of orientation, but information on the distance at which it is effective is still lacking.

In another attempt to identify cellular transcription activators, viral enhancers from SV40 or BK were used as probes to screen human genomic DNA libraries for possible sequence homologies and, indeed, human sequences homologous to the SV40 72-bp repeat (Conrad and Botchan 1982) and the human BK virus 68-bp repeat (Rosenthal et al. 1983) have been identified by this approach. In addition to their sequence homologies, both these sets of sequences also resembled their viral counterparts in their functional activity, although the former had somewhat lower activity than the latter. Similarly, sequences homologous to a mouse retroviral gene coding for reverse transcriptase were found in a λ-library of the monkey genome (Martin et al. 1981). It was suspected that retroviral LTR sequences would be present upstream (5') and downstream (3') of the monkey reverse transcriptase gene and indeed functional analysis showed enhancer elements downstream of the reverse transcriptase gene. When the functionally active fragment was subjected to sequence analysis control elements such as TATA box, polyadenylation signal (AATAAA) and a 32-bp repeat (M. Kessel, personal communication) were found.

Selection pressure was another approach used to obtain cellular enhancers (Fried et al. 1983). A polyoma mutant was generated which lacked the enhancer region but retained the remaining transcriptional control signals and tumor genes and which transformed mouse fibroblasts with low efficiency. It can be assumed that the sequences in the vicinity of the integrated polyoma virus genome contributed to the transcriptional activity and thus these cellular sequences at the immediate 5' boundary have been cloned, characterized and tested for their functional activity. Certainly transcriptional activation is dependent on these sequences (Fried et al. 1983), but it is not yet known whether they can also function if positioned at the 3' end of polyoma early genes. Recently an interesting approach has been taken by Schaffner and colleagues to select for enhancer elements. SV40 mutant DNA lacking the 72-bp repeats was cotransfected with DNA carrying potential enhancers. By recombination processes mediated by the host cell viable recombinants were selected now carrying enhancer activity. By this method some recombinants were rescued by insertion of known enhancers. Surprisingly, also by duplication of a fragment located on the late side of the SV40 genome others generated an enhancer de novo (Weber et al. 1984). Unfortunately, no cellular enhancers have yet been isolated using this protocol.

Using a shotgun method with a suitable test plasmid we were able to show the presence of enhancer-like elements in the mouse genome. This enhancer-like activity is, in one case studied in detail, present on a fragment 300 bp in length, most of which is made up of a stretch of polypurines/polypyrimidines. It occurs many times in the mouse genome (U. Drescher et al., manuscript in preparation).

Lastly, well-characterized genes have also been shown to contain cellular enhancer elements. A number of clues led Gillies et al. (1983), Banerji et al. (1983) and Queen and Baltimore (1983) to take a closer look at transcriptional control elements required for immunoglobulin gene expression. They found that, in the heavy chain genes, the enhancer element is in the JH-C region and can stimulate transcription from either homologous VH promotors or heterologous promoters from SV40 or β-globin (Gillies et al. 1983, Banerji et al. 1983) and that, even when its orientation is reversed or it is placed at the distal end of genes, this enhancer element remains active. Unlike viral enhancers, however, which have a relatively wide host range, the immunoglobulin enhancer seems to act in tissue-specific manner. This is also the case with the activity of a kappa light chain gene control element (Queen and Baltimore 1983).

One notable exception of a viral enhancer with a very limited host range has been recently discovered in our laboratory (L. Mosthaf et al., in preparation). The transcriptional control region of a lymphotropic PaPoVavirus was found to contain an enhancer element active only in cells of lymphoblastoid origin. Both cells of B and T cell origin were found to activate this particular enhancer element.

The apparent host and tissue specificity of some enhancers was a strong indication that cellular molecules interact with activator sequences. Using an in vivo competition assay we recently tested for involvement of cellular molecules in enhancer function (Schöler and Gruss 1984). We found that enhancers were involved in specific competition but that other SV40 promoter elements such as the 21-bp repeats or TATA box were not. The fact that host specificity of different enhancers is reflected in the competition assay suggests that activator/enhancer interacting molecule(s) can determine the host specificity of enhancer elements.

The LTR of the mouse mammary tumor virus (MMTV) is a good example of a genetic element regulated by a DNA/protein hormone/receptor interaction. It is possible to separate a hormone responsive fragment, which stimulates the expression of a heterologous HSV-TK promoter in the presence of hormone, from the MMTV promoter element. As the glucocorticoid response element does not appear to be severely restricted in its location and orientation, it is possible that there is an inducible enhancer element within the hormone responsive region (Chandler et al. 1983). In general the expression of MMTV is weak and so in several experiments enhancer elements were used to boost the activity of the hormone responsive MMTV LTR (Lee et al. 1981, Huang et al. 1981). Experiments to investigate mutants which successively deleted 5' sequences of the MMTV LTR while retaining an intact heterologous enhancer attracted particular interest. The activity remaining in the presence and absence of hormone showed that there was low gene activity when the hormone was absent as long as the hormone receptor binding region was preserved. When the hormone receptor binding region was deleted the gene activity was turned on constitutively (M. Kessel, personal communication). The results of these experiments would fit with the idea that the MMVT LTR are repressed in the absence of inducer. The mechanism of this repression possibly resembles the autoregulation of SV40 early genes by T antigen (Reed et al. 1976, for review, see Tijian 1981) in which T antigen surpresses SV40 enhancer/promoter activity by binding to two distinct regions situated downstream of the SV40 72-bp repeats (Tegtmeyer et al. 1983). When dealing with the question of how exactly the enhancer functions, one must consider both specific DNA sequences and the cellular molecules that interact with them in order to account for observations made so far. It would appear from the enhancers' host and tissue specificity that specific factors play a part in the recognition of enhancer sequences and it is possible that the specificity is mediated not by RNA polymerase II, but by a specific molecule, perhaps comparable to the bacterial σ factor. Moreover, enhancers would appear to have bidirectional activity from their relative independence of orientation. It has not yet been established whether this is determined by a defined structural feature of enhancer elements or a defined sequence requirement. The enhancer's independence of distance can be explained by a "tracking" mechanism by which the complex searches the DNA for promoter elements upon the initial contact of the transcriptional complex with enhancer elements (Wasylyk et al. 1983). It is thought that these processes occur at the nuclear matrix (Jackson et al. 1981), but it is also possible that enhancers organize the chromatin structure in such a way as to make it accessible to transcriptional factors. This would be an explanation for the DNase I hypersensitive region (see Moyne et al. 1981 for references), which is possibly induced by enhancer elements (Jonstra et al. 1984). Before the enhancer function can be studied further, in vitro systems that exactly reproduce the in vivo activity of enhancers must be developed. A first step in this direction has been taken by Chambon's laboratory (Sassone-Corsi et al. 1984). It seems possibly to establish an in vitro assay mimicking the SV40 enhancer activity.

We can reasonably suppose that more cellular genes have such transcriptional control elements besides the few cellular enhancers discovered up to now and thus we hope that future findings of cellular and viral enhancer-like elements will contribute to our understanding of the mechanism of the activator-enhancer function.

Acknowledgements. I thank Rosemary Franklin for her invaluable help in the preparation of the manuscript.

References

Banerji J, Rusconi S, Schaffner W (1981) Expression of a β-globin gene is enhanced by remote SV40 DNA sequences. Cell 27:299-308

Banerji J, Olson L, Schaffner W (1983) A lymphocyte-specific cellular enhancer is located downstream of the joining region in immunoglobulin heavy chain genes. Cell 33:729-740

Benoist C, Chambon P (1981) In vivo sequence requirements of the SV40 early promoter region. Nature (Lond) 290:304-310

Blair DG, McClements WL, Oskarsson MK, Fischinger PJ, Vande Woude GF (1980) Biological activity of cloned Moloney sarcoma virus DNA: Terminally redundant sequences may enhance transformation efficiency. Proc Natl Acad Sci USA 77:3504-3508

Campo MS, Spandidos DA, Lang J, Wilkie NM (1983) Transcriptional control signals in the genome of bovine papilloma virus type 1. Nature (Lond) 303:77-80

Capecchi MR (1980) High efficiency transformation by direct microinjection of DNA into cultured mammalian cells. Cell 22:479-488
Chandler VL, Maler BA, Yamamoto KR (1983) DNA sequences bound specifically by glucocorticoid receptor in vitro render a heterologous promoter hormone responsive in vivo. Cell 33:489-499
Conrad SE, Botchan MR (1982) Isolation and characterization of human DNA fragments with nucleotide sequence homologies with the SV40 regulatory region. Mol Cell Biol 2:949-965
De Villiers J, Schaffner W (1981) A small segment of polyoma virus DNA enhances the expression of a cloned β-globin gene over a distance of 1400 base pairs. Nucleic Acids Res 9:6251-6264
De Villiers J, Olson L, Banerji J, Schaffner W (1983) Studies of the transcriptional "enhancer" effect. Cold Spring Harbor Symp Quant Biol 47:911-919
De Villiers J, Olson L, Tyndall C, Schaffner W (1982) Transcriptional enhancers from SV40 and polyoma virus show a cell type preference. Nucleic Acids Res 10:7965-7976
Des Groseillers L, Rassart E, Jolicoeur P (1983) Thymotropism of murine leukemia virus is conferred by its long terminal repeat. Proc Natl Acad Sci USA 80:4203-4207
Even J, Anderson SJ, Hampe A, Galibert F, Lowy D, Khoury G, Sherr CJ (1983) Mutant feline sarcoma proviruses containing the viral oncogene (v-fes) and either feline or murine control elements. J Virol 45:1004-1016
Fried M, Griffiths M, Davies B, Bjursell G, La Mantia G, Lania L (1983) Isolation of cellular DNA sequences that allow expression of adjacent genes. Proc Natl Acad Sci USA 80:2117-2121
Fromm M, Berg P (1982) Deletion mapping of DNA regions required for SV40 early region promotor function in vivo. J Mol Appl Genet 1:457-481
Fromm H, Berg P (1983) SV40 early and late region promoter function are enhanced by the 72 base pair repeat inserted at distant locations and inverted orientation. J Mol Cell Biol 3:991-999
Fujimura FK, Deininger PL, Friedmann T, Linney E (1981) Mutation near the polyoma DNA replication origin permits productive infection of F9 embryonal carcinoma cells. Cell 23:809-814
Gheysen D, Van de Voorde A, Contreras R, Vanderheyden J, Duerinck F, Fiers W (1983) Simian virus 40 mutants carrying extensive deletions in the 72-base-pair repeat region. J Virol 47:1-14
Gillies SD, Morrison SL, Oi VT, Tonegawa S (1983) A tissue-specific transcription enhancer element is located in the major intron of a rearranged immunoglobulin heavy chain gene. Cell 33:717-728
Grosschedl R, Birnstiel ML (1980) Spacer DNA sequences upstream of the TATAATA sequence are essential for promotion of H2A histone gene transcription in vivo. Proc Natl Acad Sci USA 77:7102-7106
Gruss P, Dhar R, Khoury G (1981) Simian virus 40 tandem repeated sequences as an element of the early promoter. Proc Natl Acad Sci USA 78:943-947
Hearing P, Shenk T (1983) The adenovirus type 5 E1A transcriptional control region contains a duplicated enhancer element. Cell 33:695-703
Huang AL, Ostrowski MC, Berard D, Hager GL (1981) Glucocorticoid regulation of the HaMuSV p21 gene conferred by sequences from mouse mammary tumor virus. Cell 27:245-255
Humphries RK, Ley T, Turner P, Mouston AD, Nienhuis AW (1982) Differences in human α-, β-, and δ-globin gene expression in monkey kidney cells. Cell 30:173-183
Jackson DA, McCready SJ, Cook PR (1981) RNA is synthesized at the nuclear cage. Nature (Lond) 292:552-555
Jolly DJ, Esty AC, Rubramani S, Friedmann T, Verma IM (1983) Elements in the long terminal repeat of murine retroviruses enhance stable transformation by thymidine kinase gene. Nucleic Acids Res 11:1855-1872
Katinka M, Vasseur M, Montreau N, Yaniv M, Blangy D (1981) Polyoma DNA sequences involved in control of viral gene expression in murine embryonal carcinoma cells. Nature (Lond) 290:720-722
Kriegler M, Botchan M (1983) Enhanced transformation by a simian virus 40 recombinant virus containing a Harvey murine sarcoma virus long terminal repeat. Mol Cell Biol 3:325-339

Laimins L, Khoury G, Gorman C, Howard B, Gruss P (1982) Host-specific activation of transcription by tandem repeats from simian virus 40 and Moloney murine sarcoma virus. Proc Natl Acad Sci USA 79:6453-6457
Lee F, Mulligan R, Berg P, Ringold G (1981) Glucocorticoids regulate expression of dihydrofolate reductase cDNA in mouse mammary tumor virus chimaeric plasmids. Nature (Lond) 294:228-232
Levinson B, Khoury G, Vande Woude G, Gruss P (1982) Activation of the SV40 genome by the 72 base-pair tandem repeats of Moloney sarcoma virus. Nature (Lond) 305: 568-572
Luciw PA, Bishop JM, Varmus HE, Capecchi MR (1983) Location and function of retroviral and SV40 sequences that enhance biochemical transformation after microinjection of DNA. Cells 33:705-716
Lusky M, Weiher H, Botchan M (1983) Bovine papilloma virus contains an activator of gene expression at the distal end of the early transcription unit. Mol Cell Biol 3:1108-1122
Martin MA, Bryan T, McCutchan TF, Chan H (1981) Detection and cloning of murine leukemia virus-related sequences from African green monkey liver DNA. J Virol 39:835-844
Mellon P, Parker V, Gluzman Y, Maniatis T (1981) Identification of DNA sequences required for transcription of the human α1-globin gene in a new SV40 host-vector system. Cell 27:279-288
Moreau P, Hen R, Wasylyk B, Everett R, Gaub MP, Chambon P (1981) The SV40 base pair repeat has a striking effect on gene expression in both SV40 and other chimeric recombinants. Nucleic Acids Res 9:6047-6068
Moyne G, Freeman R, Saragosti S, Yaniv M (1981) A high-resolution electron microscopy study of nucleosomes from simian virus 40 chromatin. J Mol Biol 149:735-744
Nordheim A, Lafer EM, Peck LJ, Wang JC, Stollar BD, Rich A (1982) Negatively supercoiled plasmids contain left-handed Z-DNA segments as detected by specific antibody binding. Cell 31:309-318
Payne GS, Bishop JM, Varmus HE (1982) Multiple arrangements of viral DNA in an activated host oncogene in bursal lymphomas. Nature (Lond) 295:209-213
Queen C, Balitmore D (1983) Immunoglobulin gene transcription is activated by downstream sequence elements. Cell 33:741-748
Reed SI, Stark GR, Alwine JC (1976) Autoregulation of simian virus 40 gene A by T antigen. Proc Natl Acad Sci USA 73:3083-3087
Rosenthal N, Kress M, Gruss P, Khoury G (1983) The Bk viral enhancer element and a human cellular homology. Science (Wash DC) 222:749-755
Sassone-Corsi P, Hen R, Borelli E, Leff T, Chambon P (1983) Far upstream sequences are required for efficient transcription from the adenovirus-2 E1A transcription unit. Nucleic Acids Res 11:8735-8745
Sassone-Corsi P, Dougherty JP, Wasylyk B, Chambon P (1984) Stimulation of in vitro transcription from heterologous promoters by the simian virus 40 enhancer. Proc Natl Acad Sci USA 81:308-312
Schöler H, Gruss P (1984) Specific interaction between enhancer-containing molecules and cellular components. Cell 36:403-411
Sekikawa K, Levine AJ (1981) Isolation and characterization of polyoma host range mutants that replicate in nullipotential embryonal carcinoma cells. Proc Natl Acad Sci USA 78:1100
Shenk T (1981) Transcriptional control regions: nucleotide sequence requirements for initiation by RNA polymerase II and III. Curr Top Microbiol Immunol 93:25-46
Tegtmeyer P, Lewton BA, DeLucia AL, Wilson VG, Ryder K (1983) Topography of simian virus 40 A protein-DNA complexes: arrangement of protein bound to the origin of replication. J Virol 46:151-161
Tjian R (1981) Regulation of viral transcription and DNA replication by the SV40 large T antigen. Curr Top Microbiol Immunol 93:5-24
Tooze J (ed) (1980) Molecular biology of tumor viruses. DNA tumor viruses. Cold Spring Harbor Laboratory, Cold Spring Harbor, New York
Tyndall C, La Mantia G, Thacker C, Favaloro J, Kamen R (1981) A region of the polyoma virus genome between the replication origin and late protein coding sequences is required in cis for both early gene expression and viral DNA replication. Nucleic Acids Res 9:6231-6250

Wasylyk B, Wasylyk C, Augereau P, Chambon P (1983) The SV40 72 bp repeat preferentially potentiates transcription starting from proximal natural or substitute promoter elements. Cell 32:503-514

Weber F, de Villiers J, Schaffner W (1984) An SV40 "Enhancer-Trop" incorporates exogenous enhancers or generates enhancers from its own sequences. Cell 36:983-992

Weeks DL, Jones NC (1983) E1A control of gene expression is mediated by sequences 5' to the transcriptional starts of early viral genes. Mol Cell Biol 3:1222-1234

Weiher H, König M, Gruss P (1983) Multiple point mutations affecting the simian virus 40 enhancer. Science (Wash DC) 219:629-631

Weiher H, Botchan M (1984) An enhancer sequence from bovine papilloma virus DNA consists of two essential regions. Nucleic Acids Res 12:2917-2928

Z-DNA: Conformational Flexibility of the DNA Double Helix

A. NORDHEIM[1]

1 Introduction

During the last 5 years new insights into the molecular structure of the DNA double helix have been obtained. An enhanced awareness of the conformational flexibility of the double-helical molecule has resulted (Cantor 1981; Pohl 1983; Rich 1983). The spectrum of the structural DNA polymorphism ranges from subtle, but distinct variations (e.g., Drew and Travers 1984; Frederick et al. 1984) of the classical right-handed B-DNA (Watson and Crick 1953) to the radically different left-handed Z-DNA (Wang et al. 1979). Two methodological developments allowed the recent progress in our understanding of DNA structure: first, rapid chemical synthesis of DNA oligonucleotides with defined nucleotide sequence and, second, recombinant DNA technology. The first technique greatly facilitated DNA X-ray crystallography by ensuring the supply of sufficient quantities of pure DNA material. The second method, in vitro recombination and cloning of DNA sequences, allowed biochemical characterization of individual DNA segments of natural origin.

In this article, selected aspects of the structure and characterization of left-handed Z-DNA will be discussed. I will mainly focus on questions relevant to the search for natural occurrence and potential biological function of Z-DNA. The interrelationship between DNA topology (DNA supercoiling) and DNA double-helical structure will be described in order to emphasize the effectiveness of torsional stress for stabilizing Z-DNA. For a more comprehensive treatise on Z-DNA the reader is referred to the recent review articles by Rich et al. (1984) and Jovin et al. (1983), as well as a collection of articles on various aspects of DNA structure and function (Cold Spring Harbor Symp. Quant. Biol., Vol. 47, 1983).

2 Formation and Stabilization of Z-DNA

Z-DNA is a left-handed double-helical DNA molecule whose molecular structure was determined at atomic resolution by single crystal X-ray diffraction analysis (Wang et al. 1979; Drew et al. 1980; Fig. 1). Spectral changes associated with the B-Z transition were observed earlier using DNA oligonucleotides in elevated salt conditions (Pohl and Jovin 1972). Laser Raman studies established the identity of solid-state (crystal) and solution (4 M NaCl) forms of Z-DNA (Thamann et al. 1981).

Z-DNA is formed by Watson-Crick-type base-pairing of two complementary, anti-parallel DNA single strands. Major differences in the helical

[1]Zentrum für Molekulare Biologie der Universität Heidelberg, Im Neuenheimer Feld 364, 6900 Heidelberg, FRG

35. Colloquium - Mosbach 1984
The Impact of Gene Transfer Techniques in Eukaryotic Cell Biology

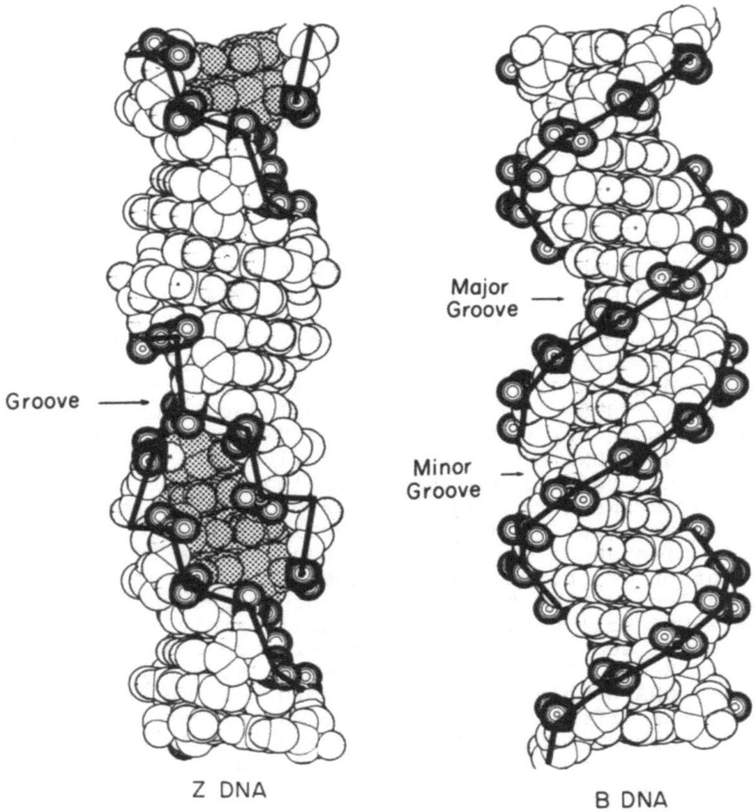

Fig. 1. Van der Waal's diagrams of left-handed Z-DNA and right-handed B-DNA. (After Wang et al. 1979). *Heavy black lines* connect positions of phosphates along the sugar phosphate DNA backbone

geometry of B and Z DNA include helical sense, helical pitch, diameter, base-pair tilt, position of helical axis, distances of phosphates from helical axis, glycosidic torsion angles of bases, and types of sugar pucker (Wang et al. 1979). Residues normally located within the major groove of B-DNA are part of the convex outer surface of Z-DNA. Consequently, reactive groups like N7 or C8 of guanine are more exposed in Z-DNA and may thereby contribute to new chemical reactivities associated with the left-handed structure.

With respect to base-sequence requirements underlying the formation of Z-DNA it is the alternating purine-pyrimidine segment that can adopt Z-DNA readily. This is based on the capability of purine residues to easily adopt either of the alternate syn/anti rotations around glycosidic bonds. Within this framework, a hierarchy of increasing Z-DNA potential is observed with sequences $d(TA/AT)_n$, $d(CA/GT)_n$, and $d(CG/GC)_n$. It should be pointed out that sequences of nonalternating purine/pyrimidine base composition appear not to be precluded from formation of Z-DNA (Nordheim et al. 1982a; Wang et al., manuscript in preparation, A. Nordheim, unpublished data). The effects of base composition on left-handed helix stability and geometry can only be ascertained with the availablity of additional X-ray crystallographic information.

The zig-zag array of phosphate positions along the sugar-phosphate backbone (hence Z-DNA) illustrates the existence in Z-DNA of shorter phosphate-phosphate distances than present in B-DNA. This is assumed to cause reduced stability of Z-DNA, which affects the equilibrium between B- and Z-DNA.

An understanding of factors that determine the equilibrium between B and Z forms of DNA is desired for physico-chemical as well as biological considerations.

In addition to nucleotide composition, there are four main effectors of the B/Z equilibrium: (A) solvent conditions, (B) chemical DNA modification, (C) protein binding, and (D) DNA supercoiling.

Unusual *solvent conditions* of high ionic strengths led to the first recognition of a left-handed DNA conformation by Pohl and Jovin (1972). The effectiveness of cations in inducing B-Z transitions increases with ion valency whereby cobalt hexamine and polyamines are especially potent (Behe and Felsenfeld 1981). Agents that change the dielectric constant of water exhibit additional effects (ethanol, methanol, polyethylene glycol) (Pohl 1976; van de Sande and Jovin 1982; Zacharias et al. 1982).

Chemical DNA modifications can very effectively reduce the salt requirements for stabilization of Z-DNA. Covalent DNA modifications of C5-cytosine or C8,N7-guanosine that affected the B-Z equilibrium of the poly(dG-dC)polymer included: methylation (Behe and Felsenfeld 1981; Möller et al. 1981), bromination (Möller et al. 1984), 5-azacytosine substitution (Jovin et al. 1983), adduction with acetylaminofluorene or aflatoxin B1 (Sage and Leng 1980; Santella et al. 1981; Nordheim et al. 1983), complexation with platinum compounds (Malfoy et al. 1981; Ushay et al. 1982), chemical cross-linking (Castleman et al. 1983) or phosphate group-thio substitution (Jovin et al. 1983). Methylation (McIntosh et al. 1983) and adduction by acetylaminofluorene (Wells et al. 1982) also facilitated Z-DNA formation of the poly(dG-dT) • poly(dC-dA)polymer.

DNA supercoiling, i.e., helix unwinding in a topologically constrained DNA segment proved to be one of the strongest driving forces for Z-DNA formation. This important mechanism will be dealt with separately in Section 5.

The effects of *protein binding* on the B-/-DNA equilibrium will be discussed in the description of Z-DNA/protein interactions in Section 6.

3 Identification of Z-DNA

The structural transition of right-handed B-DNA to left-handed Z-DNA is accompanied by alterations of both physical helix parameters and chemical reactivities. As a result, the in vitro characterization of the Z-DNA double helix is well advanced and some of the techniques that have been used successfully are summarized in the following.

The most detailed information on the Z-DNA helix can be obtained by single crystal X-ray crystallography (Wang et al. 1979; Drew et al. 1980; Wang et al. 1981). X-ray diffraction studies on DNA fibers have also yielded useful information (Arnott et al. 1980). Spectroscopic techniques of great value for Z-DNA identification include UV absorbance changes (Pohl and Jovin 1972), circular dichroism (Pohl and Jovin 1972; Sutherland et al. 1981), optical rotatory dispersion (Pohl and Jovin 1972), Laser Raman Scattering (Pohl et al. 1973; Thamann et al. 1981), and laser light scattering (Thomas and Bloomfield 1983). Nuclear magnetic resonance (NMR) studies of both the ^{31}P- and ^{1}H-type (Patel et al. 1979, 1982) also contributed greatly. Analysis of hydrogen-deuterium exchanges (Ramstein and Leng 1980) were of further value.

Finally, Z-DNA formation within supercoiled DNA circles can be detected by the altered hydrodynamic behavior of such plasmids during gel electrophoresis (Klysik et al. 1981; Peck et al. 1982) or sedimentation (Peck et al. 1982).

With respect to chemical/biochemical identification of Z-DNA, cleavage by nuclease S1 of junction regions between B- and Z-DNA has been detected (Singleton et al. 1982). Also, the finding of differential binding to B- and Z-DNA by chiral phenanthroline-ion complexes (Barton et al. 1984) is of additional significance.

Immunological approaches for detection of Z-DNA could be initiated with the generation of specific anti-Z-DNA antibodies (Lafer et al. 1981; Malfoy and Leng 1981). A detailed description of the use of these antibodies for the characterization and identification of Z-DNA is given in the following chapter (Sect. 4).

In contrast to the wealth of in vitro information on Z-DNA formation, no *direct* evidence for Z-DNA occurrence in vivo has been reported up to now. This lack of in vivo evidence may be due to the requirement for Z-DNA stabilizing factors under physiological ionic conditions. Attempts to purify Z-DNA structures out of their genomic environment necessarily lead to removal of stabilizing factors and subsequent loss of the unstable left-handed conformation. Consequently, only such methodological approaches appear promising that detect or irreversibly "freeze" Z-DNA segments in situ without major rearrangement of intracellular conditions. Obviously, to achieve this is not a trivial task. The tools currently at hand to approach this problem include specific antibodies (see below) and Z-DNA-specific binding ligands (Barton et al. 1984). Additionally, the differential chemical reactivities of B- and Z-DNA may allow specific chemical modification of Z-DNA as detectable by altered nucleotide cleavage sensitivities during genomic DNA-sequencing experiments (Church and Gilbert 1984). Finally, the photofootprinting approach developed by Becker and Wang (1984) carries some promise for in vivo identification of Z-DNA and other changes in DNA conformation.

4 Immunological Probes for Z-DNA

Should Z-DNA fulfill distinct genetic functions in the genome of a living cell it would nevertheless be expected to represent only a transiently existing, minor fraction of the total genomic DNA. Physicochemical techniques can therefore not easily be expected to provide the necessary resolution for low Z-DNA contents in biological preparations.

The finding that Z-DNA carries high immunogenic potential (Lafer et al. 1981; Malfoy and Leng 1981) led to the production of highly specific antibody probes for the identification of Z-DNA. Different chemical modifications of the poly d(CG) DNA polymer were used to generate low-salt-stabilized antigens: bromination (Lafer et al. 1981; Pohl et al. 1982; Zarling et al. 1983), methylation (Lafer et al. 1983), modification with Pt-dien (Malfoy et al. 1982), or adduction with acetylaminofluorene (AAF) (Hanau et al. 1984). In addition to polyclonal anti-Z-DNA preparations from sera of rabbits, mice, and goats, monoclonal anti-Z-DNA probes were obtained more recently from mouse hybridoma cells (Möller et al. 1982; Thomas et al. 1983; Zarling et al. 1984).

Unequivocal demonstration of true specificity for recognition of Z-DNA is the most important issue in developing antibody probes for DNA conformation. It could be shown that the chemically modified bases present in the antigens used were not required for Z-DNA recognition (Lafer et al. 1981; Nordheim et al. 1981). Antibodies specific for Z-DNA could react with chemically unmodified poly(dG-dC) DNA in 4 M NaCl when the polymer was present as Z-DNA (Nordheim et al. 1981). No significant cross-reactivity with other forms of nucleic acid could be detected.

Furthermore, certain monoclonal antibody probes can precisely distinguish selected features of the Z helix. These include backbone geometry and base positions (Möller et al. 1982). Different monoclonal antibodies also show nucleotide-sequence preferences as an additional level underlying their Z-DNA structure specificity (A. Nordheim et al., manuscript in preparation). In conclusion, it can be stated that antibodies are highly specific probes for Z-DNA identification, although the minimal recognition element along the Z helix has not yet been precisely determined with any of the available antibodies.

The use of Z-DNA-specific antibodies proved valuable for the identification of Z-DNA in natural DNA preparations. Indirect immunofluorescence studies on polytene chromosomes of *Drosophila* gave the first indirect evidence for Z-DNA-forming potential in DNA sequences of natural origin (Nordheim et al. 1981). Work in several laboratories (Lemeunier et al. 1982; Arndt-Jovin et al. 1983; Hill and Stoller 1983; Pardue et al. 1983; Robert-Nicoud et al. 1984) drew additional attention to the fact that fixation conditions during cytological preparation of specimens can greatly determine the degree to which Z-DNA structures are available for antibody recognition. Clearly, fixation of chromosomes for cytological investigation involves dramatic alterations of cellular conditions (i.e., protein removal, introduction of topological stress, changes in pH), which are expected to either unmask left-handed structures or contribute to Z-DNA formation. Antibody staining observed on cytological chromosome preparations can therefore only indicate *potential* for Z-DNA formation within a chromosomal segment. The degree to which this potential is realized at certain points in time within the living cell requires further, more direct in situ analysis.

Additional indications for the functional importance of Z-DNA in living organisms were derived from immunofluorescence in fixed mammalian nuclei (Morgenegg et al. 1983), mitotic chromosomes (Viègas-Pequiquot et al. 1983), and macronuclei of ciliated protozoa (Lipps et al. 1983).

Antibody probes for Z-DNA contributed to the finding that Z-DNA can be stabilized in negatively supercoiled DNA circles (see below; Nordheim et al. 1982a) as well as in form V DNA (Pohl et al. 1982; Lang et al. 1982). The first identification of a natural DNA sequence to assume a left-handed conformation was also achieved using anti-Z-DNA antibodies (Nordheim et al. 1982a). Likewise, these antibodies allowed the demonstration that DNA supercoiling could induce the B-Z transition within a chemically unmodified $[d(C\text{-}A)\text{-}d(G\text{-}T)]_n$ stretch of DNA (Nordheim and Rich 1983a). The latter conclusion was independently arrived at by Haniford and Pulleyblank (1983) using the technique of two-dimensional gel electrophoretic separation of supercoiled plasmids. The data obtained with these two unrelated techniques are in very good qualitative and quantitative agreement with respect to the negative superhelical densities required for the B-Z transition.

Finally, anti-Z-DNA antibodies identified the potential for Z-DNA formation within the control segment of the supercoiled SV40 genome (Nordheim and Rich 1983b), possibly suggesting that structural changes in DNA conformation contribute to biological regulatory phenomena.

5 DNA Supercoiling and Z-DNA

In a negatively supercoiled, covalently closed DNA circle the DNA exists in an underwound state. That is, the *actual* linking number α (indicating how often the two DNA strands revolve around each other) is smaller than α_0 (the linking number of two DNA strands in the relaxed situation). The degree by which α and α_0 differ is described by the number of superhelical turns, τ ($\tau = \alpha - \alpha_0$; Wang 1980; Crick 1976). To compare the

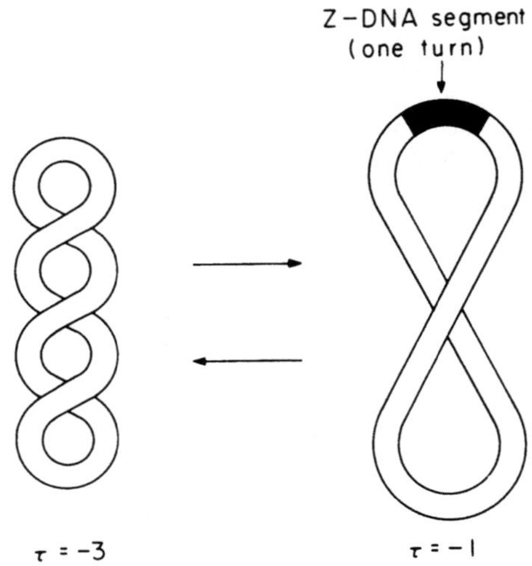

Fig. 2. Schematic demonstration of supercoil-induced B-DNA to Z-DNA transition within a defined segment of the plasmid genome. Z-DNA formation removes negative superhelical turns and thereby changes the hydrodynamic properties (e.g., sedimentation behavior or migration velocity during gel electrophoresis) of the plasmid circle

degrees of supercoiling between plasmids of different size, the number of superhelical turns are normalized over α_0 as a measure of the superhelical density (σ) of a given plasmid ($\sigma = \frac{\tau}{\alpha_0} = \frac{\alpha-\alpha_0}{\alpha_0}$). The unfavorable free energy, ΔG, that is stored in every negatively supercoiled template facilitates all processes that reduce the number of superhelical turns. Consequently, the generation and stabilization of Z-DNA is energetically favored in negatively supercoiled plasmids as could be demonstrated experimentally (Singleton et al. 1982; Peck et al. 1982; Nordheim et al. 1982a). Many examples of supercoil-induced B-Z transitions have been reported and are reviewed by Rich et al. (1984). Formation of Z-DNA segments with negatively supercoiled, right-handed plasmid genomes allowed identification and characterization of junction regions between B-DNA and Z-DNA (Peck et al. 1982; Singleton et al. 1983; Kilpatrick et al. 1983).

Negative DNA supercoiling was found to be highly potent in stabilizing Z-DNA under physiological ionic conditions. DNA sequences that could not be stabilized as Z-DNA in a chemically unmodified form at high salt concentrations were nevertheless stabilized as Z-DNA by topological stress. This included d(CA) • d(GT) sequences as well as more random compositions of natural alternating purine/pyrimidine sequences (Nordheim et al. 1982a; Nordheim and Rich 1983a,b; Haniford and Pulleyblank 1983). An inverse correlation could be found between ease of Z-DNA formation and length of the potential Z-DNA segment. Such studies also indicated that binding of anti-Z-DNA antibodies to supercoil-induced structural DNA alterations is not limited to segments of strictly alternating purines and pyrimidines (Nordheim et al. 1982a; Azorin et al. 1983; Nordheim and Rich 1983b). A careful analysis of free energy parameters involved in B-Z transitions was made possible with supercoiled plasmids (Peck and Wang 1983).

An extreme degree of topological stress is generated in form V DNA (Stettler et al. 1979) where complementary single-stranded DNA circles are allowed to anneal. The generation of significant Z-DNA segments was indicated under these conditions (Brahms et al. 1982; Lang et al. 1982; Pohl et al. 1982).

The efficiency of Z-DNA formation upon DNA supercoiling is of conceptual importance with respect to the potential biological relevance of Z-DNA. DNA supercoiling has been shown in vitro to have profound effects on DNA recombination, DNA replication, and initiation of transcription. In vivo topological stress has been documented in prokaryotic

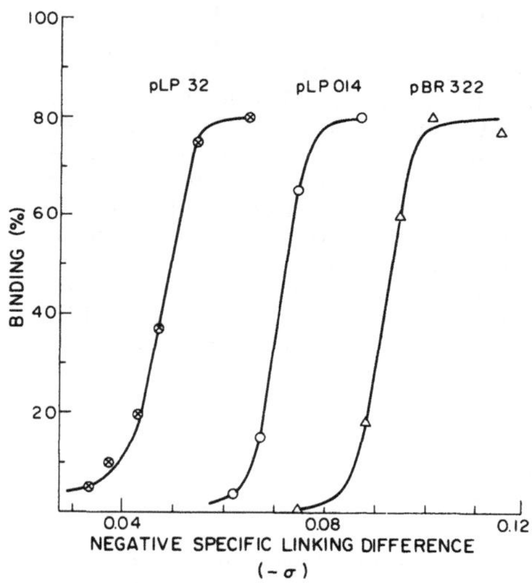

Fig. 3. Binding of anti-Z-DNA antibodies to supercoiled plasmid DNA's at increasing superhelical densities (≅ increasing negative specific linking differences). (For details see Nordheim et al. 1982a)

cells, where the maintenance of negative superhelicity is ensured by concerted action of DNA topoisomerases and DNA gyrase (Menzel and Gellert 1983). It remains to be established whether the bacterial intracellular negative superhelical densities are sufficiently high to stabilize Z-DNA segments.

Even less is known about the existence of toplogical stress in chromatin of eukaryotic cells. Although it is generally assumed that most topological stress is absorbed upon DNA wrapping around nucleosome core histones, some recent experiments suggested the presence of relaxable superhelical turns in eukaryotic cells (Luchnik et al. 1982; Ryoji and Worcel 1984). Thus, the possibility of Z-DNA stabilization by topological stress in pro- and eukaryotic cells remains a highly speculative but very intriguing idea.

For analytical purposes, on the other hand, characterized Z-DNA segments may, when introduced into cells, serve as valuable probes for the determination of intracellular superhelical densities.

6 Interaction of Z-DNA and Proteins

Polymorphic DNA conformations may either be induced by binding of proteins or could represent recognition signals for specific protein attachment. Therefore, the analysis of specific interactions between Z-DNA and proteins is expected to yield insight into potential biological functions of Z-DNA.

As outlined in the previous chapter, the recognition of Z-DNA by conformation-specific antibodies represents a model system for the study of left-handed DNA/protein interactions. Interestingly, different isolates of monoclonal antibody species can distinguish between various structural components of Z-DNA (e.g., chemical modifications; Möller et al. 1982) or nucleotide composition of Z helices (A. Nordheim et al., manuscript in preparation).

Polypeptides (poly-L-arginine, poly-L-lysine) were shown to exhibit differential effects on the equilibrium of Z-DNA versus B-DNA (Klevan and Schumaker 1982). Similarly, an investigation by Russel et al. (1983) demonstrated stabilization of Z-DNA by protamine and the nucleosome core histones H3, H4, H2a, and H2b, whereas histones H1 and H5 caused Z-DNA to revert to a right-handed double helix. Furthermore, it appears as if Z DNA is generally more difficult to package into a nucleosome configuration (Nickol et al. 1982), although successful nucleosome-core assembly has been reported at low yield (Miller et al. 1983). The use of Z-DNA affinity chromatography yielded isolation of protein preparations that exhibit selective binding to left-handed DNA polymers (Nordheim et al. 1982b). The functional analysis of such proteins by biochemical, immunochemical, and molecular genetic approaches is expected to help elucidate the biological importance of Z-DNA.

The postulated formation of Z-DNA during establishment of recombinational paranemic joints by the *Ustilago* rec 1 protein (Kmiec and Holloman 1984; see Sect. 7) may suggest Z-DNA binding properties to be associated with rec 1.

7 Conclusion and Biological Considerations

A detailed structural analysis of left-handed Z-DNA has generated insights into the conformational flexibility that is inherent in the double-helical DNA molecule (Cantor 1981; Zimmerman 1982; Rich et al. 1984). The degree to which a given DNA segment can exist in a structurally polymorphic form appears strongly dependent on the base sequence composition of the segment (Wang et al. 1979; Frederick et al. 1984; Wu and Crothers 1984). Accordingly, Rich (1983) pointed out that DNA contains conformational information which is expressed by certain nucleotide sequences in adopting altered double helical conformations. In the case of Z-DNA, the distribution of purine/pyrmidine sequences determines the Z-forming potential, although more will have to be learned to understand the nucleotide sequence "code" for Z-DNA formation (A.H-J. Wang et al., manuscript in preparation).

Although well characterized in vitro, Z-DNA still cannot be directly linked to any biological role; the in vivo occurrence of Z-DNA is still not proven. However, natural DNA sequences have been demonstrated to carry the potential to adopt Z-DNA in vitro under conditions that are expected to occur in vivo. In this regard, DNA supercoiling and specific protein binding suggest themselves as the most likely factors for stabilizing Z-DNA in vivo.

Z-DNA in itself may represent a recognition signal for certain proteins to bind at specific DNA sites. Such Z-DNA/protein complexes – or Z-DNA alone – may exert an effect on nucleosome formation (Nickol et al. 1982; Miller et al. 1983) and eukaryotic chromatin structure. Support for this speculation comes from in vitro identification of anti-Z-DNA antibody binding sites within the regulatory segment of the viral SV40 genome (Nordheim and Rich 1983b; this segment exists as a nucleosome-free region in the chromatin of SV40 minichromosomes.

There exists an intimate relationship between DNA topology and double helical DNA conformation. DNA supercoiling is a very strong driving force for the induction and stabilization of Z-DNA. This fact raises the intriguing possibility that Z-DNA formation may function to transiently releave superhelical stress within a toplogical domain. At the same time much Z-DNA formation would represent a mechanism for storage of superhelical energy. This concept of Z-DNA acting as an "energy sink" would have direct bearing on cellular processes, since DNA supercoiling is an important factor in controlling prokaryotic gene expression. However, whether intracellular superhelical densities are high enough to stabilize Z-DNA is still unclear for prokaryotic and eukaryotic systems.

Certain biological processes, like formation of paranemic joints during homologous DNA recombination, are expected to generate a localized torsional stress. Such phenomena might be accompanied by the formation of Z-DNA, which was recently investigated by Kmiec and Holloman (1984). A concerted action between topological stress and a DNA-binding protein, the rec 1 protein, may determine the B-Z equilibrium in this situation.

New advances in the development of methods for detecting Z-DNA will help elucidate questions related to the in vivo occurrence and biological significance of Z-DNA.

Acknowledgements. The author's research at the ZMBH is funded by grants from the BMFT and the Fonds der Chemischen Industrie. The help of R. Franklin in preparing this manuscript is greatly appreciated.

References

Arndt-Jovin DJ, Robert-Nicoud M, Zarling DA, Greider C, Weimer E, Jovin TM (1983) Left-handed Z-DNA in bands of acid-fixed polytene chromosomes. Proc Natl Acad Sci USA 80:4344-4348

Arnott S, Chandrasekaran R, Birdsall DL, Leslie AGW, Ratliff RL (1980) Left-handed DNA helices. Nature (Lond) 283:743-745

Azorin F, Nordheim A, Rich A (1983) Formation of Z-DNA in negatively supercoiled plasmids is sensitive to small changes in salt concentration within the physiological range. EMBO J 2:649-655

Barton JK, Basile LA, Danishefsky A, Alexandrescu A (1984) Chiral probes for the handedness of DNA helices: Enantiomers of tris (4,7-diphenyl phenanthroline) ruthenium (II). Proc Natl Acad Sci USA 81:1961-1965

Becker MM, Wang JC (1984) Use of light for footprinting DNA in vivo. Nature (Lond) 309:682-687

Behe W, Felsenfeld G (1981) Effects of methylation on a synthetic polynucleotide: The B-Z transition in poly(dG-m^5dC)$\cdot$poly(dG-m^5dC). Proc Natl Acad Sci USA 78: 1619-1623

Brahms S, Vergue J, Brahms JG, Di Capua E, Bucher P, Koller T (1982) Natural DNA sequences can form left-handed helices in low salt solution under conditions of topological constraint. J Mol Biol 162:473-493

Cantor CR (1981) DNA choreography. Cell 25:293-295

Castleman H, Hanau LH, Erlanger BF (1983) Stabilization of $(dG\text{-}dC)_n \cdot (dG\text{-}dC)_n$ in the Z conformation by a crosslinking reaction. Nucleic Acids Res 11:8421-8429

Church G, Gilbert W (1984) Genomic sequencing. Proc Natl Acad Sci USA 81:1991-1995

Crick FHC (1976) Linking numbers and nucleosomes. Proc Natl Acad Sci USA 73:2639-2643

Drew HR, Travers AA (1984) DNA structural variations in the E. coli tyr T promoter. Cell 37:491-502

Drew HR, Takano T, Tanaka S, Itakura K, Dickerson RE (1980) High salt d(CpGpCpG), a left-handed Z-DNA double helix. Nature (Lond) 286:567-573

Frederick CA et al. (1984) Kinked DNA in crystalline complex with EcoRI endonuclease. Nature (Lond) 309:327-331

Hanau LH, Santella R, Grunberger D, Erlanger BF (1984) An immunochemical examination of acetylaminofluorene-modified poly(dG-dC)$\cdot$poly(dG-dC) in the Z-conformation. J Biol Chem 259:173-178

Haniford DB, Pulleyblank DE (1983) Facile transition of poly [d(TG)$\cdot$d(CA)] into a left-handed helix in physiological conditions. Nature (Lond) 302:632-634

Hill RJ, Stollar BD (1983) Dependence of Z-DNA antibody binding to polytene chromosomes on acid fixation and DNA torsional strain. Nature (Lond) 305:338-340

Jovin TM, McIntosh LP, Arndt-Jovin DJ, Zarling DA, Robert-Nicoud M, van de Sande JH, Jorgenson KF (1983) Left-handed DNA from synthetic polymers to chromosomes. J Biomol Struct Dynam 1:21-57

Klevan L, Schumaker VN (1982) Stabilization of Z-DNA by poly-arginine near physiological ionic strength. Nucl Acids Res 10:6809-6817

Kilpatrick MW, Wei CF, Gray HR, Wells RD (1983) Bal 31 nuclease as a probe in concentrated salt for the B-Z DNA junction. Nucl Acids Res 11:3811-3822

Klysik J, Stirdivant SM, Larson JE, Hart PA, Wells RD (1981) Left-handed DNA in restriction fragments and a recombinant plasmid. Nature (Lond) 290:672-677

Kmiec EB, Holloman WK (1984) Synapsis promoted by Ustillago Rec I protein. Cell 36: 593-598

Lafer EM, Möller A, Nordheim A, Stollar BD, Rich A (1981) Antibodies specific for left-handed Z-DNA. Proc Natl Acad Sci USA 78:3546-3550

Lafer EM, Möller A, Valle RPC, Nordheim A, Rich A, Stollard BD (1983) Antibody recognition of Z-DNA. Cold Spring Harbor Symp Quant Biol 47:155-162

Lang MC, Malfoy B, Freund AM, Daune M, Leng M (1982) Visualization of Z sequences in form V of pBR322 by immuno-electron microscopy. EMBO J 1:1149-1153

Lemeunier F, Derbin C, Malfoy B, Leng M, Taillandier E (1982) Identification of left-handed Z-DNA by indirect immunofluorescence in polytene chromosomes of Chironomus thummi. Exp Cell Res 141:508

Lipps HJ, Nordheim A, Lafer EM, Ammermann D, Stollar BD, Rich A (1983) Antibodies against Z DNA react with the macronucleus but not the micronucleus of the hypotrichous ciliate Stylonychia mytilus. Cell 32:435-441

Luchnik AN, Bakayev VV, Zbarsky IB, Georgiev GP (1982) Elastic torsional strain in DNA within a fraction of SV40 minichromosomes: relation to transcriptionally active chromatin. EMBO J 1:1353-1358

Malfoy B, Hartmann B, Leng M (1981) The B to Z transition of poly(dG-dC)·poly(dG-dC) modified by some Platinum derivatives. Nucleic Acids Res 9:5659-5669

Malfoy B, Leng M (1981) Antiserum to Z-DNA. FEBS Lett 132:45-48

Malfoy B, Rousseau N, Leng M (1982) Interaction between antibodies to Z-form deoxyribonucleic acid and double-stranded polynucleotides. Biochemistry 21:4563-5467

McIntosh LP, Grieger I, Eckstein F, Zarling DA, van de Sande JH, Jovin TM (1983) Left-handed helical conformation of poly[d(A-m^{5}C)·d(G-T)]. Nature (Lond) 304: 83-86

Menzel R, Gellert M (1983) Regulation of the genes for E. coli DNA gyrase: homeostatic control of DNA supercoiling. Cell 34:105-113

Miller FD, Rattner JB, van de Sande JH (1983) Nucleosome-core assembly on B and Z forms of poly[d(G-m^{5}C)]. Cold Spring Harbor Symp Quant Biol 47:571-575

Möller A, Nordheim A, Nichols SR, Rich A (1981) 7-methylguanine residue in poly(dG-dC)·poly(dG-dC) facilitates Z-DNA formation. Proc Natl Acad Sci USA 78:4777-4781

Möller A, Gabriels JE, Lafer EM, Nordheim A, Rich A, Stollar BD (1982) Monoclonal antibodies recognize different parts of Z-DNA. J Biol Chem 257:12081-12085

Möller A, Nordheim A, Kozlowski S, Patel DJ, Rich A (1984) Bromination stabilizes poly(dG-dC) in the Z-DNA form under low-salt conditions. Biochemistry 23:54-62

Morgenegg G, Celio MR, Malfoy B, Leng M, Kuenzle CC (1983) Z-DNA immunoreactivity in rat tissues. Nature (Lond) 303:540-543

Nickol J, Behe M, Felsenfeld G (1982) Effect of the B-Z transition in poly(dG-m^{5}dC)·poly(dG-m^{5}dC) on nucleosome formation. Proc Natl Acad Sci USA 79:1771-1775

Nordheim A, Rich A (1983a) The sequence $(dC\text{-}dA)_n \cdot (dG\text{-}dT)_n$ forms left-handed Z-DNA in negatively supercoiled plasmids. Proc Natl Acad Sci USA 80:1821-1825

Nordheim A, Rich A (1983b) Negatively supercoiled simian virus 40 DNA contains Z-DNA segments within transcriptional enhancer sequences. Nature (Lond) 303:674-679

Nordheim A, Pardue ML, Lafer EM, Möller A, Stollar BD, Rich A (1981) Antibodies to left-handed Z-DNA bind to interband regions of Drosophila polytene chromosomes. Nature (Lond) 294:417-422

Nordheim A, Lafer EM, Peck LJ, Wang JC, Stollar BD, Rich A (1982a) Negatively supercoiled plasmids contain left-handed Z-DNA segments as detected by specific antibody binding. Cell 31:309-318

Nordheim A, Tesser P, Azorin F, Kwon YH, Möller A, Rich A (1982b) Isolation of Drosophila proteins that bind selectively to left-handed Z-DNA. Proc Natl Acad Sci USA 79:7729-7733

Nordheim A, Hao WM, Wogan GN, Rich A (1983) Salt-induced conversion of B-DNA to Z-DNA inhibited by aflatoxin B1. Science (Wash DC) 219:1434-1436

Pardue ML, Nordheim A, Lafer EM, Stollar BD, Rich A (1983) Z-DNA and the polytene chromosome. Cold Spring Harbor Symp Quant Biol 47:171-176

Patel DJ, Channel LL, Pohl FM (1979) "Alternating B-DNA" conformation for the oligo (dG-dC) duplex in high-salt solution. Proc Natl Acad Sci USA 76:2508-2511

Patel DJ, Kozlowski SA, Nordheim A, Rich A (1982) Right-handed and left-handed DNA: Studies of B- and Z-DNA by using proton nuclear Overhauser effect and pNMR. Proc Natl Acad Sci USA 79:1413-1417

Peck LJ, Wang JC (1983) Energetics of B-to-Z transition in DNA. Proc Natl Acad Sci USA 80:6206-6210

Peck LJ, Nordheim A, Rich A, Wang JC (1982) Flipping of cloned $d(pCpG)_n \cdot d(pCpG)_n$ DNA sequences from right- to left-handed helical structure by salt, co(III), or negative supercoiling. Proc Natl Acad Sci USA 79:4560-4564

Pohl FM (1976) Polymorphism of a synthetic DNA in solution. Nature (Lond) 260:365-366

Pohl FM (1983) Allosteric DNA. In: Sund, Vecger (eds) Mobility and recognition in cell biology. De Gruyter, New York

Pohl FM, Jovin TM (1972) Salt-induced co-operative conformational change of a synthetic DNA: Equilibrium and kinetic studies with poly(dG-dC). J Mol Biol 67:375-396

Pohl FM, Ranade A, Stockburger M (1983) Laser Raman scattering of two double-helical forms of poly(dG-dC). Biochim Biophys Acta 335:85-92

Pohl FM, Thomae R, DiCapua E (1982) Antibodies to Z-DNA interact with form V DNA. Nature (Lond) 300:545-546

Ramstein J, Leng M (1980) Salt-dependent dynamic structure of poly(dG-dC)·poly(dG-dC). Nature (Lond) 288:413-414

Rich A (1983) Right-handed and left-handed DNA: conformational information in genetic material. Cold Spring Harbor Symp Quant Biol 47:1-12

Rich A, Nordheim A, Wang AH-J (1984) The chemistry and biology of left-handed Z-DNA. Annu Rev Biochem 53:751

Robert-Nicoud M, Arndt-Jovin DJ, Zarling DA, Jovin TM (1984) Immunological detection of left-handed Z-DNA in isolated polytene chromosomes. Effects of ionic strength, pH, temperature and topological stress. EMBO J 3:721-731

Russel WC, Precious B, Martin SR, Bayley PM (1983) Differential promotion and suppression of Z-B transitions in poly[d(G-C)] by histone subclasses, polyamino acids and polyamines. EMBO J 2:1647-1653

Ryoji M, Worcel A (1984) Chromatin assembly in Xenopus oocytes: in vivo studies. Cell 37:21-32

Sage E, Leng M (1980) Conformation of poly(dG-dC)·poly(dG-dC) modified by the carcinogens N-acetoxy-N-acetyl-2-aminofluorene and N-hydroxy-N-2-aminofluorene. Proc Natl Acad Sci USA 77:4597-4601

Santella RM, Grunberger D, Weinstein IB, Rich A (1981) Induction of the Z conformation in poly(dG-dC)·poly(dG-dC) by binding of N-2-acetylaminofluorene to guanine residues. Proc Natl Acad Sci USA 78:1451-1455

Singleton CK, Klysik J, Stirdivant SM, Wells RD (1982) Left-handed Z-DNA is induced by supercoiling in physiological ionic conditions. Nature (Lond) 299:312-316

Singleton CK, Klysik J, Wells RD (1983) Conformational flexibility of junctions between contiguous B- and Z-DNAs in supercoiled plasmids. Proc Natl Acad Sci USA 80:2447

Stettler UH, Weber H, Koller T, Weissmann C (1979) Preparation and characterization of form V DNA, the doublex DNA resulting from association of complementary, circular single-stranded DNA. J Mol Biol 131:21-40

Sutherland JC, Griffin KP, Keck PC, Takacs PZ (1981) Z-DNA: vacuum ultraviolet circular dichroism. Proc Natl Acad Sci USA 78:4801-4804

Thamann TJ, Lord RC, Wang AH-J, Rich A (1981) The high-salt form of poly(dG-dC)·poly(dG-dC) is left-handed Z-DNA: Raman spectra of crystals and solutions. Nucleic Acids Res 9:5443-5457

Thomae R, Beck S, Pohl FM (1983) Isolation of Z-DNA-containing plasmids. Proc Natl Acad Sci USA 80:5500-5553

Thomas TJ, Bloomfield VA (1983) Chain flexibility and hydrodynamics of the B and Z forms of poly(dG-dC)·poly(dG-dC). Nucleic Acids Res 11:1919-1930

Ushay HM, Santella RM, Caradonna JP, Grunberger D, Lippard SJ (1982) Binding of [(dien)PtCl]Cl to poly(dG-dC)·poly(dG-dC) facilitates the B → Z conformational transition. Nucleic Acids Res 10:3573-3588

van de Sande JH, Jovin TM (1982) Z* DNA, the left-handed helical form of poly[d(G-C)] in $MgCl_2$-ethanol, is biologically active. EMBO J 1:115-120

Viegas-Péquignot E, Derbin C, Malfoy B, Taillander E, Leng M, Dutrillaux B (1983) Z-DNA immunoreactivity in fixed metaphase chromosomes of primates. Proc Natl Acad Sci USA 80:5890-5894

Wang AH-J, Quigley GJ, Kolpak FJ, Crawford JL, van Boom JH, van der Marel G, Rich A (1979) Molecular structure of a left-handed double helical DNA fragment at atomic resolution. Nature (Lond) 282:680-686

Wang AH-J, Quigley GJ, Kolpak FJ, van der Marel G, van Boom JH, Rich A (1981) Left-handed double helical DNA: variations in the backbone conformation. Science (Wash DC) 211:171-176

Wang JC (1980) Superhelical DNA. Trends Biochem Sci 5:219-222

Watson DJ, Crick FH (1953) Molecular structure of nucleic acid: a structure for deoxyribose nucleic acid. Nature (Lond) 171:737-738

Wells RD, Miglietta JJ, Klysik J, Larson JE, Stirdivant SM, Zacharias W (1982) Spectroscopic studies on acetylaminofluorene-modified $(dT\text{-}dG)_n \cdot (dC\text{-}dA)_n$ suggest a left-handed conformation. J Biol Chem 257:10166-10171

Wu H-M, Crothers DM (1984) The locus of sequence-directed and protein-induced DNA bending. Nature (Lond) 308:509-513

Zacharias W, Larson JE, Klysik J, Stirdivant SM, Wells RD (1982) Conditions which cause the right-handed to left-handed DNA conformational transitions. J Biol Chem 257:2775-2782

Zarling DA, McIntosh LP, Arndt-Jovin DJ, Robert-Nicoud M, Jovin TM (1984) Interaction of Anti-poly [d(G-Br5C)] IgG with synthetic, viral and cellular Z-DNA. J Biomol Struct Dynam 1:1081-1107

Zarling DA, Arndt-Jovin DJ, Robert-Nicoud M, McIntosh LP, Thomae R, Jovin TM (to be published 1984) Immunological recognition of synthetic and natural left-handed Z-DNA conformations and sequences. J Mol Biol

Zimmermann SB (1982) The three-dimensional structure of DNA. Annu Rev Biochem 51:395-427

DNA Rearrangements in Varions Organisms

Antigenic Variation in African Trypanosomes

H. EISEN, S. LONGACRE, and G. BUCK[1]

1 Introduction

African trypanosomes, which are extracellular parasitic hemoflagellates, are responsible for several diseases in humans and domestic animals. The most widely known of these diseases is human sleeping sickness which is found in both East and West Africa. Like most mammalian parasites, the African trypanosomes have evolved a mechanism for escaping the host immune system, thus allowing them to persist in the immune competent mammal. For the African trypanosomes this is done by a mechanism known as antigenic variation, which consists of periodically changing the predominant and perhaps unique antigen presented on the surface of the parasite. These variable antigens are composed of immunologically distinct glycoproteins and are called variable surface glycoproteins (VSG's). Although the total repertoire of different VSG's which a given trypanosome species is capable of expressing is not known, over 100 have been characterized in a single clone of *T. equiperdum* (Capbern et al. 1977). Individual trypanosomes express only one VSG at a time, and the frequency of switching has been estimated to be 10^{-4} - 10^{-6} per cell per generation (Doyle 1977).

There are several interesting questions concerning antigenic variation: with respect to the VSG repertoire; are all the different VSG's encoded by individual genes in the trypanosome genome, or can VSG genes be created by combinatorial processes analogous to those involved in the formation of immunoglobulin genes? If there is a combinatorial process for VSG gene generation how does it work? Concerning the regulation of VSG gene expression; how are the VSG genes activated and inactivated and what is the mechanism which allows expression of only a single gene at any given time? Finally, is the variation itself a stochastic process or a regulated process?

T. equiperdum. T. equiperdum is a parasite of horses. Unlike other African trypanosomes, it is not transmitted by the tsetse fly, but has evolved into a venereal disease (Capbern et al. 1977). The parasite can be grown in laboratory animals. It causes an acute infection in mice and rats, thus allowing the isolation of parasite clones and the production of large quantities of material. In rabbits, it causes a chronic infection which lasts 2 - 3 months with the constant appearance of different variant antigen types (VAT's). *T. equiperdum* has two interesting properties for the laboratory (Capbern et al. 1977). First, in infected rabbits, the different VAT's appear in a loosely defined order and have been arbitrarily divided into early, middle, and late groups. The appearance of the different VAT's within each group does not appear to be ordered, but the time of appearance of a given VAT in an individual rabbit is more or less constant. Secondly, infection of a naîve rabbit

[1]Unité d'immunoparasitologie, Institut Pasteur, 28, rue du Dr. Roux, 75015 Paris, France

35. Colloquium - Mosbach 1984
The Impact of Gene Transfer Techniques in Eukaryotic Cell Biology

with any cloned VAT results in the rapid appearance of VAT-1 followed by the other early VAT's. It thus appears that, in a nonimmune environment, the parasites have a mechanism for resetting their VSG gene clock to zero. We have used these properties to study the molecular mechanism of antigenic variation in *T. equiperdum*.

2 Results and Discussion

Genomic Rearrangements and VSG Gene Activation. Using cDNA clones prepared from VSG mRNA's several laboratories have examined the genomic rearrangements which accompany activation of VSG genes in African trypanosomes. It was shown quite early that in the majority of cases, expression of a given VSG gene is accompanied by the duplicative transposition of a silent basic copy gene (BC) to form an expression-linked copy of the gene (ELC) (see Borst and Cross 1982, for review). Furthermore DNAase I digestion of the chromatin of the parasites results in the selective degredation of the ELC, suggesting that it is this copy of the VSG gene which is transcribed (Pays et al. 1981; Longacre et al. 1983). All of the ELC's thus far described are located near chromosome telomeres in the orientation 5'-3'-telomere. The DNA situated between the VSG gene and the telomere shows peculiar properties. Although these regions can be up to 30 kilobasis (kb) long, they do not contain sites for restriction endonucleases (Williams et al. 1982; De Lange and Borst 1982; Raibaud et al. 1983). In *T. equiperdum* the deoxycytidines in these regions possess an as yet unidentified modification which renders them extremely dense and appears to destabilize the hydrogen bonding between dC and dG (Raibaud et al. 1983).

The BC's of VSG genes may or may not be located near telomeres. When they are near telomeres they can be in either orientation and may be activated without being duplicated (see below) (Williams et al. 1979; Laurent et al. 1983). Those VSG BC genes which are not situated near telomeres invariably form ELC's when they are activated, suggesting that only the telomerically located copy of the gene can be transcribed. In *T. brucei*, there appears to be a unique site into which the duplicated VSG gene copies are inserted. This finding gave rise to a simple model for VSG gene activation which also accounts for the finding that only a single gene is expressed in a given trypanosome (Borst and Cross 1982). This model for VSG gene regulation is similar to the cassette model proposed to explain mating type regulation in yeast (Hicks et al. 1977). It proposes that for expression it is necessary and sufficient for a copy of a VSG gene to be inserted into a unique telomeric expression site. If there is only one such site per genome then only one gene can be expressed at a time.

Several results are difficult to explain by the simple single expression site model. First, in *T. brucei* some BC genes are situated in a 5'-3' orientation near telomeres. These genes can be activated either via a duplicative transposition or *without* apparent rearrangements (Williams et al. 1979; Laurent et al. 1984). In *T. equiperdum* all of the VSG genes thus far examined give rise to ELC's when they are expressed, even when the BC is at a telomere. However, examination of the VSG-1 ELCs in eight independent VSG-1 expressors has revealed four different expression sites (Longacre et al. 1983 and unpublished results). Furthermore, more than one expression site can be occupied at a time even though only one appears to be active. This was seen when BoTat (Bordeaux Trypanozoon antigen type)-1 of *T. equiperdum* was allowed to vary to other variants. In 50% of these VAT's the previously expressed VSG-1 ELC was still present in the new VAT's (Buck et al. 1984). Although no rearrangements could be detected in the 10 kb situated 5' to the VSG-1 gene in the residual ELC (R-ELC) it was clearly inactivated as judged by Northern blot analysis and DNAase I sensitivity. BoTat-2, one of the early VATs which contains an R-ELC of the VSG-1 gene was allowed to vary back to VAT-1 in order to determine whether R-ELC could be reactivated. Five independent BoTat-2 derived VSG-1 expressing trypanosome clones were examined and three were found to have reactivated the VSG-1 R-ELC. Thus, the presence of an ELC is not sufficient for transcription of a VSG gene. Therefore other regulatory mechanisms must be implicated.

It is possible to rationalize the apparent multiple expression sites in *T. equiperdum* and the in situ regulation of some VSG genes with the finding that only one gene is active at a time. It has been proposed that an "activating element" situated either 5' or 3' to an expression site is necessary for the site to be active (Buck et al. 1984). If there were only one such element per genome, expression sites could be regulated by chromosome and exchange, thus activating one VSG gene by placing the duplicated gene either 3' or 5' to the element. This would also result in the creation of an R-ELC of the previously expressed VSG gene. To test this model we examined the fate of the VSG-2 gene in the BoTat-2 derived VSG-2 expressors in which the VSG-1 R-ELC had been reactivated. If the VSG-1 R-ELC were reactivated by a reciprocal and exchange a VSG-2 R-ELC should be created. However, this was not the case for the three trypanosome clones examined. It thus seems unlikely that expression sites are activated by simple chromosome and exchanges. It appears more likely that the different expression sites are regulated without obvious secondary rearrangements.

Mode of Duplication, Hybrid Genes and Order of Expression. It has been demonstrated that some ELC's are not faithful copies of their respective BC's but are rather hybrids (Bernards et al. 1981; Pays et al. 1983). In most cases the 5' portion of the ELC is derived from the BC while the 3' segment comes from another gene. However, in *T. brucei* an ELC composed of segments derived from three genes has been reported (Pays et al. 1983). We have examined the formation of hybrid genes in *T. equiperdum* using the S1 protection method (S. Longacre, in preparation). This consists of hybridizing a given VSG mRNA with denatured DNA from either a homologous or a heterologous VAT under conditions where only RNA/DNA hybrids are stable. The reactions are then treated with S1 endonuclease to digest remaining single-stranded DNA, the RNA degraded and the protected fragments of DNA analyzed by Southern blot analysis using, as probe, cDNA clones corresponding to the mRNA used. Thus, if the ELC is a faithful copy of the BC, a messenger-sized fragment will be protected independently of the origin of the genomic DNA. On the other hand, if the ELC is a hybrid gene a messenger-sized fragment will only be protected when homologous DNA is used. DNA which only contains the BC will give a shorter protected fragment. Such an analysis was performed on seven trypanosome clones expressing early antigens and three clones expressing middle and late antigens.

For all of the early VSG genes, the ELC appears to be identical to the BC, suggesting that the basic copy is entirely duplicated. On the other hand all of the ELC's for the middle and late VSG genes are hybrid. The formation of the hybrid ELC appears to be a specific event since two independently isolated VAT's expressing VSG-78 (a middle antigen) have indistinguishable hybrid ELC's. Thus the 3' ends of these VSG-78 ELC's are derived from the same gene which is not the BC from which the 5' ends are generated.

The results presented above suggest a model to explain how the order of VSG gene expression is determined in *T. equiperdum*. This is shown in Figure 1. It is proposed that the BC's of the early VSG genes have homology with the expression sites, both 5' and 3' to the coding regions. Thus, these genes can be inserted into an expression site independent of which gene is occupying the site. However the BC's of

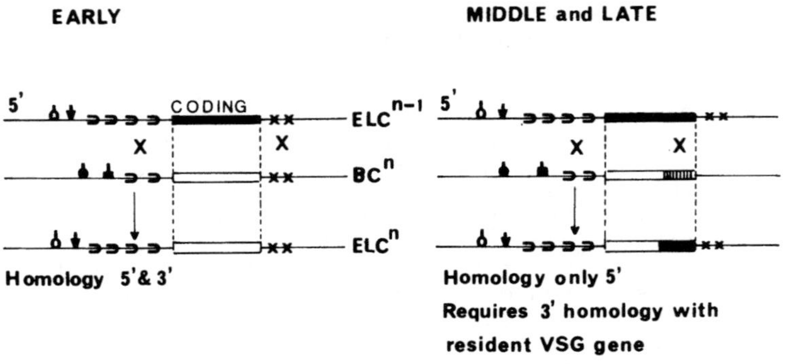

Fig. 1. Model for VSG gene integration into expression sites in *T. equiperdum*. *Symbols* represent the presence or absence of homologies between VSG gene flanking regions in the BC and ELC

the middle and late VSG genes have lost their homology with the expression sites either on one side or the other or both of the coding regions. If insertion of a VSG gene into an expression site requires homology, they could not be inserted unless they found homology with the resident VSG gene itself. Thus, the middle and late VSG genes would not be able to be duplicated and inserted into an expression site until that site was occupied by a gene with which they had homology. This would lead to an ordered expression of the middle and late genes. It would also explain why the early VSG genes are expressed early.

The mechanism of VSG gene duplication is not yet understood. Pays et al. (1983) have suggested that the duplicative transposition which creates an ELC is a form of gene conversion. Formally, this must be the case since the result is a new copy of the gene. However, the conversion event must span large distances of nonhomology (in some cases more than 3 kb). It has not been ruled out that a copy of the BC is made and that this duplicate then recombines with a telomeric expression site. These sorts of model suppose that the duplication event requires DNA homology between the expression sites and the flanking regions of the BC's. Such homology has recently been demonstrated for a *T. brucei* VSG gene where similar repeats of approximately 76 base pairs were observed 5' to both the ELC and the BC of a VSG gene (Campbell et al. 1984).

References

Bernards A, Van der Ploeg L, Frasch A, Borst P, Boothroyd J, Coleman S, Cross G (1981) Cell 27:497

Borst P, Cross GAM (1982) Cell 29:291

Buck G, Longacre S, Raibaud A et al. (1984) Nature (Lond) 307:563

Campbell D, van Bree M, Boothroyd J (1984) Nucleic Acid Res 12:26

Capbern A, Giroud C, Baltz T, Mattern P (1977) Exp Parasitol 42:6

De Lange T, Borst P (1982) Nature (Lond) 299:451

Doyle JJ (1977) Antigenic variation in the salivarian trypanosomes. In: Miller L, Pino J, Mekelvey J (eds) Immunity to blood parasites in animals and man. Plenum, New York, p 27

Hicks J, Strathern J, Herskowitz I (1977) The cassette model for mating type interconversion. In: Bukhari A et al. (eds) DNA issertion, elements, plasmids and episomes. Cold Spring Harbor, New York, p 457

Laurent M, Pays E, Magnus E, Van Meirvenne N, Matthyssens G, Williams R, Steinert M (1983) Nature (Lond) 302-263

Laurent M, Pays E, Delinte K, Magnus E, Van Meirvenne N, Steinert M (1984) Nature (Lond) 308:370

Longacre S, Hibner U, Raibaud A, Eisen H, Baltz T, Giroud C, Baltz D (1983) Mol Cell Biol 3:399

Pays E, Lheureux M, Steinert M (1981) Nature (Lond) 292:262

Pays E, Van Assel S, Laurent M, Darville M, Vervoort T, Van Meirvenne N, Steinert M (1983) Cell 34:371

Raibaud A, Gaillard C, Longacre S, Hibner U, Buck G, Bernardi G, Eisen H (1983) Proc Natl Acad Sci USA 80:4603

Williams R, Young J, Majiwa F (1979) Nature (Lond) 282:847

Williams R, Young J, Majiwa F (1982) Nature (Lond) 299:417

The Plant Transposable Elements *Tam1*, *Tam2* and *Spm-I8*

H. SAEDLER, U. BONAS, A. GIERL, B. J. HARRISON[1], R. B. KLÖSGEN, E. KREBBERS, P. NEVERS, P. A. PETERSON[2], ZS. SCHWARZ-SOMMER, H. SOMMER, K. UPADHYAYA[3], and U. WIENAND[4]

1 Introduction

Genetic instability of loci affecting a directly observable property such as pigmentation is often responsible for the strikingly variegated appearance of many plant parts. Variegated plants of this kind may also frequently produce phenotypically wild-type or nearly wild-type progeny due to reversion of mutable allele in germinal tissue. Although variegated plants have been a source of fascination for centuries, as documented by the detailed description of multi-colored maize kernels as early as 1588 in Jacob Thedor von Bergzabern's herbal, an understanding of the physical basis of mutability began to emerge only 35 years ago as a result of the pioneer work of McClintock on maize. She proposed (1950) that discrete, transposable genetic elements are responsible for the instability of certain mutable alleles in maize and referred to them as "controlling elements" (1956a,b) because of their ability to influence the expression of loci at which they integrate. At the same time mutable alleles with characteristics similar to those of maize were described by classical genetic means in innumerable other plants including *Antirrhinum majus* (see Nevers et al. 1984 for review). With the help of molecular cloning procedures we are now able to isolate as physical entities the elements previously proposed by classical genetic means, three of which will be described below.

2 Alleles of *Antirrhinum majus* Affecting Flower Pigmentation

In a series of experiments involving pollen mutagenesis Stubbe and Baur isolated a number of mutants of *A. majus* that exhibit unusual red and white variegated flowers (described in Kuckuck 1936). One of the mutations seemed to be due to instability of the *nivea* locus and was named *nivea-recurrens* (Kuckuck 1936). The *niv-rec* mutant was further characterized genetically by Harrison and Carpenter (1973a) and analyzed at a molecular level in our laboratory. We shall refer to this mutant as *niv-53*.

We selected the *niv-53* mutant for detailed analysis since the *nivea* locus is molecularly well characterized. It has been shown to encode the enzyme chalcone synthase (CHS), a key enzyme in anthocyanin biosynthesis (Spribille and Forkman 1982). A heterologous cDNA clone of the chalcone synthase gene of parsley enabled us to determine the molecular nature of the *niv-53* mutation (Wienand et al. 1982). The *niv-53*

[1]John Innes Institute, Colney Lane, Norwich NR4 7UH, UK
[2]Iowa State University, Agronomy Dept., Ames, Iowa 50010, USA
[3]School of Life Sciences, Jawaharlal Nehru University, New Delhi-110067, India
[4]Max-Planck Institut für Züchtungsforschung, Egelspfad, 5000 Köln 30, FRG

35. Colloquium - Mosbach 1984
The Impact of Gene Transfer Techniques in Eukaryotic Cell Biology

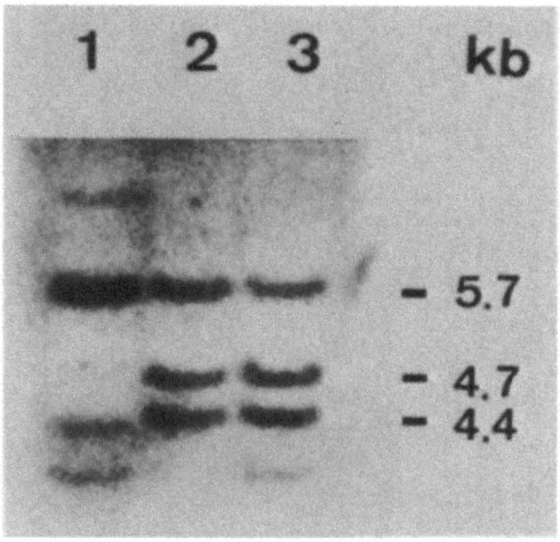

Fig. 1. Excision of *Tam1* in leaves and flowers of the *A. majus* mutant *niv-53*. Genomic DNA from *niv*+ and *niv-53* plants was digested with *EcoRI* and hybridized with ^{32}P-labelled Am3 DNA, which contains almost the entire sequence of the chalcone synthase gene. *Lane 1*, *Niv*+ DNA; *lane 2*, DNA from *niv-53* flowers; *lane 3*, DNA from *niv-53* leaves. (Bonas et al. 1984a)

mutant was found to harbor a 17 kb element called *Tam1* (Transposon *Antirrhinum majus*) at the *nivea* locus (Bonas et al. 1984a).

Two other *nivea* alleles have been used in this study, *niv-44* and *niv-45*, both of which are stable, recessive alleles that result in colorless flowers. The *niv-44* mutant contains a 5 kb insertion called *Tam2* at the *nivea* locus (Upadhyaya et al. 1985; E. Krebbers, unpubl. results).

^{32}P-labeled DNA of a genomic clone, Am3, was used to probe the DNA of *niv-53* and *niv-44* lines in Southern hybridization experiments (Wienand et al. 1982; Upadhyaya et al. 1984). The Am3 clone contains two thirds of the chalcone synthase gene (H. Sommer, unpubl. results). The results of one such experiment are shown in Figure 1. With EcoRI digested *Niv*+ DNA a characteristic 5.7 kb band hybridizes to the Am3 probe. In contrast, two new bands 4.4 and 4.7 kb in size are observed in addition to the 5.7 kb band when EcoRI digested DNA from the *niv-53* mutant is used (Wienand et. al. 1982; Bonas et al. 1984a). Subsequent cloning of *Tam1* revealed that this element contains two EcoRI restriction sites, which give rise to the 4.4 and 4.7 kb fragments observed in Southern blots when *niv-53* is cleaved with EcoRI. The 5.7 kb band observed with genomic DNA from the *niv-53* mutant is thought to result from excision of *Tam1* from the *nivea* locus in some tissue of the plant. Thus somatic variegation in *niv-53* seems to be based on occasional excision of *Tam1* in some cells of the plant.

When genomic DNA from the *niv-44* line is cleaved with EcoRI and tested with Am3 DNA, two bands 2.0 and 9.0 kb in size hybridize to the probe that cannot be seen with wild type DNA (Upadhyaya et al. 1985; E. Krebbers, unpubl. results). However, no 5.7 kb signal is observed, indicating that *Tam2* is not excised in somatic tissue of the plant (ibid.).

3 Molecular Cloning of *Tam1* and *Tam2*

To clone the Tam1 element from the niv-53 mutant, Sau3A digested plant DNA was ligated to BamHI cleaved λ1059 vector DNA (Bonas et al. 1984a). Two clones were isolated that together include the entire sequence of *Tam1* with some overlap plus adjacent sequences of the *nivea* locus. Analysis of hetero-duplexes formed between DNA from these two clones and Am3 DNA was used to determine the size of *Tam1* as 16.9 kb (Bonas et al. 1984a). For sequence analysis of the termini of *Tam1* and the ad-

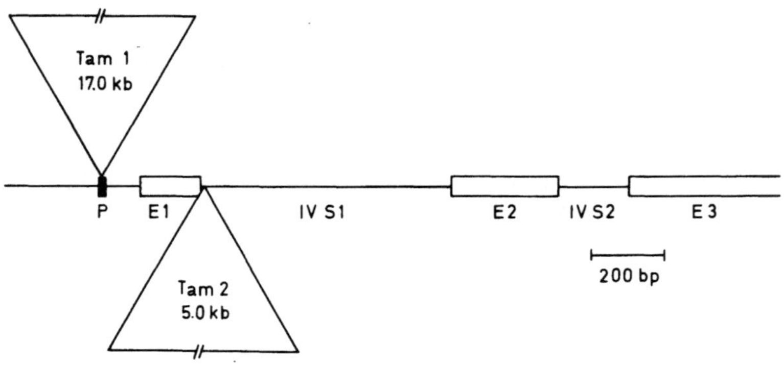

Fig. 2. Integration sites of *Tam1* and *Tam2* at the *nivea* locus of *A. majus*

jacent *nivea* sequences, the 4.4 and 4.7 kb EcoRI fragments described above were cloned (Bonas et al. 1984b).

Since *Tam2* apparently contains only one EcoRI cleavage site (Upadhyaya et al. 1984), the 9.0 and 2.0 kb EcoRI fragments from *niv-44* DNA observed in Southern blots probably contain the entire sequence of the element. These fragments were cloned in λgtWES and subcloned in pUC9 (Upadhyaya et al. 1984). The complete sequence of *Tam2* plus adjacent sequences of the *nivea* target site have been determined (E. Krebbers, unpubl. results; Upadhyaya et al. 1985).

4 Physical Characteristics of *Tam1* and *Tam2* and Their Positions in the *nivea* Locus

The integration sites of *Tam1* and *Tam2* are shown in Figure 2. DNA sequence analysis revealed that *Tam1* is located 17 bp upstream of the putative TATA box of the CHS gene and 125 bp upstream of the translational initiation codon ATG (H. Sommer, personal communication; Bonas et al. 1984b). *Tam2*, on the other hand, is integrated within the first intron of the chs gene 3 bp away from the first exon/intron junction (Upadhyaya et al. 1985).

Examination of the sequences adjacent to *Tam1* and *Tam2* showed that both elements generate a 3 bp duplication of the target site upon integration (Bonas et al. 1984b; Upadhyaya et al. 1985). Furthermore, each element terminates in an inverted repeat sequence, whereby 13 bp of the terminal 14 bp inverted repeat of *Tam2* are identical to the endmost 13 bp inverted repeat of *Tam1* (Bonas et al. 1984b; Upadhyaya et al. 1985). Additional sequence homology is observed between at least the first 200 bp of the left ends of the elements (Bonas et al. 1984b; Upadhyaya et al. 1985; E. Krebbers, unpubl. results). These structural similarities suggest that there may be transposition functions capable of operating with both elements. However, the sequences of the rest of the elements share little or no homology, as shown by Southern hybridization experiments (U. Bonas, unpubl. results), which suggests that the amino acid sequences of any functions encoded by these elements may be quite dissimilar.

As shown in Figure 3, the termini of each element contain several stretches of inverted repeat sequence homology that can be paired to form elaborate stem-and-loop structures. Similar structures have been

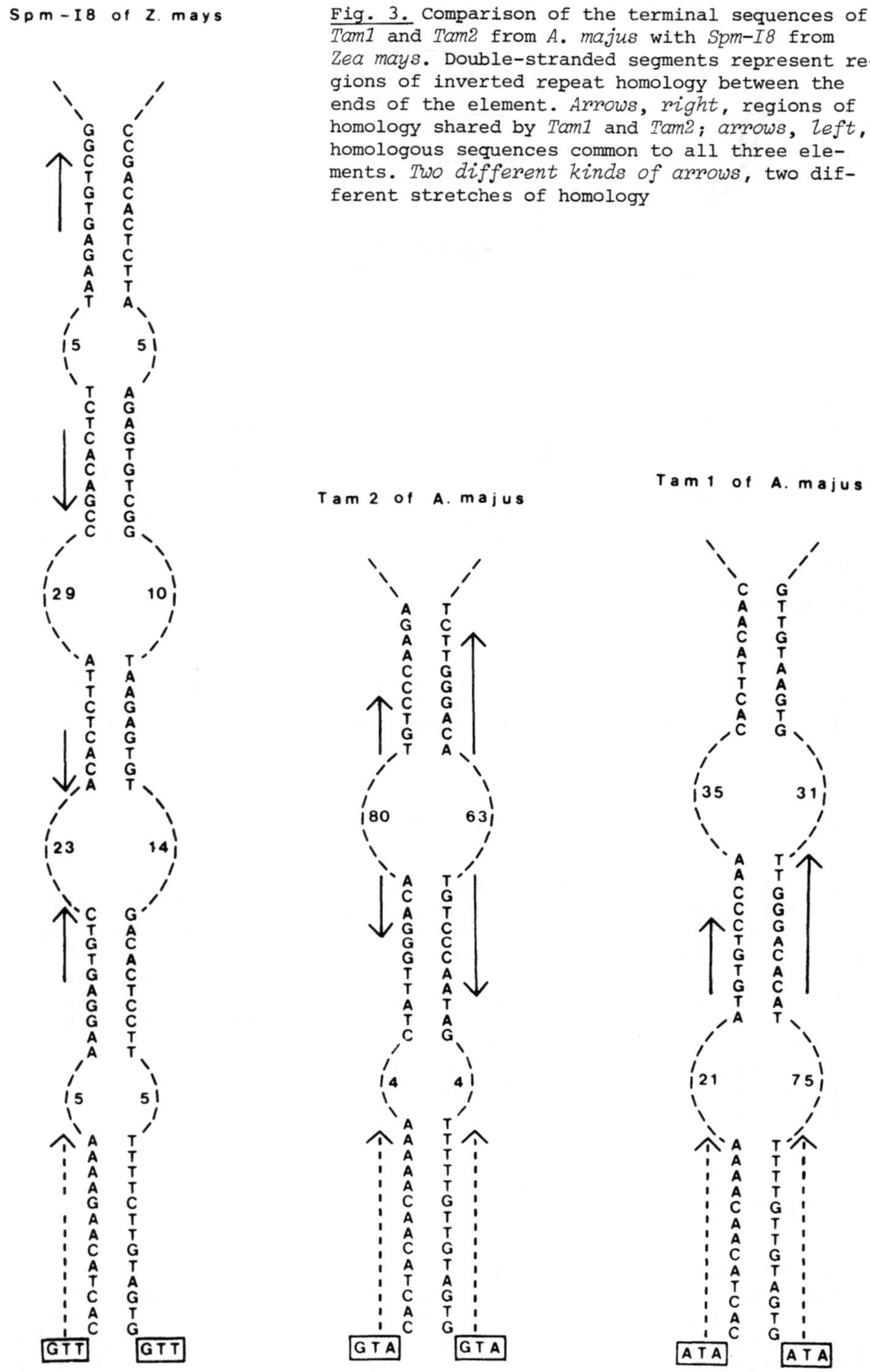

Fig. 3. Comparison of the terminal sequences of *Tam1* and *Tam2* from *A. majus* with *Spm-I8* from *Zea mays*. Double-stranded segments represent regions of inverted repeat homology between the ends of the element. *Arrows, right*, regions of homology shared by *Tam1* and *Tam2*; *arrows, left*, homologous sequences common to all three elements. *Two different kinds of arrows*, two different stretches of homology

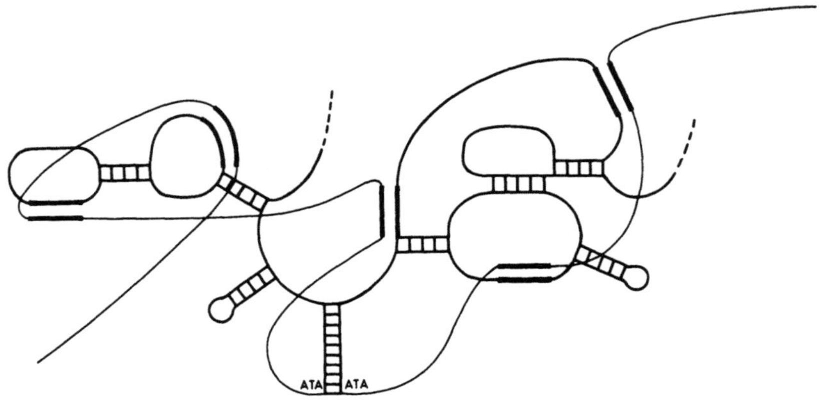

Fig. 4. Possible basis for integration specificity of *Tam1*. A single strand of the *Tam1* element plus flanking sequences of the *nivea* locus is shown. The thin line represents *nivea* sequences and the middle-sized line designates the *Tam1* element. Extending upward from the ATA target-site duplication, the terminal inverted repeat sequences of a single strand of the element have been paired. Further pairing between inverted repeat sequences within each terminus is shown by double-stranded stem structures. The loops generated by such pairing contain 8 - 9 bp sequences that are homologous to sequences of the *nivea* locus, as illustrated by the *thick lines*. It is conceivable that this structure is stabilized by proteins and serves as an intermediate during excision and/or integration

reported for the maize transposable elements *Spm-I8* (see Fig. 3 and Schwarz-Sommer et al. 1984) and *Ac* (Courage et al. 1984). The possible biological significance of such structures is presently unknown.

5 Possible Basis of Integration Specificity of *Tam1*

Inverted repeat sequence homology can be found not only *between* the ends of *Tam1* but also *within* each terminus of the element (Bonas 1984). If the endmost inverted repeats of a single strand of the element are matched, and if further pairing of homologous sequences occurs within each terminus, a structure reminiscent of a tRNA can be formed, as shown in Figure 4. The loops exposed in this structure contain stretches of 8 or 9 bp that are homologous to sequences of the target site. Pairing between such sequences in the element and the target site, as illustrated by the thick lines in Figure 4, may play a part in the process of integration and/or excision. Perhaps such structures are further stabilized by specific proteins.

6 Characterization of a Revertant of *niv-53*

As discussed previously, excision of *Tam1* in somatic tissue seems to be responsible for restoration of chalcone synthase activity in the pigmented sectors of variegated *niv-53* flowers. To determine whether excision also occurs in germinal tissue, the DNA sequence corresponding to the *Tam1* integration site was examined in a homozygous *Niv*$^+$ revertant of *niv-53* (Bonas et al. 1984b). The *Tam1* element is indeed no longer

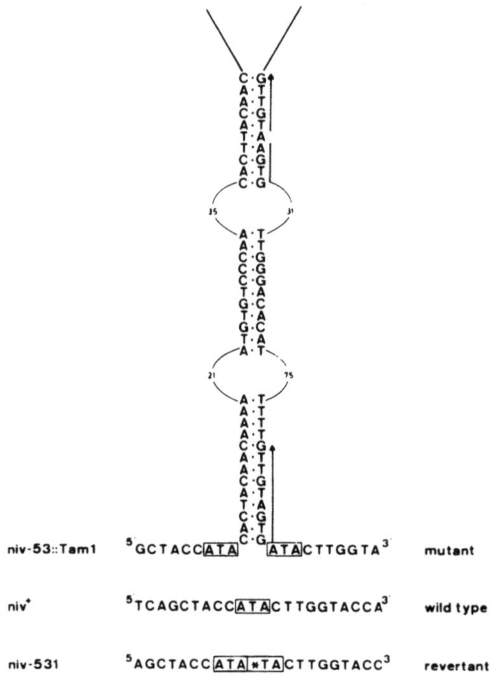

Fig. 5. Comparison of the sequences at the target site of *Tam1* integration in the *niv-53* mutant, in wild type and in revertant plants. (Bonas et al. 1984b)

present at the *nivea* locus in this revertant. However, in this case loss of the element is not accompanied by exact restoration of the wild type *nivea* sequence. Instead 2 bp of the target site duplication seem to have been retained in the revertant following excision, as illustrated in Figure 5. Since flowers of the revertant exhibit phenotypically wild type pigmentation, these additional 2 bp do not seem to interfere with expression of the *nivea* locus. Similar results have been reported for revertants of transposable element induced mutants in maize (Sachs et al. 1983; Weck et al. 1984; Pohlmann et al. 1984). This suggests that imprecise excision may be a general feature of transposable elements in plants. If an element is integrated in the coding sequence of a gene, imprecise excision events of this kind may generate new protein sequences and thus help to promote variability of the organism. Altered proteins have indeed been reported for a number of revertants of mutants caused by transposable elements in maize (Dooner and Nelson 1979; Echt and Schwartz 1981; Tuschall and Hannah 1982; Shure et al. 1983).

7 Evidence for Mobility of *Tam1* and *Tam2*

By definition, a transposable element is a DNA element which can alter its position in the genome. By classical genetic means this has not been demonstrated for the two elements of *Antirrhinum* described here. However, there is molecular evidence that suggests that each belongs to a family of elements that is truly transposable, irrespective of whether the particular elements we have cloned are themselves mobile.

First of all it should be recalled that transposition in plants appears to occur primarily via excision and reintegration of the element (for summary of evidence see Fedoroff 1983 and Nevers et al. 1984).

Thus it can be inferred that since *Tam1* can undergo excision, as discussed above, it is probably also capable of reintegration. Although revertants of the *Tam2*-induced mutant *niv-44* have not been found, recent results indicate that under certain circumstances *Tam2* can indeed be excised (E. Krebbers, unpubl. results). The same arguments therefore also hold true for *Tam2*.

More direct evidence for transposition of *Tam1* and *Tam2* derives from Southern hybridization analysis of the *niv-53* and *niv-44* mutants. First genomic DNA from the mutants was digested with EcoRI, which cleaves twice within the *Tam1* sequence and once within *Tam2*. This DNA was then probed with a unique fragment from one end of each element. In this manner a particular pattern of hybridization signals is observed. Each band corresponds to a particular copy of the element at a specific position in the genome. When the banding pattern of individual plants of the *nivea-recurrens* mutant *niv-53* are compared, they are found to differ (Bonas 1984). Although several explanations for such shifts in banding pattern are possible (e.g., restriction site polymorphisms, rearrangements in sequences adjacent to a given transposable element), at least some of these alterations may indeed reflect changes in position of the *Tam1* element. Differences in banding pattern were also observed between *niv-44* and *niv-53* plants using a *Tam2* specific probe (E. Krebbers, unpubl. results). This suggests that *Tam2* transposed after divergence of these two mutants of *A. majus*.

8 Possible Interaction Between *Tam1* and *Tam2*

In plants a number of cases are known in which one allele of a given locus is capable of inducing a heritable alteration of another allele of the same locus when the two alleles are combined in a heterozygote. This phenomenon has been termed *paramutation*, as opposed to classical mutation, since it occurs with a very high frequency in appropriate heterozygotes, and since the change that occurs is always in one particular direction (for reviews see Brink 1983; Kermicle 1983; Hagemann and Berg 1977; Gavazzi 1977).

In *Antirrhinum*, a phenomenon that strongly resembles paramutation has also been reported (Harrison and Carpenter 1973a,b). When the highly variegated *niv-53* line is crossed to the white-flowering *niv-44* stock, the majority (66 - 88%) of F1 heterozygotes do not have variegated flowers, as would be expected, but rather uniformly pink-colored ones. These plants fail to segregate the *niv-53* allele in two subsequent generations of selfings and backcrosses (Harrison and Carpenter 1973b). This effect is not observed in a cross between *niv-53* and another stable, recessive mutant, *niv-45*. Thus it seems that the unstable *niv-53* allele is frequently converted to a new, stable allele by exposure to the *niv-44* genomic background. Whether the *niv-44* allele itself is responsible for this change, or whether the genomic background of the *niv-44* stock is decisive is unknown.

Since both the *niv-53* and *niv-44* alleles are induced by transposable elements, namely by *Tam1* and *Tam2*, and since these elements share certain important structural features, it is tempting to speculate that *Tam1* somehow interacts with *Tam2* in the process of paramutation. It is conceivable that a kind of gene conversion occurs between the two *nivea* homologs in *niv-53/niv-44* heterozygotes, converting the *niv-53* allele to a new one. Alternatively some product of *Tam2* may influence transcription and/or excision of *Tam1*. Elucidation of this problem can be expected from the molecular analysis of paramutant F1 plants currently in progress in our laboratory.

9 The *Spm(En)* Controlling Element System in *Zea mays*

The transposable elements can indeed interact is clearly demonstrated by the two-component systems of *Zea mays*. One component of such a system is an autonomous transposable element capable of producing in trans certain functions required for transposition and other element-induced activities. This component has been termed the "regulator" (McClintock 1961a) or autonomous component (Fedoroff 1983). The second element in such a two-component system is defective with respect to transposition and other activities. It can inhibit expression of a locus simply by virtue of its presence, but it is stable unless it is transactivated by an appropriate regulator component. This component has been termed the "receptor" (Peterson 1965; Fincham and Sastry 1974) or nonautonomous component (Fedoroff 1983).

The interaction between the two components is specific since a given receptor component will respond to some regulator elements but not to others (Peterson 1980). In some cases the two components are also known to be structurally related (Fedoroff et al. 1983; Pohlman et al. 1984; Courage et al. 1984), that is, the receptor component seems to represent an altered regulatory component.

The *Spm* (suppressor-mutator) system (McClintock 1954), also called the *En* enhancer system (Peterson 1953), is the genetically best-characterized two-component system in maize (for reviews see Fincham and Sastry 1974; Nevers and Saedler 1977; Fedoroff 1983). Since both the *Spm* and *En* regulatory components can transactivate the same receptor element, the two systems are considered to be equivalent (Peterson 1965). The receptor component was given no particular name by McClintock but designated *I* (inhibitor) by Peterson (1953), and we shall refer to it as such here.

10 Molecular Cloning of the *Spm-I8* Receptor Component

For cloning purposes we selected the *wx-m8* mutant since it is known to harbor the *I* component of the *Spm(En)* system at the *wx* (waxy) locus. Furthermore, the product of *wx* locus can be readily isolated, which allows standard cDNA cloning procedures to be employed. The *wx* locus encodes the enzyme UDP-glucose starch transferase, the major starch granule bound protein in the endosperm (Nelson and Rines 1962; Tsai 1974). This enzyme is involved in amylose synthesis in the endosperm (Sprague et al. 1943). Assuming that the *I* element is structurally related to the regulatory component, it should be possible to isolate the latter in subsequent experiments once the *I* element has been cloned.

Using a cDNA clone as a probe a library of Line C (wild type) maize DNA cloned into the λEMBL4 vector was screened. A recombinant phage, λWx-45, with a 14 kb insert was isolated and subcloned. Both λWx-45 and the cDNA clone were used as probes to isolate the *wx* locus from *wx-m8* DNA. In this case MboI partially digested genomic DNA from *wx-m8* plants was ligated into the BamH1 sites of λEMBL4. A clone carrying a 19 kb insert, wx-m8-2, was identified and used for further experiments (Schwarz-Sommer et al. 1984).

11 Physical Characteristics of the *I* Element and the *wx-m8* Locus

A comparison of the DNA sequence of the wild type *wx* locus with that of the *wx-m8* mutant revealed the presence of three DNA insertions, 6.1,

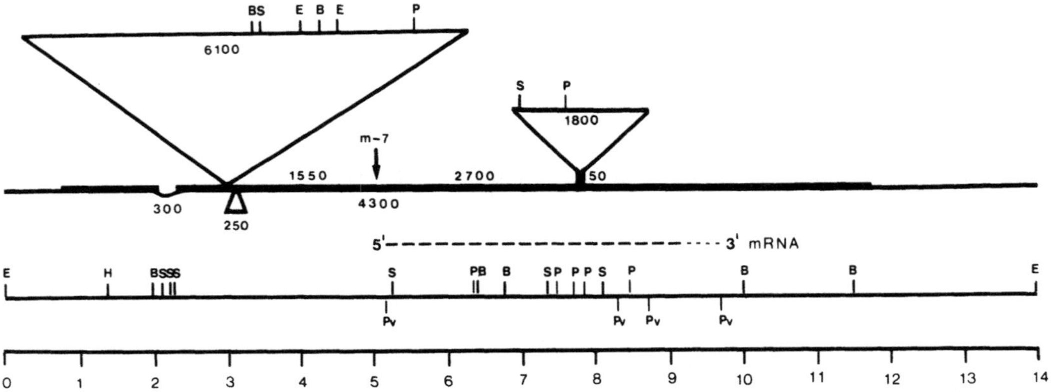

Fig. 6. Integration site of *Spm-I8* in the *wx* locus of *wx-m8* plants. Data are derived from heteroduplex analysis of wild type and *wx-m8* DNA. *Thin lines* indicate single-stranded stretches of DNA within the heteroduplex. *Numbers* designate the distances and sizes of various DNA intervals. *Middle line* represents the physical map of the 14 kb EcoRI insert of the wild type clone λwx-45. *Lower line* indicates the scale in kilobases. *Dashed line* above the physical map represents the transcribed region of the *wx*-gene. (Schwarz-Sommer et al. 1984)
Abbreviations for restriction enzymes: E = EcoRI; H = HindIII; B = BamHI; S = SalI; P = PstI; PV = PvuI

2.0 and 0.25 kb long, within the *wx-m8-2* clone (Schwarz-Sommer et al. 1984). These are shown in Figure 6. To determine which of these insertions corresponds to *I*, an *En(Spm)* regulatory element was crossed into the *wx-m8* line, and the behavior of the insertions was then analyzed in Southern hybridization experiments. The results of these experiments show that the 6.1 kb insertion is stable in response to *En(Spm)* while the 2.0 kb insertion is excised in part of the somatic tissue. Thus the 2.0 kb insertion is regarded as *I*. No conclusions could be drawn concerning the 0.25 kb insertion (Schwarz-Sommer et al. 1984).

As shown in Figure 6 the 2.0 kb *I* element (termed *Spm-I8*) is located within the transcribed region of the *wx* gene. The stable 6.1 kb insertion, on the other hand, appears to be integrated approximately 2 kb upstream of the putative 5' end of the structural gene (Schwarz-Sommer et al. 1984).

The sequence of the first 100 - 200 bp of the ends of *Spm-I8* plus the flanking sequences of the target site were determined and compared with the corresponding sequence of the target site in wild type DNA. This revealed that *Spm-I8* generates a 3 bp duplication upon integration (see Fig. 3). Furthermore the termini of the *Spm-I8* element exhibit an imperfect inverted repeat structure at least 50 bp long that can be envisioned as forming the stem-and-loop structure shown in Figure 3. Four out of the five double-stranded regions contain inverted or direct repetitions of the same motif: CTCACA.

12 Structural Similarities Between Transposable Elements from Different Plant Species

As discussed previously, the terminal inverted repeat sequences of the *Antirrhinum* elements *Tam1* and *Tam2* are almost identical, and both elements generate a 3 bp duplication of the target site sequence. Surprisingly the maize element *Spm-I8* also induces a 3 bp duplication upon integration, and 12 bp of the terminal 13 bp inverted repeat sequence of this element are homologous to the endmost sequences of *Tam1* and *Tam2*. Still another element, the *Tgm1* insertion responsible for the Le1 mutation in *Glycine max*, possesses similar structural features (Vodkin et al. 1983). It also is flanked by a 3 bp duplication of the target site and has a terminal 13 bp inverted repeat sequence, 11 bp of which are homologous to the terminal 13 bp inverted repeat of *Spm-I8*. All four elements, *Tam1*, *Tam2*, *Spm-I8* and *Tgm1*, share the sequence CACTA at their very ends. On the basis of the structural similarities listed above these elements may be considered members of a common "family" of transposable elements.

The fact that transposable elements from such widely divergent species as *Antirrhinum*, maize and soybean share prominent structural properties raises the question of their origin. Perhaps these structures reflect convergent evolution. Alternatively one could speculate that a common progenitor element existed prior to divergence of these species, or that horizontal spread of a common progenitor element occurred at a later stage in evolution. If and when appropriate interspecific transformation systems are available, it will also be interesting to examine whether these elements can interact functionally.

References

Bonas U (1984) In-vitro Klonierung eines transponierbaren Elements im Gen der Chalkon-Synthase von *Antirrhinum majus*. Ph D thesis, University of Cologne

Bonas U, Sommer H, Harrison BJ, Saedler H (1984a) The transposable element Tam1 of *Antirrhinum majus* is 17 kb long. Mol Gen Genet 194:138-143

Bonas U, Sommer H, Saedler H (1984b) The 17 kb Tam1 element of *Antirrhinum majus* induces a 3 bp duplication upon integration into the chalcone synthase gene. EMBO J 3:1015

Brink RA (1973) Paramutation. Annu Rev Genet 7:129-152

Courage U, Döring H-P, Frommer W-B et al. (1984) Transposable elements *Ac* and *Ds* at the *shrunken*, *waxy* and *alcohol dehydrogenase* loci in *Zea mays* L. Cold Spring Harbor Symp Quant Biol (Manuscript submitted)

Dooner HK, Nelson OE (1979) Heterogeneous flavonoid glucosyltransferase in purple derivatives from a controlling element-suppressed *bronze* mutant in maize. Proc Natl Acad Sci USA 76:2369-2371

Echt CS, Schwartz D (1981) Evidence for the inclusion of controlling elements whithin the structural gene at the *waxy* locus in maize. Genetics 99:275-284

Fedoroff N (1983) Controlling elements in maize. In: Shapiro JA (ed) Mobile genetic elements. Academic Press, New York, London, pp 1-63

Fedoroff N, Wessler S, Shure M (1983) Isolation of the transposable maize controlling elements *Ac* and *Ds*. Cell 35:235-242

Fincham JRS, Sastry GRK (1974) Controlling elements in maize. Annu Rev Genet 8:15-50

Friedemann P, Peterson PA (1982) The Uq controlling element system in maize. Mol Gen Genet 187:19-29

Gavazzi G (1977) The genetic complexity of the R locus in maize. Stadler Genet Symp 9:38-60

Hagemann R, Berg W (1977) Vergleichendc Analyse der Paramutationssysteme bei höheren Pflanzen. Biol Zentralbl 96:257-301

Harrison BJ, Carpenter R (1973a) A comparison of the instabilities at the *nivea* and *pallida* loci in *Antirrhinum majus*. Heredity 31:309-323

Harrison BJ, Carpenter R (1973b) Paramutation in *Antirrhinum majus*. John Innes Annual Report, Norwich, England, p 52

Kermicle JL (1973) Organization of paramutational components of the R locus in maize. Brookhaven Symp Biol 25:262-280

Kuckuck H (1936) Über vier neue Serien multipler Allele bei *Antirrhinum majus*. Z indukt Abstammungs- u Vererbungsl 71:429-440
McClintock B (1950) The origin and behavior of mutable loci in maize. Proc Natl Acad Sci USA 36:344-355
McClintock B (1954) Mutations in maize and chromosomal aberrations in *Neurospora*. Carnegie Inst Wash Year Book 53:254-260
McClintock B (1956a) Intranuclear systems controlling gene action and mutation. Brookhaven Symp Biol 8:58-74
McClintock B (1956b) Controlling elements and the gene. Cold Spring Harbor Symp Quant Biol 51:197-216
McClintock B (1961a) Some parallels between gene control systems in maize and in bacteria. Am Nat 95:265-277
McClintock B (1961b) Further studies of the suppressor-mutator system of control of gene action in maize. Carnegie Inst Wash Year Book 60:469-476
Nelson OE, Rines HW (1962) The enzymatic deficiency in the *waxy* mutant of maize. Biochem Biophys Res Commun 9:297-300
Nevers P, Saedler H (1977) Transposable genetic elements as agents of gene instability and chromosomal rearrangements. Nature (Lond) 268:109-115
Nevers P, Shepherd N, Saedler H (1984) Plant transposable elements. Adv Bot Res (Manuscript submitted)
Peterson PA (1953) A mutable pale green locus in maize. Genetics 38:682-683
Peterson PA (1965) A relationship between the *Spm* and the *En* control systems in maize. Am Nat 99:391-398
Peterson PA (1976) Basis for the diversity of states of controlling elements in maize. Mol Gen Genet 149:5-21
Peterson PA (1980) Instability among the components of a regulatory element transposon in maize. Cold Spring Harbor Symp Quant Biol 45:447-455
Pohlman RF, Fedoroff NV, Messing J (1984) The nucleotide sequence of the maize controlling element activator. Cell (Manuscript submitted)
Sachs MM, Peacock WJ, Dennis ES, Gerlach WL (1983) Maize *Ac/Ds* controlling elements: a molecular viewpoint. Maydica 28:289-303
Schwarz-Sommer Z, Gierl A, Klösgen RB, Wienand U, Peterson PA, Saedler H (1984) The *Spm/En* transposable element controls the excision of a 2 kb DNA insert at the *wx-m8* locus of *Zea mays*. EMBO J 3:1021-1028
Shure M, Wessler S, Fedoroff N (1983) Molecular identification and isolation of the waxy locus. Cell 35:225-233
Sprague GF, Brimhall B, Hixon RM (1943) Some effects of the waxy gene in corn on the properties of the endosperm starch. J Am Soc Agron 35:817-822
Spribille R, Forkmann G (1982) Genetic control of chalcone synthase activity in flowers of *Antirrhinum majus*. Phytochemistry (Oxf) 21:2231-2234
Tsai CY (1974) The function of the *waxy* locus in starch synthesis in maize endosperms. Biochem Genet 11:83-96
Tuschall DM, Hannah LC (1982) Altered maize endosperm ADP-glucose pyrophorylase from revertants of a shrunken-2-dissociation allele. Genetics 100:105-111
Upadhyaya KC, Sommer H, Krebbers E, Saedler H (1985) Chalcone synthase gene from a paramutagenic line of *Antirrhinum majus* contains the insertion sequence Tam2. (Manuscript in preparation)
Vodkin LO, Rhodes PR, Goldberg RB (1983) A lectin gene insertion has the structural features of a transposable element. Cell 34:1023-1031
Weck E, Courage U, Döring H-P, Fedoroff N, Starlinger P (1984) Analysis of sh-m6233, a mutation induced by transposable element *Ds* in the sucrose synthase gene of *Zea mays*. EMBO J (Manuscript submitted)
Wienand U, Sommer H, Schwarz Z et al. (1982) A general method to identify plant structural genes among genomic DNA clones using transposable element induced mutations. Mol Gen Genet 187:195-201

Mobile Genetic Elements and Their Use for Gene Transfer in Drosophila

W. J. Gehring[1]

1 Mobile Genetic Elements in Drosophila

Following the pioneering work on "controlling elements" in maize by McClintock (1957), mobile genetic elements were also discovered in *Drosophila* on the basis of their genetic properties. The first mobile genetic element was described by Green (1967, 1969a,b) and found to be associated with the *white* locus. The *white (w)* mutation was the first mutant which was isolated in *Drosophila* by Morgan in 1910. Since the w^1 mutation lacks the red and brown eye pigments present in the wild type, it shows a *white* eye color. Green found an unstable mutant *white-crimson* (w^c) which confers a light reddish-orange eye color and mutates with an unusually high frequency ($\sim 10^{-3}$) to lighter or darker eye pigmentation. It also generates deletions which extend from the *white* locus over varying distances, and most importantly it gives rise to transpositions of parts of the *white* locus to other chromosomes. On the basis of the genetic data, Green postulated the existence of a genetic element responsible for the high mutability and the generation of chromosomal rearrangements and transposition, and this was later confirmed by gene cloning experiments (Collins and Rubin 1982; Paro et al. 1983; see below).

Following Green's discovery, Ising and Ramel found a remarkable "transposing element", designated as TE, which not only includes the *white* locus but also the neighboring *roughest* gene (Ising and Ramel 1976). The TE will be described in more detail in the following section.

When gene cloning techniques became available a large number of transposable elements were discovered on the basis of their chromosomal distribution. They were first isolated as moderately repetitive DNA sequences, and subsequent mapping by in situ hybridization to polytene giant chromosomes indicated that they occurred at many different sites in the genome, which differ widely from stock to stock. In several cases it has been shown conclusively that this variable chromosomal distribution is due to transposition.

A large fraction, on the order of 10%, of the *Drosophila melanogaster* genome is composed of mobile genetic elements. They can be considered as intragenomic parasites which can be detected by their ability to integrate into the *Drosophila* (host) genome. They consist of DNA segments, a few kilobases (kb) long, that are usually characterized by direct or inverted repeats at their ends which are necessary for transposition. They may encode transposition functions, and upon integration they generate short direct repeats of defined length in the host DNA. Mobile genetic elements give rise to a major proportion of spontaneous mutations which they either cause directly by inserting into the host genome or by causing secondary chromosomal rearrangements. They carry regulatory signals which control gene expression both within the ele-

[1]Abteilung Zellbiologie, Biozentrum der Universität Basel, Klingelbergstrasse 70, CH-4056 Basel, Schweiz

35. Colloquium - Mosbach 1984
The Impact of Gene Transfer Techniques in Eukaryotic Cell Biology

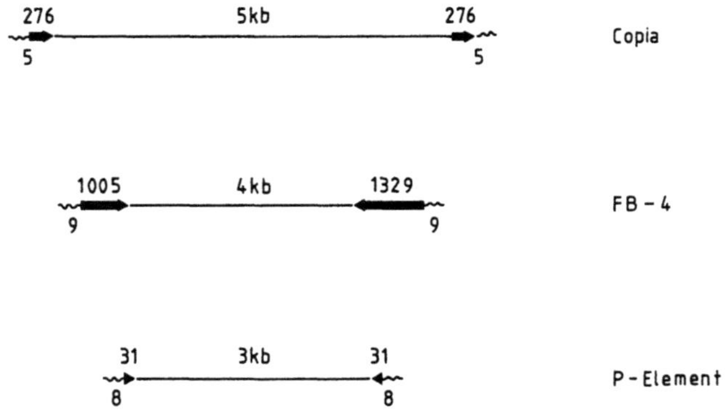

Fig. 1. Diagrammatic representation of the structure of three types of transposable elements found in *Drosophila melanogaster*. *Copia*-like elements: *copia* (Levis et al. 1980); Fold-back elements: FB-4 (Potter 1982a); P elements: P factor (O'Hare and Rubin 1983). *Black arrows* indicate the terminal repeats of the elements and their polarity. *Wavy lines* indicate the short direct duplications in the flanking host DNA generated during insertion of the transposable elements. *Numbers* refer to basepairs, kb = kilobase (pair)

ment and in flanking host DNA. Mobile genetic elements can also transpose host sequences.

The three classes of mobile genetic elements which have been studied in *Drosophila* most extensively are the *copia-like elements*, the *fold-back DNA elements* and the *P-transposon* (see Rubin 1983, for a review). Their basic structural features are diagrammatically shown in Figure 1.

The *copia-like elements* share several characteristics with those of retroviruses. They are between 5 and 10 kb long, flanked by direct long terminal repeats of a few hundred nucleotides and generate a short direct duplication of 3 to 5 nucleotides (depending on the element) at the target site in the host DNA. There are approximately 30 different families of *copia*-like elements in *Drosophila melanogaster*, each represented by 10 to 100 copies per genome. The original *copia* element (Fig. 1) is 5 kb long and flanked by long terminal repeats (LTR) of 276 bp, which show some homology to the LTR's of spleen necrosis virus, a retrovirus which also generates a 5 bp. duplication at the target site (Levis et al. 1980). Furthermore, circular DNA copies of *copia* have been identified similar to those in retroviruses which are thought to be intermediates in the integration of the virus into the host genome (Flavell and Ish-Horowicz 1981). Finally, virus-like particles containing *copia* RNA and reverse transcriptase activity (Shiba and Saigo 1983) have been found in nuclei of *Drosophila* tissue culture cells. This evidence taken together strongly suggests that *copia*-like elements are evolutionarily related to retroviruses. A large fraction of the spontaneous mutations at the *white* locus (Gehring and Paro 1980; Zachar and Bingham 1982) or in the *bithorax* complex (Bender et al. 1983) is due to the insertion of *copia*-like elements, indicating that they are a major cause of spontaneous mutation. They can also give rise to unequal crossing-over, and thereby generate gene duplications and deletions (Goldberg et al. 1983).

The second class of mobile genetic elements, the *fold-back elements*, were first discovered on the basis of their property to form fold-back structures upon renaturation of denatured DNA. One member of this family, FB-4, is illustrated in Figure 1. Fold-back elements are characterized by inverted long terminal repeats of varying length, which in turn are composed of short tandem repeats (Potter et al. 1980; Truett et al.

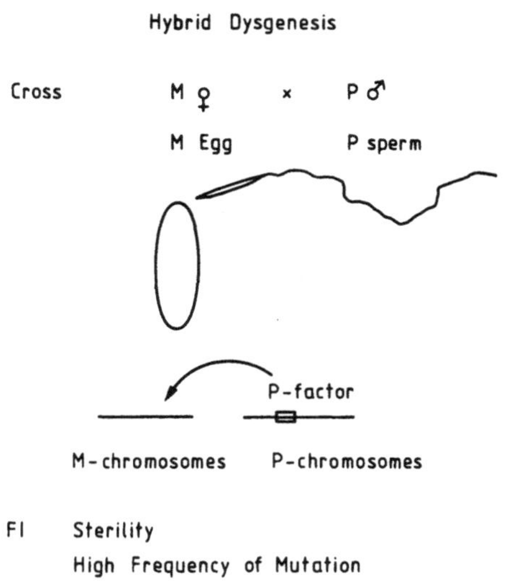

Fig. 2. Hybrid dysgenesis (PM-type). The P-factors introduced by the P-sperm into an M-egg become mobilized and insert into the chromosomes of the germ-line generating partial sterility and a high frequency of mutations in the F1 generation

1981). The internal core segment is highly variable in length and DNA sequence, and in some cases no DNA has been found between the inverted repeats. However, there is suggestive evidence that the intact functional fold-back element is FB-NOF which is characterized by a specific 4.5 kb internal segment (Paro et al. 1983). This element is associated with the *white-crimson* (w^c) mutation described above (Collins and Rubin 1982) and a w^c derived transposon (Paro et al. 1983) as well as with the TE (see next section). Upon integration into the host genome, FB-4 elements generates a 9 bp duplication. The variable chromosomal distribution of FB elements in different *Drosophila* stocks indicates that they are mobile.

The third class of mobile genetic elements, the *P-elements* were discovered on the basis of a genetic syndrome designated as *hybrid dysgenesis* (Kidwell et al. 1977; Rubin et al. 1982; O'Hare and Rubin 1983). The principle of PM-hybrid dysgenesis is outlined in simplified form in Figure 2. *Drosophila melanogaster* stocks can be classified into two types, M-stocks (m = maternal) and P-stocks (p = paternal). P-stocks carry the intact P-factor, a 3 kb transposon characterized by short inverted ends of 31 bp (Fig. 1). In contrast to P-stocks, M-stocks lack active P-factors although they may carry defective ones. In a cross of an M-female with a P-male, the P-transposons carried by the sperm enter the M-cytoplasm and become mobilized. In the germ-line of the hybrid embryo, the P-elements transpose to multiple new chromosomal sites and generate insertion mutations and chromosomal rearrangements. Therefore, the F1 progeny of these hybrid flies is largely sterile and shows a high frequency of mutations. The P-factor is thought to encode at least two functions, a "transposase activity" catalyzing transposition and an immunity or repressor function, which prevents the continued transposition of the P-elements in P-stocks. The M-cytoplasm apparently lacks the repressor function and allows the P-factors, introduced by the sperm, to be mobilized. A large fraction of the P-elements in P-stocks is defective and carries internal deletions. However, such defective P-elements can be mobilized *in trans* by intact P-factors, if the defective elements have intact ends.

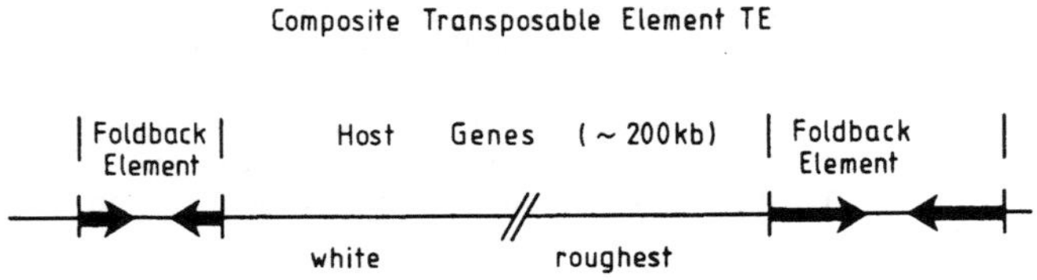

Fig. 3. Diagram of the structure of composite transposable element TE. The host genes, whose dimensions are not known precisely, are flanked by the fold-back elements. (After Paro et al. 1983)

2 Transposons Carrying the *White* Locus

The transposable elements do not only transpose by themselves but they can also mobilize host genes. In the case of *white-crimson* (see above) a portion of the *white* locus was shown to transpose from the X chromosome to different autosomal sites. A large transposable element (TE) carrying at least two genes, *white* and *roughest*, was discovered by Ising and Ramel (1976). It arose spontaneously in an X chromosome carrying the *white-aprociot* mutation, which produces apricot-colored eyes, and transposed first to the second chromosome. Subsequently, the TE carrying *white-apricot* and *roughest* was found to transpose to well over 100 insertion sites scattered over the *Drosophila* genome (Ising and Block 1981). The frequency of transposition is about 10^{-4} per generation. The TE consists of two or more chromosomal bands which can be visualized in the giant polytene chromosomes. The transposition events are accompanied by considerable genetic variation: *white-apricot* may revert to near wild-type eye color, it may become duplicated or lost from the element, and genes adjacent to the TE insertion site may "hitchhike" with the TE during subsequent transpositions. Molecular cloning of the TE borders from three different insertion sites indicates that the TE is flanked by fold-back elements on both sides (Fig. 3; Paro et al. 1983). The association of the FB elements with the ends of the TE's suggests that they are involved in the mobilization of the host genes which they enclose. Thus, a composite transposable element may be formed by two-fold-back elements enclosing a segment of host DNA. The function of the inverted terminal repeats may be tested by constructing artificial transposons and introducing them into the germ-line of different recipient stocks. The combination of short tandem repeats within large inverted repeats makes the FB elements highly recombinogenic (Potter 1982b). Such a structure allows for unequal crossing-over within a chromosome or even between different chromosomes which may lead to the genetic variability which is observed. Recombination between FB elements may also provide a mechanism for transposition of the TE. In situ hybridization experiments to polytene chromosomes indicate that two other transposable elements carrying *white*, the *white-crimson* derived *Tp* $w^{c\text{-}1}$ (Green 1969a) and the *Tp* w^{+IV} (Rasmusson et al. 1980), are also associated with a fold-back element, FB-NOF (Paro et al. 1983).

3 P-Factor Mediated Gene Transfer: "Synthetic" Transposons

Since transposons have efficient mechanisms for the insertion of defined DNA segments into the chromosomes, they may be used as vectors for gene transfer. For this purpose, the P-factor has some important

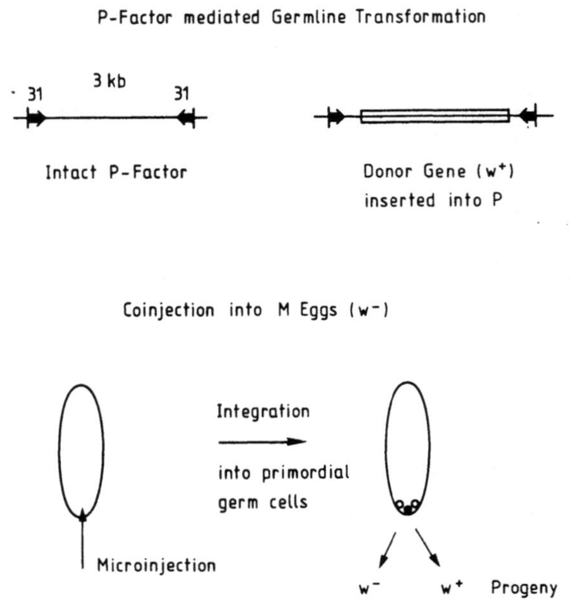

Fig. 4. P-factor mediated germline transformation. Intact P-factor DNA and a P-factor carrying an insert of donor DNA are coinjected into M-eggs. The complete P-factor can complement the defective one carrying the donor DNA, presumably by providing the "transposase" activity

advantages: the P-factor shows a very high frequency of transposition in germ cells, it can be mobilized by PM-interaction, defective P-elements can be complemented by active ones *in trans*, and the P-factor is only 3 kb in size, which is convenient for cloning experiments. An elegant technique for P-factor mediated germ-line transformation has been developed by Spradling and Rubin (1982; Rubin and Spradling 1982). It is based on the PM-interaction (Fig. 2) which can be induced by microinjection of cloned intact P-factor DNA into M-eggs. In a large fraction of the injected eggs (approximately 50%) the P-factor becomes integrated into the chromosomes of the germ cells of the recipient. Since active P-factors can mobilize defective ones, a donor gene can be inserted into a P-element which is used as a vector (Fig. 4). Coinjection of this DNA with intact P-factor DNA into M-recipient eggs can lead to the integration of the donor DNA into the germ cells of the recipient and the transformants can be isolated on the basis of genetic markers. We have used this procedure to construct a "synthetic" *white* transposon, $P(w^+)$ by insertion of a 12 kb DNA segment spanning the w^+ locus into a P-element carrying an internal deletion (Gehring et al. 1984). Upon coinjection of this construct and intact P-factor DNA into fertilized *white* mutant (w^-) eggs, several red eyed transformants have been isolated. The analysis of the eye pigments and the interaction of the transfered w^+ gene with the *zeste* mutation indicates that all the information needed for functional expression of the w^+ locus is contained in the transfered 12 kb segment which includes the RNA coding region and approximately 3 kb of flanking DNA on either side. Similar results have been obtained by Hazelrigg et al. (1984). Using a newly developed method for in situ hybridization of cloned DNA to RNA transcripts in tissue sections (Hafen et al. 1983), we have shown that the tissue specific accumulation of w^+ transcripts in the transformants is the same as in the wild-type controls (Fjose et al. 1984). We have also demonstrated that the transfered DNA has become integrated as a functional $P(w^+)$ transposon that can be mobilized again by a second PM-interaction. Therefore, it may be considered as a "synthetic" transposon. At least for other *Drosophila* genes, xanthine dehydrogenase *(ry)*, alcoholdehydrogenase *(adh)*, dopadecarboxylase *(ddc)* and the salivary gland protein gene *Sgs-3* (Spradling and Rubin 1983; D.A. Goldberg et al. 1983; Scholnick et al. 1983; Richards et al. 1983) appear to be expressed normally after P-factor mediated transformation at most of the insertion sites. Therefore, transposons can provide a highly efficient means for gene transfer. Methods for site-specific transformation in *Drosophila* remain to be developed.

Acknowledgements. I would like to acknowledge the contribution of R. Paro, M. Goldberg, R. Klemenz, A. Fjose, U. Weber, and R. Kloter to this work. I thank Iain Mattaj for critical reading of the manuscript and Erika Wenger-Marquardt for typing it. Our research has been supported by the Kanton Basel-Stadt and the Swiss National Science Foundation.

References

Bender W, Spierer P, Hogness DS (1983) Science (Wash DC) 221:23-29
Collins M, Rubin GM (1982) Cell 30:71-79
Fjose A, Polito LC, Weber U, Gehring WJ (1984) EMBO J 3:2078-2094
Flavell AJ, Ish-Horowicz D (1981) Nature (Lond) 292:591-595
Gehring WJ, Paro R (1980) Cell 19:897-904
Gehring WJ, Klemenz R, Weber U, Kloter U (1984) EMBO J 3:2077-2085
Goldberg ML, Paro R, Gehring WJ (1982) EMBO J 1:93-98
Goldberg ML, Sheen J-Y, Gehring WJ, Green MM (1983) Proc Natl Acad Sci USA 80:5017-5021
Goldberg DA, Posakony JW, Maniatis T (1983) Cell 34:59-73
Green MM (1967) Genetics 56:467-482
Green MM (1969a) Genetics 61:423-428
Green MM (1969b) Genetics 61:429-441
Hazelrigg T, Levis R, Rubin GM (1984) Cell 36:469-481
Hafen E, Levine M, Garber RL, Gehring WJ (1983) EMBO J 2:617-623
Ising G, Ramel C (1976) In: Ashburner M, Novitski E (eds) Genetics and Biology of *Drosophila*, Vol 1b. Academic Press, New York, pp 947-954
Ising G, Block K (1981) Cold Spring Harbor Symp Quant Biol 45:527-551
Kidwell MG, Kidwell JF, Sved JA (1977) Genetics 86:813-833
Levis R, Dunsmuir P, Rubin G (1980) Cell 21:581-588
McClintock B (1957) Cold Spring Harbor Symp Quant Biol 21:197-216
Morgan TH (1910) Science (Wash DC) 32:120-122
O'Hare K, Rubin GM (1983) Cell 34:25-35
Paro R, Goldberg ML, Gehring WJ (1983) EMBO J 2:853-860
Potter SS (1982a) Nature (Lond) 297:201-204
Potter S (1982b) Mol Gen Genet 188:107-110
Potter S, Truett M, Philipps M, Maher A (1980) Cell 20:639-647
Rasmuson B, Montell I, Rasmuson A, Suahlin H, Westerberg BM (1980) Mol Gen Genet 177:567-570
Richards G, Cassab A, Bourouis M, Jarry B, Dissous C (1983) EMBO J 2:2137-2142
Rubin GM (1983) In: Shapiro JA (ed) Mobile genetic elements. Academic Press, New York, pp 329-361
Rubin GM, Spradling AC (1982) Science (Wash DC) 218:348-353
Rubin GM, Kidwell MG, Bingham PM (1982) Cell 29:987-994
Scholnick SB, Morgan BA, Hirsh J (1983) Cell 34:37-45
Shiba T, Saigo K (1983) Nature (Lond) 302:119-124
Spradling AC, Rubin GM (1982) Science (Wash DC) 218:341-347
Spradling AC, Rubin GM (1983) Cell 34:47-57
Truett MA, Jones RS, Potter SS (1981) Cell 24:753-763
Zachar Z, Bingham PM (1982) Cell 30:529-541

Genetic Engineering of Plants

Genetic Engineering of Plants

J. SCHELL[1,2], L. HERRERA-ESTRELLA[1], P. ZAMBRYSKI[1], M. DE BLOCK[1], H. JOOS[1], L. WILLMITZER[2], P. ECKES[2], S. ROSAHL[2], and M. VAN MONTAGU[1]

1 Introduction

The soil bacterium *Agrobacterium tumefaciens* can infect almost all dicotyledonous plants (Braun 1978, 1982). As a result of the infection, the wound tissue proliferates as a neoplastic growth, commonly referred to as a crown gall tumor. Once induced, the tumor no longer requires the presence of the bacteria to grow, and can be cultivated in vitro as an axenic culture (Braun 1943). Two main properties characterize crown galls: their ability to grow in vitro without the supplement of hormones required by normal plant cells, and their ability to synthesize a set of new metabolites termed opines, which are not present in normal cells. Opines are amino acid or sugar derivatives which can be used as a carbon and nitrogen source by the bacteria responsible for inciting the tumor (Tempé and Petit 1982).

Agrobacterium first attracted the attention of molecular biologists when it was discovered that a large plasmid, the tumor-inducing or Ti-plasmid, was responsible for the oncogenic capacity of the bacterium (Zaenen et al. 1974; Van Larebeke et al. 1974; Watson et al. 1975), and that tumor formation is the direct result of the transfer and stable integration of a defined segment of the Ti-plasmid (T-DNA) into the nuclear genome of plant cells (Chilton et al. 1977, 1980; Willmitzer et al. 1980).

In this paper we shall discuss some of the more recent advances in the systematic design of Ti-plasmid derivatives efficient at introducing foreign DNA sequences to plant cells capable of regenerating to form whole plants and some examples of expression of foreign genes in plants will be described.

2 Genetic and Functional Organization of the Ti-Plasmid

Ti-plasmids can be classified according to the type of opines they induce in transformed plant cells. To date, most studies have concentrated on the Ti-plasmids which induce tumors that produce nopaline or octopine. Crown gall induced by nopaline-type Ti-plasmids also produce agrocinopine A and B (Firmin and Fenwick 1978; Ellis and Murphy 1981), while octopine-induced tumors also produce agropine and mannopine (Tate et al. 1982).

Genetic studies defined two regions of the Ti-plasmid essential for tumor formation. These results are diagrammed in Figure 1. Mutants in one of these regions, called the Vir regions, have a nononcogenic phenotpye, while mutants in the other region of the Ti-plasmid important

[1]Laboratorium voor Genetika, Rijksuniversiteit Gent, B-9000 Gent
[2]Max-Planck-Institut für Züchtungsforschung, 5000 Köln 30, FRG

35. Colloquium - Mosbach 1984
The Impact of Gene Transfer Techniques in Eukaryotic Cell Biology

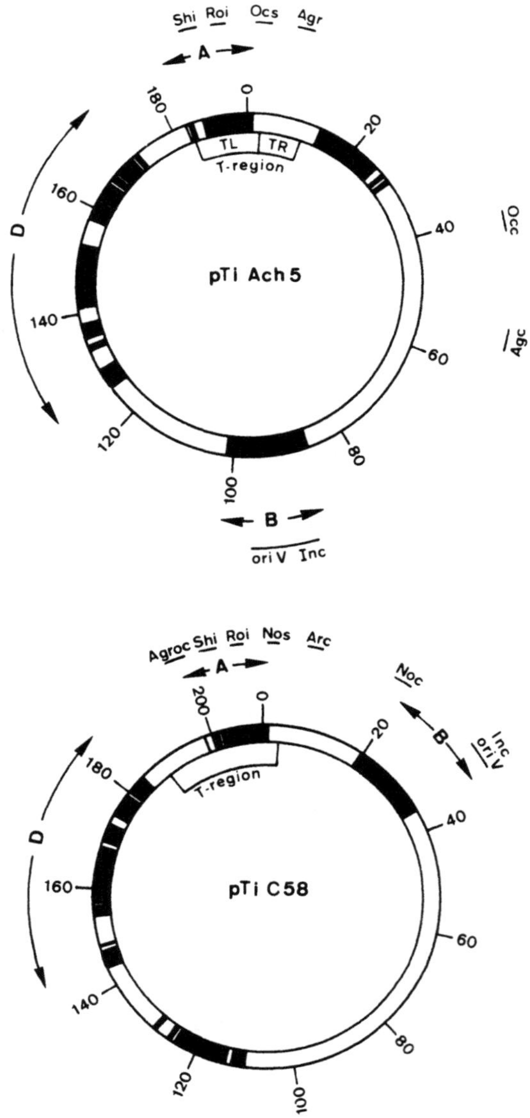

Fig. 1. Functional organization of the octopine pTiAch5 and the nopaline pTiC58 plasmids. *Numbers* indicate the size in kb of each plasmid. *Black bars* represent the regions of homology between the two plasmids as described by Engler et al. (1981). *Shi* and *roi* indicate the position of the loci involved in shoot or root inhibition, respectively. The genes responsible for opine synthesis are indicated as *ocs* (octopine synthase), *agroc* (agrocinopine synthase), *agr* (agropine synthase), and *nos* (nopaline synthase). The region containing the genes involved in octopine or nopaline catabolism are indicated *occ* and *noc* respectively. The D or Vir region has large stretches of homology between both plasmids and is the region essential for the virulence of *Agrobacterium*. A more detailed map and description of these plasmids is given by De Greve et al. (1981) and Holsters et al. (1980)

for tumor induction (the T-region) either have an attenuated oncogenic phenotype or induce tumors with altered morphology (Holsters et al. 1980; De Greve et al. 1981; Garfinkel et al. 1981; Leemans et al. 1982; Joos et al. 1983; Inzé et al. 1984).

Cot and Southern blot hybridization analysis demonstrated that the T-region contains all of the Ti-plasmid sequences which are transferred and integrated into the nuclear genome of tumor lines (Chilton et al. 1977, 1980; Thomashow et al. 1980; Lemmers et al. 1980). In some octopine plasmids, e.g., pTiA6NC, pTiAch5, and pTiB6S3, the T-region is devided into two adjacent DNA segments, one of 13 kb (left T-DNA), and one of approximately 7 kb (right T-DNA) (Thomashow et al. 1980; De Beuckeleer et al. 1981). These two DNA segments can either be trans-

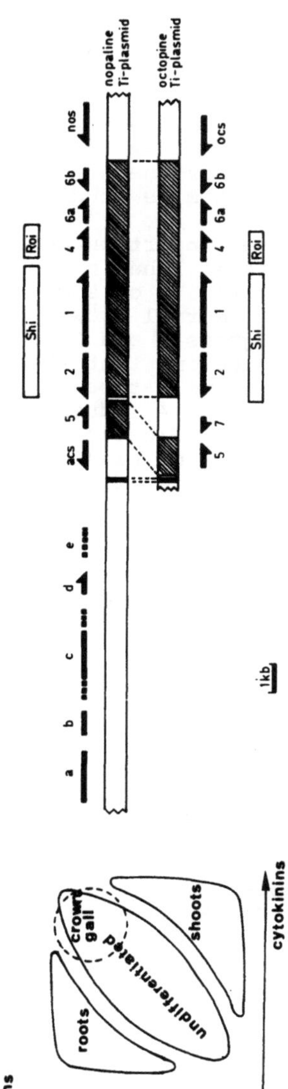

Fig. 2. Transcriptional map of the octopine and nopaline T-DNA's. The regions of homology between the two T-DNA's (common region) are marked as *hatched areas*. The position, length, and direction of the T-DNA transcripts is as described by Willmitzer et al. (1982, 1983). Transcripts *1*, *2*, *3*, *4*, *5*, *6a* and *6b* are common to both T-DNA's. *Shi* indicates transcripts which control shoot inhibition; *roi* indicates the transcript which controls root inhibition; *nos*, nopaline synthase; *ocs*, octopine synthase; *acs*, agrocinopine synthase. The auxin/cytokinin diagram is a schematic representation of how the hormone balance can influence callus morphology as first described by Skoog and Miller (1957)

ferred to the plant genome independently of each other or else as a continuous stretch of DNA colinear with the T-region present in the Ti-plasmid. Nopaline plasmids, like pTiC58 and pTiT37, transfer a single DNA segment of approximately 23 kb (Lemmers et al. 1980; Hepburn et al. 1983a; Ursic et al. 1983) (see Fig. 1).

Octopine and nopaline T-DNA's are expressed in plant cells, encoding several polyadenylated transcripts, which are transcribed by the RNA polymerase II of the plant (Willmitzer et al. 1981, 1982, 1983; Bevan and Chilton 1982; Gelvin et al. 1982). The position of these transcripts was mapped by hybridization with different T-region probes and the direction of transcription was identified by hybridzation to both strands of the corresponding T-region fragment. Six of the transcripts (5, 2, 1, 4, 6a, and 6b) are homologous to both T-DNA's, and are encoded by a 9 kb DNA segment common to both octopine and nopaline T-DNA's (Chilton et al. 1978; Depicker et al. 1978; Engler et al. 1981; Willmitzer et al. 1983). The T-DNA transcripts contribute to no more than 10^{-5} - 10^{-6} of the total poly(A)$^+$ RNA population; nevertheless the expression of the T-DNA genes has a dramatic effect on the physiology and phenotype of the transformed tissue. At least three of these transcripts seem to be directly responsible for the tumorous mode of growth of transformed cells.

An analysis of plant cells transformed by transposon insertion and deletion mutants of the T-DNA region of Ti-plasmids established a correlation between the phenotype of the tumor and the activity of some of the genes encoding the different T-DNA transcripts. Normal crown gall tumors grow as unorganized tissue. Mutations in genes *1* and *2* induce tumors which produce an abundance of shoots, indicating that these genes encode functions that suppress shoot formation in wild-type tumors (Shi; see Fig. 2). Ti-plasmids with mutations in gene *4* induce tumors which characteristically show root proliferation indicating that gene *4* encodes a function which blocks the normal ability of plant cells to undergo differentiation to form roots (Roi; see Fig. 2) (Garfinkel et al. 1981; Leemans et al. 1982; Joos et al. 1983a).

Other T-DNA genes have been shown to encode the enzymes responsible for opine synthesis. Such is the case for the nopaline (Holsters et al. 1980) and agrocinopine (Joos et al. 1983a) synthases in pTiC58, and octopine and agropine synthases in pTiAch5 and pTiB6S3 (De Greve et al. 1981; Salomon et al. 1984). Although there are several other genes which are transcribed, it has not been possible thus far to correlate their mutants with any easily detectable change in the phenotype of the corresponding tumor.

One of the main conclusions from the studies using deletion mutants is that none of the genes encoded by the T-DNA are essential for transfer or integration of the T-DNA. T-DNA transfer occurred with all mutants tested, including those that were unable to induce tumor formation. In the latter case, using Ti-plasmid mutants with large deletions removing most of the internal part of the T-DNA and containing only one of the genes encoding opine synthesis and the T-DNA borders, it was shown that cells at the wound site were still able to synthesize opines, indicating that T-DNA transfer and integration had occurred (Leemans et al. 1982; Joos et al. 1983a; Zambryski et al. 1983).

3 T-DNA Integration

The mechanism by which *Agrobacterium* transfers the T-region segment of its Ti-plasmid to the genome of plant cells, is as yet still unknown. The interaction of *Agrobacterium* with the competent plant cells initiates a series of events, activating specific genes in both the bacteria (G. Nester, personal communication) and the host. To understand how T-DNA transfer occurs, regions of the Ti-plasmid which are integrated into the plant genome were identified.

Four nopaline Ti-plasmid-induced tumor lines have been analyzed in detail (Zambryski et al. 1982). The data obtained from this analysis show that there is little variation in the T-DNA integrated into the genome of the different tumor lines analyzed. Some of these tumor lines appear to have a single copy of the T-DNA; others have multiple copies, several of which are organized in tandem arrays. Molecular cloning of the ends of the T-DNA from the genome of transformed plant cells showed that some DNA fragments contain sequences derived only from the right or left end of the T-DNA, while others contain sequences derived from both, confirming the presence of tandem copies of the T-DNA in trans-

G C T G G	T G G C A G G A T A T A T T G	T G	G T G T A A A C	A A A T T	Nopaline L
G T G T T	T G A C A G G A T A T A T T G	G C	G G G T A A A C	C T A A G	Nopaline R
A G C G G	C G G C A G G A T A T A T T C	A A	T T G T A A A T	G G C T T	Octopine A
C T G A C	T G G C A G G A T A T A T A C	C G	T T G T A A T T	T G A G C	Octopine B
A A A G G	T G G C A G G A T A T A T C G	A G	G T G T A A A A	T A T C A	Octopine C
A C T G A	T G G C A G G A T A T A T G C	G G	T T G T A A T T	C A T T T	Octopine D

Fig. 3. Comparison of the 25 bp terminal sequence flanking the T-DNA's of octopine and nopaline Ti plasmids. *Boxes* indicate the regions with higher degrees of homology between these terminal repeats. The two base pairs in between the boxes are not conserved. Nopaline L *(left)* and R *(right)* are the repeats flanking the T-DNA of nopaline-type plasmids pTiC58 of pTiT37. Octopine *A/B* and *C/D (left/right)* represent the terminal repeats flanking the TL or TR regions, respectively, present in the octopine-type Ti plasmids pTiAch5 and pTi15955. Data from Zambryski et al. 1982; Yadav et al. 1982; Simpson et al. 1982; Holters et al. 1983; Barker et al. 1983)

formed cells (Zambryski et al. 1980, 1982; Yadav et al. 1982; Holters et al. 1983).

Determination of the nucleotide sequence of the junction fragments has revealed the precise ends of the T-DNA present in the genome. In all cases, the homology between the sequences in the tumor DNA, and those present in the Ti-plasmid end within, or close to a 25 bp sequence which flanks the T-region in the Ti-plasmid as direct-repeats (Simpson et al. 1982; Yadav et al. 1982; Zambryski et al. 1982; Holters et al. 1983). Within the 25 bp of these repeats, twelve base pairs show a perfect homology and the rest imperfect homology. These terminal repeated sequences are not only present at the ends of the T-region of nopaline plasmids, but also at the ends of the TL- and TR-regions of the octopine plasmid (Holsters et al. 1983; Barker et al. 1983). Figure 3 shows the nucleotide sequence of the 25 bp sequence present in nopaline and octopine T-DNA's. Analysis of octopine TL-borders, recloned from transformed cells, showed that the TL integration is quite analogous to that found for nopaline T-DNA, including the presence of multiple T-DNA copies in tandem array. Detailed analysis of one of the junctions between two copies of the T-DNA in tandem showed that the junction consisted of a 40 bp unit of plant origin repeated six times. The presence of plant DNA between the tandem T-DNA copies suggests that the generation of the tandem in this particular tumor took place during or after the insertion of an original copy of the T-DNA in the plant genome (Holsters et al. 1983).

A search in two nucleotide sequence data banks revealed that the 12 bp of perfect homology of the 25 bp T-DNA flanking repeats is only present in the Ti-plasmid flanking the T-DNA, but not in any other DNA sequence known to date (Barker et al. 1983).

Although the 25 bp repeat is present at both ends of the T-DNA, genetic analysis using deletion mutants has shown that the T-DNA ends are functionally different. Deletions of the right ends of the nopaline T-DNA makes the Ti-plasmid virtually avirulent on most plant species; in contrast, however, deletions of the left end have little or no effect on the transfer and integration of the T-DNA (Joos et al. 1983a).

Recently, a series of experiments has shown that the right end of the nopaline T-DNA is sufficient to direct the integration of bacterial sequences in the plant genome. It was shown that the nopaline synthase gene which is linked to the right end of the T-DNA can still be efficiently transferred when inserted in a different part of the Ti-plasmid, or when inserted in an independent replicon (Joos et al. 1983b; Caplan, Depicker, personal communication).

It has also been shown that in order to be transferred the T-DNA does not have to be physically linked to the Vir region. Tumor formation by *Agrobacterium* harboring two compatible plasmids, one containing the Vir region, and the other the T-DNA, is indistinguishable from tumor formation by *Agrobacterium* harboring the wild-type Ti-plasmid. However, *Agrobacterium* containing any of these plasmids alone are not oncogenic. This demonstrates that the Vir region is trans-acting in relation to the necessary functions for T-DNA transfer (De Framond et al. 1983; Hoekema et al. 1983). The finding that the octopine Vir region can complement the nopaline T-DNA and vice versa, demonstrates that there are no plasmid-specific virulence functions.

One can think of several models to explain T-DNA transfer and integration. One possibility is that the T-region is excised from the Ti-plasmid in *Agrobacterium* during the process of infection; in this model the T-region would be cleaved out by enzymes that are probably encoded by the Vir region, and which specifically recognize the T-DNA borders. Since non of the T-DNA genes are involved in T-DNA integration and if only the T-DNA region enter the plant cell, then the host must provide the functions necessary for integration. A second model suggests that the whole Ti-plasmid, and possibly all of the bacterial DNA enters the plant cell, and after the T-region has been inserted into a chromosome the remainder of the bacterial DNA is lost. It is conceivable that the T-region is excised in the *Agrobacterium* prior to fusion of the bacterial cells to the plant cells or that excision and integration of the T-DNA are provided by the host plant.

Most mutations in the Vir region decrease or eliminate the oncogenicity of the Ti-plasmid. Complementation studies using cosmid clones overlapping these mutations have shown that the *vir* genes are clustered in a number of complementation groups (Klee et al. 1982, 1983), and if we assume that the cosmid is not transferred to the host, this also suggests that most, if not all, the genes encoded by the Vir region are functional in the bacterium. It is known that when a plant is coinfected on the same wound with bacteria containing two different Ti-plasmids, the T-DNA of both plasmids can integrate in the same cell (Ooms et al. 1982). When different Vir mutants were used to coinfect on the same wound, they are not able to complement each other (Iyer et al. 1982).

4 Transfer and Expression of Foreign Sequences in Plant Cells

Having established that the interaction between *Agrobacterium* and wounded plant cells results in the transfer of a segment of Ti-plasmid (T-DNA) into the genome of plant cells, the question arose as to whether other DNA sequences, experimentally inserted within the T-DNA would be efficiently co-transferred. A Tn7 insertion in the nopaline synthase locus produces a Ti-plasmid mutant still capable of T-DNA transfer. Analysis of DNA from these tumors showed that the whole T-DNA region, including the inserted Tn7 sequence, had been transferred and integrated as a single 38×10^3 bp segment into the genome of tobacco cells (Hernalsteens et al. 1980).

Several different DNA sequences have since been introduced into various regions of the T-region of Ti-plasmids. Analysis of the DNA transferred by such mutated Ti-plasmids led to the conclusion that probably any DNA sequences inserted within the T-DNA will be cotransferred and integrated without any detectable rearrangement. So far, no limit has been found to the size of DNA that can be transferred and integrated when inserted between T-DNA borders. Sequences of up to 50 kb have been introduced into the genome of tobacco as a single DNA segment.

Early attempts to express foreign genes in plants, using the antibiotic resistance gene encoded by the transposons used to mutagenize the T-region, showed that prokaryotic genes are not expressed in plant cells, presumably because the prokaryotic transcriptional signals of these genes are not recognized by the transcriptional machinery of the plant. Further attempts to express heterologous genes in plants were made using genes from other eukaryotic organisms, such as the yeast alcohol dehydrogenase gene (Barton et al. 1983), or genes from mammalian cells, such as β-globin, interferon, and genes under control of the early SV40 promoter (Koncz et al. 1984). Analysis of the messenger RNA of tissues containing the genes reveal that none of them were expressed. This suggests that specific factors or signals required for their expression are present only in the cells or specific tissues of their original host and are not present in plant cells. This can be correlated with proteins which may interact with specific segments of DNA in the promoter region, like enhancers (Benoist and Chambon 1981), or sequences which determine differential or tissue-specific expression, as has been demonstrated for some animal genes (Walker et al. 1983). In contrast some foreign plant genes have been found to be expressed in heterologous systems, e.g., the phaseolin gene (Murai et al. 1983) and the Zein gene in sunflower (Matzke et al., in press).

Nevertheless, an essential requirement for systematic expression of heterologous genes in plants was to use the transcriptional signal of a gene known to be expressed in plants. Most of the plant genes cloned to date are specifically regulated and may only be expressed at a particular stage of the plant development. For instance, the gene families encoding for the zein or phaseolin storage proteins are only expressed during seed formation, and the leghemoglobins are only induced during interaction between legumes and *Rhizobium* (Geraghty et al. 1981; Brisson and Verma 1982; Wiborg et al. 1982; Pedersen et al. 1982). The best candidates for constitutively expressed genes proved to be those encoded by the T-DNA, particularly the ones encoding the nopaline or the octopine or the agropine synthase. These genes have been shown to be normally expressed both in callus tissue and in all differentiated tissues of plants regenerated from calli containing the opine genes (De Greve et al. 1982a; Wöstemeyer et al. 1983; Velten et al., in preparation).

The nucleotide sequences of the nopaline and octopine synthase genes, as well as other T-DNA genes, revealed that they have transcriptional signals that resemble the consensus sequences which have been found to be important for the start and stop of transcription of other eukaryotic genes. At the 5'-flanking region the genes contain sequences homologous to the TATA or Goldberg Hogness box, 30 to 40 nucleotides upstream of the start of transcription, and a sequence similar to the CAT box, 60 to 80 nucleotides upstream the 5' end of the transcript. In addition, they have sequences similar to the AATAAA between 20 and 100 nucleotides 5' to the polyadenylation site, which strongly resembles the consensus sequence similarly placed in animal genes (Depicker et al. 1982; De Greve et al. 1982b; Bevan et al. 1983a; Barker et al. 1983; Gielen et al. 1984).

So far, most of the information about the transcriptional signals used by plant cells has been obtained from studies using T-DNA-encoded genes. Thus, it has been shown that for the *ocs* gene, sequences upstream of position -170 from the start of transcription are essential for efficient transcription of this gene. The functional *ocs* promoter is comprised within 294 bp of the 5' region of the start point of transcription (Koncz et al. 1983).

The *ocs* gene produces transcripts that are polyadenylated at either of two positions in the 3'-untranslated region of this gene, although the longer transcript was found to be more abundant. A mutant of the *ocs* with deletion of the 3'-untranslated region was found to be still active. This deletion terminates 19 bp upstream of the major polyadenylation site of the wild-type transcript, removing the AATAAA signal preceding this site. The *ocs* mutant gene produces transcripts with a polyadenylated tail added at four different positions (Dhaese et al. 1983). The most abundant of these transcripts is polyadenylated at the same position as the minor one in the wild-type gene. All these results suggest that the AATAAA signal in plants is mainly involved in the post-transcriptional processing of the transcripts, similar to what was previously found in animal cells.

Based on this information, a series of vectors have been constructed using the promoter and terminator signals for transcription of the nopaline synthase gene *(nos)* gene. In-between these signals, restriction sites were introduced for the insertion of any desired sequence. The functionality of these vectors was first tested using the coding sequence of another T-DNA gene, the octopine synthase. It was shown that the *nos* promoter was able to direct functional expression of the octopine synthase coding sequence when transferred to plant cells (Herrera-Estrella et al. 1983a). An example of a heterogenous gene being expressed in plants was the bacterial chloramphenicol acetyl transferase *(cat)* gene from pBR325. When the coding sequence of the *cat* gene was brought under control of the *nos* promoter signal and transferred to tobacco cells, the transformed cells were shown to express this chimeric gene and to produce a functional protein (Herrera-Estrella et al. 1983a).

Since foreign genes were shown to be expressed in plant cells, and the resulting proteins were shown to be stable and active, it became feasable to construct genes that can be used as dominant selectable markers.

Bacterial genes, such as the aminoglycoside phosphotransferase APH(3')II and APH(3')I carried by transposons Tn5 and Tn903 respectively, inactivate related aminoglycoside antibiotics like neomycin, kanamycin or G418. The APH(3')II from Tn5 had been shown to provide selectable resistance to G418 in mammalian cells, when expressed under control of the SV40 promoter, and in *Dictyostelium* when expressed from the *Dictyostelium* actin 8 promoter (Southern and Berg 1982; Hirth et al. 1982). Since plant cells are sensitive to several of these aminoglycosides, expression of an APH activity could therefore be expected to determine a resistance phenotype.

Chimeric genes containing the APH(3')I and APH(3')II coding sequences under control of the *nos* transcriptional signals were constructed and inserted into the T-DNA of the Ti-plasmids (Herrera-Estrella et al. 1983b; Bevan et al. 1983b; Fraley et al. 1983). A modification of an in vitro cell transformation system, based on the cocultivation of regenerating plant cell protoplasts with *Agrobacterium* (Martón et al. 1979) was used to transfer the chimeric gene into plant cells (De Block et al. 1984). When small, fastgrowing plant cell colonies derived from protoplasts used for the cocultivation experiments were transferred to selective plates containing 50 μg ml^{-1} of kanamycin, resistant colonies were obtained with frequencies up to 50%. The resistant colonies were shown to be transformed by Southern blot hybridization and testing for linkage with the hormone independence markers of the T-DNA. Later experiments have shown that the methotrexate-insensitive dihydrofolate reductase encoded by the plasmid R67, and the chloramphenicol acetyl transferase can also be used as selectable markers, although the resistance phenotype is not as marked as in the case of the APH(3')II (Herrera-Estrella et al. 1983b; De Block et al. 1984). These experiments have shown that four different bacterial coding sequences can be properly transcribed and their messenger RNA translated into functional proteins in plants. Thus, the codon usage for the plant translation machinery may allow expression of any other bacterial, fungal or mammalian gene, including those which could confer a useful trait to plant cells.

The first chimeric *nos* Km^R gene used for transformation experiments contained an ATG codon in the 5'-untranslated sequence out of frame with the APH coding sequence. It has been shown that in most eukaryotic genes, the first ATG in the messenger RNA is the initiation codon, and that when ATGs other than the initiation codon are present in the 5' untranslated region, the proper translation of the mRNA is less efficient. To test whether the same phenomenon occurs for plant mRNA's, Steve Rogers at the Monsanto laboratory and we constructed chimeric genes such that the extra ATG in the leader of the APH(3')II gene was deleted. When plant cells transformed with this gene were tested for the level of resistance to kanamycin, they were found to grow normally at a level four times higher than that of the original construction. Since the level of transcription is the same for both genes, this suggests that at least for this gene the messenger RNA is more efficiently translated if the initiation codon is the first ATG present in the transcribed part of the gene.

5 Regeneration of Whole Plants Containing Foreign DNA Sequences

Based on our knowledge regarding which genes of the T-DNA prevent differentiation, and which functions enable the Ti-plasmid to transfer DNA sequences to plant cells, it was possible to systematically design Ti-plasmid derivatives capable of efficient transfer of DNA to plant cells, but devoid of the genes preventing normal differentiation. Based on the Ti-plasmid pTiC58, a deletion was constructed which removes most of the T-DNA, but which retains the T-DNA border sequences allowing transfer, as well as the nopaline synthase gene as a T-DNA specific marker gene. In this Ti-plasmid the internal portion of the T-DNA was replaced by the cloning vehicle pBR322. When wounded plants were inoculated with this Ti-plasmid derivative (pGV3850), it was found that although there was no tumor formation, the cells of the wound site did produce nopaline (Zambryski et al. 1983). After propagation in vitro, the wound calli were transferred to medium which stimulates shoot regeneration. Plantlets derived from these shoots were screened for nopaline production, and it was found that between 10 and 70% of these plantlets were nopaline-positive (Zambryski et al. 1983). Southern blotting analysis showed that the nopaline-positive plantlets contained one or more copies of the modified pGV3850 T-DNA. These plantlets are able to grow into mature plants which are fertile, and produce seeds. It has also been shown that the newly acquired trait is stable through the meiotic process and is transmitted normally to the progeny. It is worth mentioning that some of these plants transmitted the nopaline synthase trait to 100% of the F1 progeny. This suggests that some abnormal loss and/or duplication of chromosomes or gene-conversion occurred.

Because pBR322 is inserted in-between the T-DNA borders in pGV3850, this modified Ti-plasmid is a versatile acceptor for introducing genes into plants. Any DNA sequence cloned into pBR322, or its derivatives, can be easily introduced into the pGV3850 T-DNA, by a single homologous recombination event between the pBR present in pGV3850 and the homologous sequence present in the cloning vehicle (Fig. 4). pBR322-like plasmids can be easily mobilized from *E. coli* to *Agrobacterium*, if one provides in trans the Mob functions of Co1E1 and the Tra functions of a selftransmissible plasmid-like R64*drd*11 (Van Haute et al. 1983). As the origin of replication of pBR322 is not functional in *Agrobacterium*, the only way its resistance markers can be maintained after being transferred to *Agrobacterium* is by recombination with the homologous region in the pGV3850 plasmid carrying the genetic markers. The cointegrates can be selected for and maintained in *Agrobacterium* by including a drug

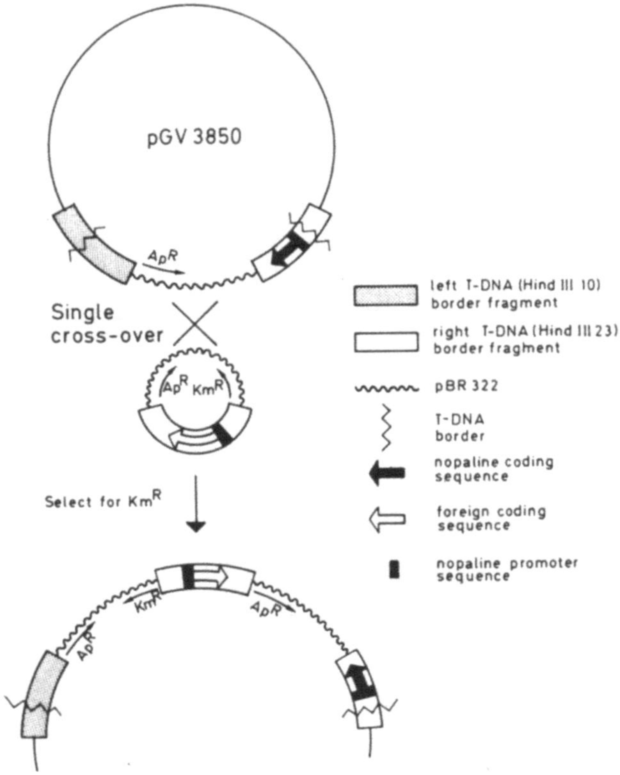

Fig. 4. Ti-plasmid-derived vector pGV3850 and its use as an acceptor plasmid for the introduction of foreign DNA sequences into plants. The pGV3850 T-DNA only contains the nopaline synthase gene and the border sequences essential for T-DNA transfer. *Wavy line* represents pBR322 used to replace the internal part of the wild-type T-DNA. As described in the text, a single recombination event between the pBR322 sequences present in the T-DNA and the homologous sequences in the cloning vehicle allows the introduction of foreign sequences into the T-DNA

resistance marker other than ampicillin in the cloning vehicle. As shown in Figures 4 and 5, the cointegrates contain a direct duplication of the pBR sequences. In *Agrobacterium*, the cointegrate can be maintained by selective growth in media containing the appropriate drugs. It is unlikely that these cointegrates will be unstable in plant cells, as the plant genome is composed of much repeated DNA, including, at times, the T-DNA itself which can be present in tandem copies. So far, we have not found any evidence for instability of the T-DNA from these cointegrates in tissue or plants have been growing for several months. In other experiments, it has been found that some tumor tissues grown in vitro for long periods of time, do not produce any detectable amount of opines. In other cases, tumors that were thought to be opine-negative produce transformed shoots that regain opine production. In both cases, the loss of opine production has been correlated with methylation of the gene encoding the corresponding opine synthase, since opine production can be restored by adding the 5-azacytidine compound that inhibits DNA methylation (Hepburn et al. 1983b; Van Slogteren et al. 1983).

6 Expression of Foreign Genes in Whole Plants

Using the pGV3850 system described before, cointegrates were constructed containing chimeric genes that confer resistance to either kanamycin,

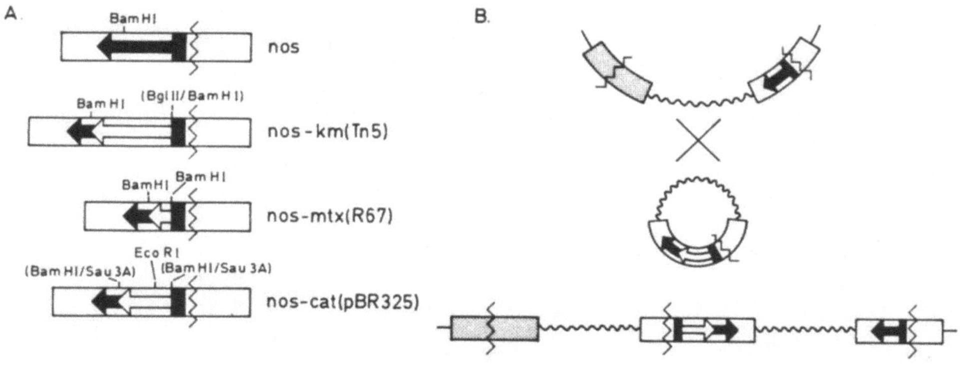

Fig. 5. Use of pGV3850 as acceptor to introduce dominant selectable markers into plant cells. A Schematic representation of the various chimeric genes containing bacterial coding sequences under control of the *nos* promoter. The wild-type *nos* gene is shown compared to the structure of the chimeric gene; B Structure of one of the cointegrates formed between one of the chimeric genes shown in A and pGV-3850. Such cointegrates contain an extra right T-DNA border sequence *(jagged line)*

methotrexate or chloramphenicol in tissue cultures of plant cells (Fig. 5). The T-DNA of these cointegrates was transferred to tobacco cells using the co-cultivation method (De Block et al. 1984). Plant cell colonies obtained after co-cultivation were plated on selective medium containing the appropriate antibiotic. Transformed colonies able to grow in the presence of the antibiotics were easily obtained. The co-cultivation method was used to obtain resistant colonies, since these colonies arise from a single or very few protoplasts, and in most of the cases represent a single T-DNA transfer event.

After being propagated in vitro, the transformed colonies were transferred to regeneration medium. After a few weeks plantlets were obtained; these plantlets were shown to grow into mature plants and to produce seeds. Using Southern blot analysis it was shown that these transformed plants contained the chimeric genes in their genome and that they express the chimeric genes: (1) pGV3850::neo-transformed shoots and control shoots were transferred to agar medium containing 100 $\mu g\ ml^{-1}$ kanamycin; after 3 weeks the transformed shoots were able to develop roots and maintain normal growth, in contrast to control shoots which did not form roots, and eventually died; (2) seeds obtained from these plants and control nontransformed plants were germinated in the presence of 100 $\mu g\ ml^{-1}$ kanamycin, although both transformed and nontransformed seeds were able to germinate, control seeds became etiolated and died after 3 weeks whereas transformed seeds were able to maintain normal growth after germination despite the presence of antibiotic; (3) plants transformed with pGV3850cat were shown to express CAT activity in all their differentiated tissues.

Using the ability of pGV3850neo-transformed seeds to germinate and to grow in kanamycin-containing medium, the genetic transmission pattern of the kanamycin resistance trait was examined by self-fertilizing the regenerated plants. In all cases a ratio close to 3:1 resistant to sensitive seeds was found (Fig. 6), suggesting that the mother plant was heterozygous for the resistance trait and that the trait is transmitted as a single dominant Mendelian factor. Similar results were recently published by Horsch et al. (1984).

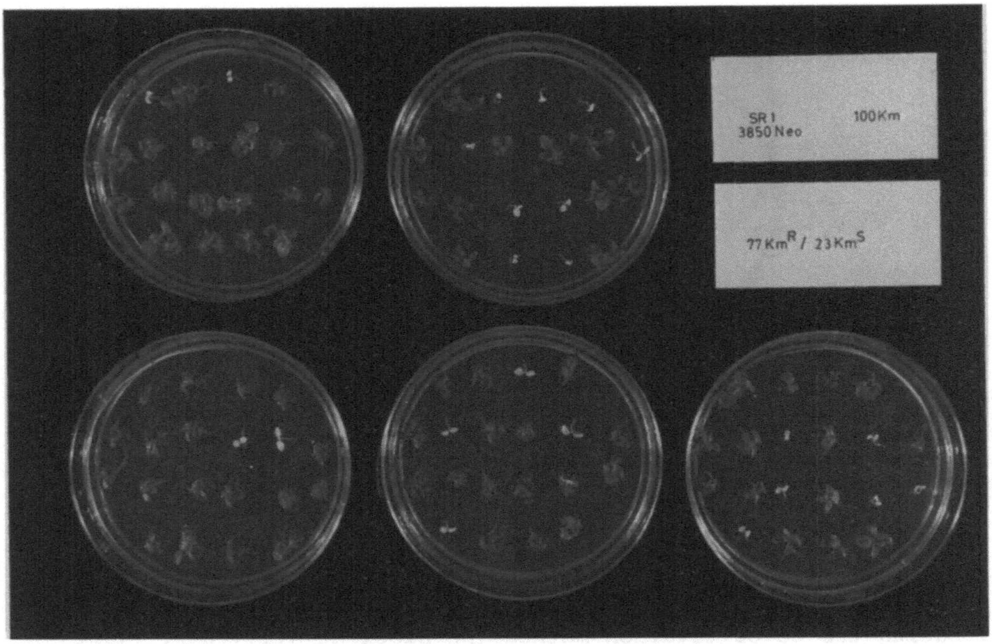

Fig. 6. Germination of seed obtained by selfing plants containing the *nos*-Km^R chimeric gene in kanamycin-containing medium. Seeds were germinated in solid medium containing Murashige and Skoog medium salts, and 100 μg ml^{-1} kanamycin sulfate. All seeds germinated, but whereas those harbouring the *nos*-Km^R gene developed normally, the sensitive ones etoilated, and eventually died. The genetic transmission pattern of the first generation (S1) from the self-fertilized plant containing the *nos*-Km^R gene is shown in the photograph. From 100 seeds tested, 77 showed resistance to kanamycin whereas 23 were sensitive. Thus, the Km^R trait was inherited as a dominant Mendelian factor, as deduced from the 3:1, resistant to sensitive ratio

7 Regulated Expression of Foreign Genes in Plant Cells

The induction by light of many of the genes involved in the photosynthetic pathway of plants is one area of particular interest in plant molecular biology. Several members of the gene families encoding the small-subunit of the ribulose bisphosphate carboxylase (RuBisCo) have been isolated from different plant species (Berry-Lowe et al. 1982; Cashmore 1983; Broglie et al. 1983). RuBisCo performs the first step in the Calvin cycle (hydrolytic cleavage of ribulose-1,5-bisphosphate to form two molecules of 3-phosphoglycerate) and is composed of two subunits, the large subunit which is encoded by the chloroplast genome and the small subunit which is encoded by the nuclear genome. The small subunit (ss) is synthesized on free cytoplasmic ribosomes as a 20,00 MW precursor. The ss is post-translationally processed during transport into chloroplasts to yield the mature polypeptide.

The light-inducible expression of the ss gene family has been shown to be regulated at the transcriptional level in green tissue (Gallagher and Ellis 1982). To determine whether this type of light-regulated expression is controlled either by the sequences 5' to the promoter, or by sequences located in another region of the gene, a chimeric gene was constructed containing 900 bp of the promoter region of a small subunit gene isolated from pea, coupled to the CAT-coding sequence from Tn9 and the 3' end sequence of the *nos* gene (Herrera-Estrella et al. 1984). The *cat* gene is an important tool to monitor the transcription of promoters, since a simple assay allows quantitation of the level of enzymatic activity in cells of different organisms which usually represents the level of transcription initiated at the promoter under study (Gorman et al. 1982).

This chimeric gene was introduced into tobacco cells and the influence of light on the CAT activity was determined. It was found that the 900 bp of the 5'-upstream region of the *ss* gene contains signals sufficient to confer light-inducible expression of the *cat* gene. Analysis of the levels of CAT mRNA confirmed that there is a 20-fold increase in the level of expression of the *ss-cat* chimeric gene, when green tissue is grown under light condition as compared to tissue grown in the dark. It was also observed that in white undifferentiated tissue, the level of expression is extremely low, even when the tissue is grown under light conditions (Herrera-Estrella et al. 1984).

Expression of endogenous wild-type ss-genes is thought to follow a certain organ-specificity, i.e., there is high expression in leaf/stem tissue and very low or nearly no expression in root tissue. However, recent experiments performed in our group with potato plants indicate that ss-genes, although silent in white roots, are very active in green roots, which are easily obtained under certain tissue culture conditions. It may therefore be that expression of ss-genes is coupled to the presence of light plus in addition to chloroplast development, as indicated by the greening of the tissue. These findings constitute evidence that sequences upstream of the start of transcription are sufficient to confer regulated expression of a plant gene, and also demonstrate that a promoter from one plant species (pea) can be functional in cells of another plant species (tobacco).

8 Organ-Specific Expression of Genes in Plants

One of the ultimate goals concerning the introduction of foreign genes into plants is their controlled and regulated expression. We have therefore recently started to clone genes from potato displaying an organ-specific expression. The reason for this is twofold: In a first step we want to understand how organ-specific expression is regulated, and subsequently we hope to use the regulatory elements of these genes to express foreign genes under their control.

Using conventional cDNA-cloning methods we obtained three cDNA-clones from each leaf-A^t or tuber-A^t RNA, which exhibited an organ-specific expression. One of the leaf-clones is a ss-clone, one of the tuber-clones represents the cDNA for patatin, a 40,000 MW protein representing the major storage protein of tubers. Run-off experiments indicate that the organ-specific expression of most of the genes is controlled at the level of transcription. Genomic clones have been isolated from a library set up using EMBL4 vectors and are presently characterized with the ultimate goal of defining what regulate their organ-specific expression.

Acknowledgement. The authors are much indebted to their colleagues at the Laboratory of Genetics of the State University, Gent; the Laboratory of Genetische Virologie, Free University Brussels and the Max-Planck-Institut für Züchtungsforschung, Köln, for their continued help and stimulation.

LHE is indebted to CONACYT Mexico for a Ph.D. fellowship, MDB is a Senior Research Assistant of the National Fund for Scientific Research (Belgium), and PZ is supported by a long-term EMBO fellowship. The research was supported by grants from the A.S.L.K.-

Kankerfonds, from the Instituut tot Aanmoediging van het Wetenschappelijk Onderzoek in Nijverheid en Landbouw (I.W.O.N.L. 3894A), from the Services of the Prime Minister (O.O.A. 12052179), from the Fonds voor Geneeskundig Wetenschappelijk Onderzoek (F.G.W.O. 3.001.82), to MVW and JS, and was carried out under Research Contract no. GVI-4-017-B (RS) of the Biomolecular Engineering Programme of the Commission of the European Communities.

References

Barker RF, Idler KB, Thompson DV, Kemp JD (1983) Nucleotide sequence of the T-DNA region from the *Agrobacterium tumefaciens* octopine Ti plasmid pTi195955. Plant Mol Biol 2:335-350

Barton KA, Binns AN, Matzke AJM, Chilton MD (1983) Regeneration of intact tobacco plants containing full length copies of genetically engineered T-DNA, and transmission of T-DNA to R1 progeny. Cell 32:1033-1043

Benoist C, Chambon P (1981) In vivo sequence requirements of the SV40 early promoter region. Nature (Lond) 290:304-310

Berry-Lowe SL, McKnight TD, Shah DM, Meagher RB (1982) The nucleotide sequence, expression, and evolution of one member of a multigene family encoding the small subunit of ribulose-1.5-bis-phosphate carboxylase in soybean. J Mol Appl Genet 1:483-498

Bevan MW, Chilton MD (1982) Multiple transcripts of T-DNA detected in nopaline crown gall tumors. J Mol Appl Genet 1:539-546

Bevan M , Barnes WM, Chilton MD (1983a) Structure and transcription of the nopaline synthase gene region of T-DNA. Nucleic Acids Res 11:369-385

Bevan MW, Flavell RB, Chilton MD (1983b) A chimaeric antibiotic resistance gene as a selectable marker for plant cell transformation. Nature (Lond) 304:184-187

Bisaro DM, Hamilton WDO, Coutts RHA, Buch KW (1982) Molecular cloning and characterization of the two DNA components of tomato golden mosaic virus. Nucleic Acids Res 10:4913-4922

Braun AC (1943) Studies on tumor inception in crown gall disease. Am J Bot 30:674-677

Braun AC (1978) Plant tumours. Biochim. Biophys Acta 516:167-191

Braun AC (1982) A history of the crown gall problem. In: Kahl G, Schell J (eds) Molecular biology of plant tumors. Academic, New York, p 155

Brisson N, Verma DP (1982) Soybean leghemoglobin gene family: normal, pseudo, and truncated genes. Proc Natl Acad Sci USA 79:4055-4059

Broglie R, Coruzzi G, Lamppa G, Keith B, Chua NH (1983) Structural analysis of nuclear genes coding for the precursor to the small subunit of wheat ribulose-1.5-bisphosphate carboxylase. Bio/technology 1:55-61

Cashmore AR (1983) Nuclear gene encoding the small subunit of ribulose-1.5-biphosphate carboxylase. In: Kosuge T, Meredith CP, Hollaender A (eds) Genetic engineering of plants - an agricultural perspective. Plenum, New York, p 29

Chilton MD, Drummond HJ, Merlo DJ, Sciaky D, Montoya AL, Gordon MP, Nester EW (1977) Stable incorporation of plasmid DNA into higher plant cells: the molecular basis of crown gall tumorigenesis. Cell 11:263-271

Chilton MD, Drummond MH, Merlo DJ, Sciaky D (1978) Highly conserved DNA of Ti-plasmids overlaps T-DNA, maintained in plant tumors. Nature (Lond) 275:147-149

Chilton MD, Saiki RK, Yadav N, Gordon MP, Quetier F (1980) T-DNA from *Agrobacterium* Ti plasmid is in the nuclear DNA fraction of crown gall tumor cells. Proc Natl Acad Sci USA 77:4060-4064

Comai L, Kosuge T (1982) Cloning and characterization of iaaM, a virulence determinant of *Pseudomonas savastanoi*. J. Bacteriol 149:40-46

De Beuckeleer M, Lemmers M, De Vos G, Willmitzer L, Van Montagu M, Schell J (1981) Further insight on the transferred DNA of octopine crown gall. Mol Gen Genet 183: 283-288

De Block M, Herrera-Estrella L, Van Montagu M, Schell J, Zambryski P (to be published 1984) Expression of foreign genes in regenerated plants. EMBO J 3

De Framond AJ, Barton KA, Chilton MD (1983) Mini-Ti: a new vector strategy for plant genetic engineering. Bio/technology 1:262-269

De Greve H, Decraemer H, Seurinck J, Van Montagu M, Schell J (1981) The functional organization of the octopine *Agrobacterium tumefaciens* plasmid pTiB6S3. Plasmid 6:235-248

De Greve H, Leemans J, Hernalsteens JP et al. (1982a) Regeneration of normal and fertile plants that express octopine synthase, from tobacco crown galls after deletion of tumour-controlling functions. Nature (Lond) 300:752-755

De Greve H, Dhaese P, Seurinck J, Lemmers M, Van Montagu M, Schell J (1982b) Nucleotide sequence and transcript map of the *Agrobacterium tumefaciens* Ti plasmid-encoded octopine synthase gene. J Mol Appl Genet 1:499-512

Depicker A, Van Montagu M, Schell J (1978) Homologous DNA sequences in different Ti-plasmids are essential for oncogenicity. Nature (Lond) 275:150-153

Depicker A, Stachel S, Dhaese P, Zambryski P, Goodman HM (1982) Nopaline synthase: transcript mapping and DNA sequence. J Mol Appl Genet 1:561-574

Dhaese P, De Greve H, Gielen J, Seurinck J, Van Montagu M, Schell J (1983) Identification of sequences involved in the polyadenylation of higher plant nuclear transcripts using *Agrobacterium* T-DNA genes as models. EMBO J 2:419-426

Ellis JG, Murphy PJ (1981) Four new opines from cronw gall tumours - their detection and properties. Mol Gen Genet 181:36-43

Engler G, Depicker A, Maenhaut R, Villaroel-Mandiola R, Van Montagu M, Schell J (1981) Physical mapping of DNA base sequence homologies between an octopine and a nopaline Ti-plasmid of *Agrobacterium tumefaciens*. J Mol Biol 152:183-208

Fedoroff N (1983) Controlling elements in maize. In: Shapiro JA (ed) Mobile genetic elements. Academic, New York, p 1

Fedoroff N, Wessler S, Shure M (1983) Isolation of the transposable maize controlling elements *Ac* and *Ds*. Cell 35:235-242

Firmin JL, Fenwick GR (1978) Agropine - a major new plasmid-determined metabolite in crown gall tumours. Nature (Lond) 276:842-844

Fraley RT, Rogers SG, Horsch RB et al. (1983) Expression of bacterial genes in plant cells. Proc Natl Acad Sci USA 80:4803-4807

Gallagher TF, Ellis RJ (1982) Light-stimulated transcription of genes for two chloroplast polypeptides in isolated pea leaf nuclei. EMBO J 1:1493-1498

Garfinkel DJ, Simpson RB, Ream LW, White FF, Gordon MP, Nester EW (1981) Genetic analysis of crown gall: fine structure map of the T-DNA by site-directed mutagenesis. Cell 27:143-153

Gelvin SB, Thomashow MR, McPherson JC, Gordon MP, Nester EW (1982) Sizes and map positions of several plasmid-DNA-encoded transcripts in octopine-type crown gall tumors. Proc Natl Acad Sci USA 79:76-80

Geraghy D, Pfeifer MA, Rubenstein I, Messing J (1981) The primary structure of a plant storage protein: zein. Nucl Acids Res 9:5163-5174

Gielen J, De Beuckeleer M, Seurinck J et al. (1984) The complete nucleotide sequence of the TL-DNA of the *Agrobacterium tumefaciens* plasmid pTiAch5. EMBO J 3:835-846

Gorman CM, Moffat LF, Howard BH (1982) Recombinant genomes which express chloramphenicol acetyltransferase in mammalian cells. Mol Cell Biol 2:1044-1051

Hepburn AG, Clarke LE, Blundy KS, White J (1983a) Nopaline Ti plasmid, pTiT37, T-DNA insertions into a flax genome. J Mol Appl Genet 2:211-224

Hepburn AG, Clarke LE, Blundy KS, White J (1983b) The role of cytosine methylation in the control of nopaline synthase gene expression in a plant tumor. J Mol Appl Genet 2:315-329

Hernalsteens JP, Van Vliet F, De Beuckeleer M et al. (1980) The *Agrobacterium tumefaciens* Ti plasmid as a host vector system for introducing foreign DNA in plant cells. Nature (Lond) 287:654-656

Herrera-Estrella L, Depicker A, Van Montagu M, Schell J (1983a) Expression of chimaeric genes transferred into plant cells using a Ti-plasmid-derived vector. Nature (Lond) 303:209-213

Herrera-Estrella L, De Block M, Messens E, Hernalsteens JP, Van Montagu M, Schell J (1983b) Chimeric genes as dominant selectable markers in plant cells. EMBO J 2: 987-995

Herrera-Estrella L, Van den Broeck G, Maenhaut R, Van Montagu M, Schell J, Timko M, Cashmore A (to be published 1984) Light-inducible expression of a bacterial *cat* gene controlled by the 5'-flanking sequence of the small subunit gene of Rubisco. Nature (Lond)

Hirth KP, Edwards CA, Firtel RA (1982) A DNA-mediated transformation system for *Dictyostelium discoideum*. Proc Natl Acad Sci USA 79:7356-7360

Hoekema A, Hirsch PR, Hooykaas PJJ, Schilperoort RA (1983) A binary plant vector strategy based on separation of *vir-* and T-region of the *Agrobacterium tumefaciens* Ti plasmid. Nature (Lond) 303:179-181

Holsters M, Silva B, Van Vliet F et al. (1980) The functional organization of the nopaline *A. tumefaciens* plasmid pTiC58. Plasmid 3:212-230

Holsters M, Villarroel R, Gielen J, Seurinck J, De Greve H, Van Montagu M, Schell J (1983) An analysis of the boundaries of the octopine TL-DNA in tumors induced by *Agrobacterium tumefaciens*. Mol Gen Genet 190:35-41

Horsch R, Fraley R, Rogers S, Sanders P, Lloyd A, Hoffmann N (1984) Inheritance if functional foreign genes in plants. Science (Wash DC) 223:496-498

Inzê D, Follin A, Van Lijsebettens M, Simoens C., Genetello C, Van Montagu M, Schell J (1984) Genetic analysis of the individual T-DNA genes of *Agrobacterium tumefaciens*; further evidence that two genes are involved in indole-3-acetic acid synthesis. Mol Gen Genet 194:265-274

Iyer VN, Klee HJ, Nester EW (1982) Units of genetic expression in the virulence region of a plant tumor-inducing plasmid of *Agrobacterium tumefaciens*. Mol Gen Genet 188:418-424

Joos H, Inzê D, Caplan A, Sormann M, Van Montagu M, Schell J (1983a) Genetic analysis of T-DNA transcripts in nopaline crown galls. Cell 32:1057-1067

Joos H, Timmerman B, Van Montagu M, Schell J (1983b) Genetic analysis of transfer and stabilization of *Agrobacterium* DNA in plant cells. EMBO J 2:2151-2160

Klee HJ, Gordon MP, Nester EW (1982) Complementation analysis of *Agrobacterium tumefaciens* Ti plasmid mutations affecting oncogenicity. J Bacteriol 150:327-331

Klee HJ, White FF, Iyer VN, Gordon MP, Nester EW (1983) Mutational analysis of the virulence region of an *Agrobacterium tumefaciens* Ti plasmid. J Bacteriol 153: 878-883

Koncz C, Dė Greve H, Andrê D, Deboeck F, Van Montagu M, Schell J (1983) The octopine synthase genes carried by Ti plasmids contain all signals necessary for expression in plants. EMBO J 2:1597-1603

Koncz C, Kreuzaler F, Kalman Z, Schell J (1984) A simple method to transfer, integrate and study expression of foreign genes, such as chicken ovalbumin and α-actin in plant tumors. EMBO J 3:1029-1037

Larkin PJ, Scowcroft WR (1981) Somaclonal variation - a novel source of genetic variability from cell cultures for improvement. Theor Appl Genet 60:197-214

Leemans J, Deblaere R, Willmitzer L, De Greve H, Hernalsteens JP, Van Montagu M, Schell J (1982) Genetic identification of functions of TL-DNA transcripts in octopine crown galls. EMBO J 1:147-152

Lemmers M, De Beuckeller M, Holsters M et al. (1980) Internal organization, boundaries and integration of Ti-plasmid DNA in nopaline crown gall tumours. J Mol Biol 144: 353-376

McClintock B (1967) Genetic systems regulating gene expression during development. Dev Biol Suppl 1:84-112

Martôn L, Wullems GJ, Molendijk L, Schilperoort RA (1979) In vitro transformation of cultured cells from *Nicotiana tabacum* by *Agrobacterium tumefaciens*. Nature (Lond) 277:129-130

Murai N, Sutton DW, Murray MG et al. (1983) Phaseolin gene from bean is expressed after transfer to sunflower via tumor-inducing plasmid vectors. Science (Wash DC) 222:476-482

Ooms G, Molendijk L, Schilperoort RA (1982) Double infection of tobacco plants by two complementing octopine T-region mutants of *Agrobacterium tumefaciens*. Plant Mol Biol 1:217-226

Pedersen K, Devereux J, Wilson DR, Sheldon E, Larkins BA (1982) Cloning and sequence analysis reveal structural variation among related zein genes in maize. Cell 29: 1015-1026

Rubin GM, Spradling AC (1982) Genetic transformation of *Drosophila* with transposable element vectors. Science (Wash DC) 218:348-353

Salomon F, Deblaere R, Leemans J, Hernalsteens JP, Van Montagu M, Schell J (1984) Genetic identification of functions of TR-DNA transcripts of octopine crown galls. EMBO J 3:141-146

Schröder G, Klipp W, Hillebrand A, Ehring R, Koncz C, Schröder J (1983) The conserved part of the T-region in Ti-plasmids expresses four proteins in bacteria. EMBO J 2:403-409

Schröder G, Waffenschmidt S, Weiler EW, Schröder J (1984) The T-region of Ti plasmids codes for an enzyme synthesizing indole-3-acetic acid. Eur J Biochem 138:387-391

Simpson RB, O'Hara PJ, Krook W, Montoya AL, Lichtenstein C, Gordon MP, Nester EW (1982) DNA from the A6S/2 crown gall tumors contains scrambled Ti-plasmid sequences near its junctions with the plant DNA. Cell 29:1005-1014

Skoog F, Miller CO (1957) Chemical regulation of growth and origin formation in plant tissues cultured in vitro. Symp Soc Exp Biol 11:118-131

Southern PJ, Berg P (1982) Transformation of mammalian cells to antibiotic resistance with a bacterial gene under control of the SV40 early region promoter. J Mol Appl Genet 1:327-341

Spradling AC, Rubin GM (1982) Transposition of cloned P elements into *Drosophila* germ line chromosomes. Sciences (NY) 218:341-347

Stanley J (1983) Infectivity of the cloned gernini virus genome requires sequences from both DNAs. Nature (Lond) 305:643-645

Stanley J, Gay RM (1983) Nucleotide sequence of cassava latent virus. Nature (Lond) 301:260-262

Tate ME, Ellis JG, Kerr A, Tempé J, Murray K, Shaw K (1982) Agropine: a revised structure. Carbohyd Res 104:105-120

Tempé J, Petit A (1982) Opine utilization by *Agrobacterium*. In: Kahl G, Schell J (eds) Molecular biology of plant tumors. Academic, New York, p 451

Thomashow MF, Nutter R, Montoya AL, Gordon MP, Nester EW (1980) Integration and organisation of Ti-plasmid sequences in crown gall tumors. Cell 19:729-739

Ursic D, Slightom JL, Kemp JD (1983) *Agrobacterium tumefaciens* T-DNA integrates into multiple sites of the sunflower crown gall genome. Mol Gen Genet 190:494-503

Van Haute E, Joos H, Maes M, Warren G, Van Montagu M, Schell J (1983) Intergenic transfer and exchange recombination of restriction fragments cloned in pBR322: a novel strategy for the reversed genetics of Ti plasmids of *Agrobacterium tumefaciens*. EMBO J 2:411-418

Van Larebeke N, Engler G, Holsters M, Van den Elsacker S, Zaenen I, Schilperoort RA, Schell J (1974) Large plasmid in *Agrobacterium tumefaciens* essential for crown gall-inducing ability. Nature (Lond) 252:169-170

Van Slogteren GMS, Hoge JHC, Hooykaas PJJ, Schilperoort RA (1983) Clonal analysis of heterogenous crown gall tumor tissue induced by wild-type and shooter mutant strains of *A. tumefaciens*. Plant Mol Biol 2:317-321

Walker MD, Edlund T, Boulet AM, Rutter WJ (1983) Cell-specific expression controlled by the 5'-flanking region of insulin and chymotrypsin genes. Nature (Lond) 306: 557-561

Watson B, Currier TC, Gordon MP, Chilton MD, Nester EW (1975) Plasmid required for virulence of *Agrobacterium tumefaciens*. J Bacteriol 123:255-264

Wiborg O, Hyldig-Nielsen J, Jensen E, Paludan K, Marcker K (1982) The nucleotide sequences of two leghemoglobin genes from soybean. Nucleic Acids Res 10:3487-3494

Willmitzer L, De Beuckeleer M, Lemmers M, Van Montagu M, Schell J (1980) DNA from Ti-plasmid is present in the nucleus and absent from plastids of plant crown-gall cells. Nature (Lond) 287:359-361

Willmitzer L, Schmalenbach W, Schell J (1981) Transcription of T-DNA in octopine and nopaline crown gall tumours is inhibited by low concentrations of α-amanitin. Nucleic Acids Res 9:4801-4812

Willmitzer L, Simons G, Schell J (1982) The TL-DNA in octopine crown gall tumours codes for seven well-defined polyadenylated transcripts. EMBO J 1:139-146

Willmitzer L, Dhaese P, Schreier PH, Schmalenbach W, Van Montagu M, Schell J (1983) Size, location and polarity of T-DNA-encoded transcripts in nopaline crown gall tumors; evidence for common transcripts present in both octopine and nopaline tumors. Cell 32:1045-1056

Wöstemeyer A, Otten L, De Greve H, Hernalsteens JP, Leemans J, Van Montagu M, Schell J (1983) Regeneration of plants from gall cells. In: Lurquin P, Kleinhofs A (eds) Genetic engineering in eukaryotes, NATO ASI Series A, Vol 61. Plenum, New York, p 137

Yadav NS, Vanderleyden J, Bennett DR, Barnes WM, Chilton MD (1982) Short direct repeats flank the T-DNA on a nopaline Ti plasmid. Proc Natl Acad Sci USA 79:6322-6326

Zaenen I, Van Larebeke N, Teuchy H, Van Montagu M, Schell J (1974) Supercoiled circular DNA in crown gall inducing Agrobacterium strains. J Mol Biol 86:109-127

Zambryski P, Holsters M, Kruger K, Depicker A, Schell J, Van Montagu M, Goodman HM (1980) Tumor DNA structure in plant cells transformed by *A. tumefaciens*. Science (Wash DC) 209:1385-1391

Zambryski P, Depicker A, Kruger K, Goodman H (1982) Tumor induction by *Agrobacterium tumefaciens*: analysis of the boundaries of T-DNA. J Mol Appl Genet 1:361-370

Zambryski P, Joos H, Genetello C, Leemans J, Van Montagu M, Schell J (1983) Ti plasmid vector for the introduction of DNA into plant cells without alteration of their normal regeneration capacity. EMBO J 2:2143-2150

The Chloroplast Genome, Its Interaction with the Nucleus, and the Modification of Chloroplast Metabolism

P. H. SCHREIER[1,2] and H. J. BOHNERT[1,3]

1 Introduction

Plastids and mitochondria are the only two cell organelles which are enclosed by two unit membranes and which possess a limited genetic and synthetic autonomy from the nucleus/cytosol compartment. This is believed to mirror the descent of these organelles from prokaryotic ancestors (Gray and Doolittle 1982). The partial autonomy is reflected by the DNA content of these organelles, their RNA- and protein-synthesizing capacity (Bohnert et al. 1982; Bottomley and Bohnert 1982; Ellis 1981; Curtis and Clegg 1984) and, in the case of mitochondria, by a different codon recognition (Barrell et al. 1980). While several mitochondrial genomes are either totally (human) (Anderson et al. 1981) or largely (yeast) sequenced, the larger chloroplast DNA molecule is not studied so extensively. The structure of several genes is, however, already known, and details of gene organization, gene content, transcription and transcriptional regulation are beginning to emerge (Whitfeld and Bottomley 1983; Crouse et al. 1984).

The size of the plastid genome is not large enough to provide the coding capacity of all plastid constituents. A rough estimate may be that approximately 70% of the proteins which make a plastid function are coded for by genes located in the nucleus (Ellis 1981; Ellis and Robinson 1984). The proteins made from the transcripts of these nuclear genes have to be transported into the plastid. Many of these proteins along with translation products of chloroplast ribosomes have to be assembled into complex structures, like the chloroplast ribosome or the photosynthetic membrane.

Plastids are the predominant producers of energy for the biosynthesis of organic matter. It is therefore obvious that chloroplasts have been the target of many biophysical and biochemical studies performed to understand this process. Now that plants can reliably be transformed it seems to be equally obvious to design experiments by which chloroplast metabolism might be modified. Such experiments might provide the mutant plants which could help to study chloroplast function further and ultimately increase chloroplast performance.

2 Plastid Chromosomes

The plastid - mainly chloroplast - chromosomes from many species have been characterized over the last years (reviewed in Bohnert et al. 1982). Without exception up to now these are circular DNA molecules ranging in size from 57 kbp in the alga *Codium* sp. to 195 kb in the alga

[1]Max-Planck Institut für Züchtungsforschung, 5000 Köln 30, FRG
[2]Bayer AG, Leverkusen, FRG
[3]Department of Biochemistry, University of Arizona, Tucson, Arizona 85721, USA

35. Colloquium - Mosbach 1984
The Impact of Gene Transfer Techniques in Eukaryotic Cell Biology

Chlamydomonas. While these appear to be the most extreme values, the majority of species investigated have a size of between 120 and 160 kbp for their plastid DNA.

2.1 Organization

All chromosomes are characterized in principle by a distinctive feature, that they contain a sequence repetition. In green algae and higher plants this repetition is present twice in inverted orientation. The two repeat units are not adjacent and divide the molecule into a small and a large single-copy DNA region. Species-specific variation in chromosome size appears to be due to insertions or deletions of DNA which can occur in all these segments (see below).

Another type of arrangement of repeated segments is found in the alga *Euglena gracilis*. Repeat units are arranged in tandem close together and they appear in "odd" numbers (e.g., 1, 3, 5) in closely related strains (Koller and Delius 1982a). The reason for this diversity is most probably that *intra* molecular recombination leads to a relatively frequent excision of repeat units followed by recombination between two molecules and integration of extra copies. Intramolecular recombination has already been demonstrated for other plastid DNA's (Bohnert and Loeffelhardt 1982, Palmer 1983).

One apparent exception to this rule appears to be some members of the legume plant family where no repeat units have been found. The explanation for this is that, in a late evolutionary event, these chloroplasts have lost one unit (or two half units!), most probably by a recombinational accident (Palmer and Thompson 1982).

2.2 Genes

Among the genes located on chloroplast DNA, the most abundantly transcribed rRNA and tRNA genes were characterized first. Whereever a repeat unit has been found, it contains the genes for 16S rRNA, 23S rRNA and 5S rRNA. These genes from several species are sequenced (see Whitfeld and Bottomley 1983) and were found to be virtually homologous to the corresponding *E. coli* genes in many regions of the molecule which are essential for ribosome function. Some chloroplast-specific evolutionary changes were also revealed. In many higher plants a molecule of species-specific variable length, 4.5S rRNA, has been found as a physically separate entity which represents the functional 3'-end of the bacterial 23S rRNA gene (Edwards et al. 1981). The rRNA genes are, as in *E. coli*, arranged as an operon (see Bohnert et al. 1982). The transcription unit is preceded by three partially overlapping putative promoters for which a regulatory control on transcription has recently been postulated (Briat et al. 1983).

Unlike in *E. coli*, the rDNA operon does not include the 5S rRNA gene which in chloroplasts is under the control of a separate promoter, although the gene is (still) close to the 3'-end of the 23S rRNA gene. Spacer DNA segments separating the 16S and 23S rRNA genes contain the genes for trnA and trnI, as in some *E. coli* rDNA operons. Several of those tRNA genes have been found to be interrupted by introns in higher plant chloroplast DNA's. Other chloroplast tRNA genes are widely distributed over the chromosome, some regions have been found where several tRNA genes are clustered (Karabin and Hallick 1983) and possibly cotranscribed. In most species studied in the order of 25 tRNA genes with several isoaccepting species for some amino acids have been found. These are sufficient for carrying 17 amino acids. As chloroplasts contain ca. 40 different tRNA species which have not yet all been identi-

Table 1. Plastid protein genes and gene control regions

Gene	Species	"-35		"-10"		References
rbcL	Maize	TTGATA	9b	TATCAT	34b	Jolly et al. 1981
		TTGATA	20b	TTAGAT	23b	
		TTGATA	21b	TAGATT	22b	
	Spinach	TTGCGC	18b	TACAAT	4b	Zurawski et al. 1981
	Tobacco	TTGCGC	18b	TACAAT	4b	Shinozaki and Sugiura 1982
	Pea	TTGCGC	18b	TATAAT	2b	Whitfeld and Bottomley 1983
	Chl.r.	TTTACA	20b	TATAAT	6b	Dron et al. 1982
atpA	Tobacco	TTGAAC	17b	TACCAT	102b	Deno et al. 1983
atpB/E	Maize	TTGACA	18b	TAGTAT	2-6b	Krebbers et al. 1982
	Spinach	TTGACA	21b	TATCCT	3b	Zurawski et al. 1982a
	Tobacco	TAGATA	17b	TATAAT	4-5b	Shinozaki and Sugiura 1982
	Pea	TTGACA	21b	AATCCT	3b	Whitfeld and Bottomley 1983
psbA	Spinach	TTGACA	18b	TATACT	4b	Zurawski et al. 1982b
	N.deb.	TTGACA	18b	TATACT	4b	Zurawski et al. 1982b
	Soybean	TTGACA	18b	TATACT	87b	Spielmann and Stutz 1983
	Mustard	TTGACA	17b	TATACT	3-6b	Link and Langridge 1984
psbB	Spinach	TTGCGT	18b	TAGAAT	6b	Morris and Herrmann 1984
rpsD	Maize	TTGCTA	21b	TAATAT	(191b)	Subramanian et al. 1983
		TTGTGT	15b	TACTCT	(109b)	
		TTGAAT	14b	TAATGT	(59b)	
		TTGTGT	15b	TAAAAT	(15b)	
rpsS	Tobacco	(TAGTAA)	19b	TAAATT	(38b)	Sugita and Sugiura 1983
tufA	*Euglena*	TTTAAA	23b	TAAGAT	6b	Montandon and Stutz 1984
Plastid consensus		TTGana	9-23b	TAtaat	2-34b	
E. coli consensus		TTGaca	16-19b	TATaat	7-10b	Hawley and McClure 1983

Numbers between the sequences indicate numbers of nucleotides, or nucleotides to the start of transcription. () not indicated as a promoter in the reference.

rbcL - large subunit of rubisco (ribulosebisphosphate carboxylase);
psbA - "photogene", subunit of photosystem II, M.-apparent 32,000;
atpA/B/E - alpha/beta/epsilon subunits of the ATPsynthase complex, CF_1;
psbB - subunit of photosystem II, M.-51,000, chlorophyll a associated;
rpsD - small subunit ribosomal protein, S 4;
rpsS - small subunit ribosomal protein, S19;
tufA - protein chain elongation factor, EF-Tu

Identified, though not sequenced, are the following plastid protein genes (see Alt et al. 1983; Westhoff et al. 1983a,b; W. Bottomley, personal communication; Willey et al. 1983):

psbC - photosystem II subunit, M.-44,000 chlorophyll a associated;
psbA - photosystem I protein, P_{700} chlorophyll a apoprotein;
petF - cytochrome f apoprotein;
petB6 - cytochrome b apoprotein;
petC4 - subunit 4 of cyt. b_6/f complex;
rp1B - large subunit ribosomal protein, L2;
atpH - subunit III of CF_O from the ATPsynthase complex

fied, it appears justified to postulate that a set specific for all amino acids will be present.

The coding capacity of this genome, taking pea chloroplast DNA as a model (120 kbp in size, no sequence repetition) might be estimated as follows. One set of rRNA genes, 7 - 8 kbp in an operon, a complete set of tRNA genes some of which present as iso-accepting species, 30 times 0.4 kbp, 12 kbp, are used for the structural RNA's of the translation machinery. This leaves approximately 100 kbp for genes coding for proteins. This would be compatible with a number of approximately 70 protein genes of an average size of 1.5 kbp, or M.W. 30,000, assuming 30% of the gene for regulatory regions.

Sixteen protein genes have been positively identified (Table 1) on chloroplast DNA up to now. Several more, like ribosomal protein genes or subunits of a chloroplast RNA polymerase, might have to be added soon. As far as a generalization can be made, the following statements are possible.

Plastid protein genes come with or without introns, depending on the species. Genes can be partially overlapping (Krebbers et al. 1982; Zurawski et al. 1982) from one strand or can be overlapping transcribed from both strands (Zurawski et al., personal communication). Genes may be transcribed mono- or polycistronically. Protein coding parts are especially highly conserved between chloroplast DNA's, highly conserved with respect to cyanobacterial genes (Nierzwicki-Bauer et al. 1984) and fairly conserved also when compared with *E. coli* genes.

Until recently, a generally accepted hypothesis had been that the genes on all different sized plastid DNA's were the same, i.e., conserved in kind and number. With the discovery of the gene for the small subunit of ribulosebisphosphate carboxylase, which is normally a nuclear gene, on an organelle chromosome (Heinhorst and Shively 1983) this view has to be abandoned. Extant plant species show a greater variety of gene content than expected, which will be helpful in the study of organelle evolution.

Differences between plant species, or forms/cultivars of one species appear to be restricted to noncoding, nonregulatory regions separating genes. Further differences exist in the presence, number, and size of introns (e.g., Koller et al. 1984). Some differences must exist also in the regulatory regions. This may be inferred from the different expression of the rbcL gene in maize bundle-sheath and mesophyll chloroplasts (Link et al. 1978) which has no equivalent in C_3 plant chloroplasts. This differential transcription might have its molecular explanation in sequence differences upstream of the transcription start point in the spinach and maize rbcL genes (McIntosh et al. 1980; Zurawski et al. 1981). Similarly, phytochrome inducibility of some plastid genes (Link 1982) might be caused by additional control regions. It is for the future to find which sequences on plastid DNA's and which additional factors, presumably of nuclear origin, have evolved for the creation and establishment of a molecular control system. Up to now the most obvious control regions appear to be of a "prokaryotic" type (Table 1). Preceding all genes which are sequenced so far a "-10" and a "-35" box is present which resemble *E. coli* promoter consensus sequences (Hawley and McClure 1983).

Considering, as outlined above, the regulatory influence of the nucleus and differential gene expression, it is likely that these putative promoters are necessary, though not sufficient, control regions. Studies dissecting the process of chloroplast transcription and its regulation during plant development and under different physiological conditions will thus be the next major task, as the identification of other genes and sequencing of those continues.

3 Chloroplast Protein Genes in Nuclear Chromosomes

Considering the low complexity of chloroplast chromosomes it is immediately suggestive that not all plastid constituents can be encoded by the unique sized plastid DNA. Using two-dimensional electrophoresis of uniformly labeled chloroplast proteins, it has been shown that at least 190 discrete spots (140 soluble, i.e., plastid stromal, proteins and 50 proteins releasable from the photosynthetic membrane) may be distinguished (Roscoe and Ellis 1982). This represents a conservative estimate, with over 300 genes expected (Bottomley and Bohnert 1982) to contribute proteins to the structure and function of the organelle. It follows then that most genes will be located in other compartments, most probably in nuclear chromosomes. Transcripts of nuclear genes have indeed been found whose products reside in the plastid. These transcripts constitute a subpopulation of the cytoplasmic poly(A)$^+$ mRNA which is found tissue-specifically expressed in leaves (Kamalay and Goldberg 1980). It has been shown by in vitro translation of these mRNAs that polypeptides arise which can specifically be recognized by antibodies raised against chloroplast proteins. From such experiments and by inhibiting the protein synthesis in either the nuclear or the chloroplast compartment the coding site of many proteins have been determined with the, apparently justified, assumption that no RNA is transported between compartments (Bottomley and Bohnert 1982).

Among those nuclear genes characterized in more detail are genes for the small subunit protein of ribulose-bisphosphate carboxylase/oxygenase and for the major chlorophyll a/b binding proteins. Both genes are coded for by small gene families, in the order of 10 - 20 genes per family, although not all of these genes may be active (Coruzzi et al. 1983; Cashmore 1979; Dunsmuir et al. 1983; Tobin 1978). The genes are induced by light and then constitute a major portion of the extractable mRNA.

4 Compartmental Interaction

4.1 Transport of Proteins

The transport mechanism of proteins entering plastids and mitochondria (Schatz and Butow 1983) is only one specialized mechanism among others. Probably the best understood of these is the co-translational transport of secretory proteins across the endoplasmic reticulum (Blobel 1980). An amino-terminal signal peptide of the protein to be transported makes contact between the ribosome and the membrane. The signal is characterized by a hydrophobicity moment, hydrophobic amino acids facing toward the membrane with charged amino acids at the opposite surface of an alpha-helix (Eisenberg 1984). Processing of this signal extension occurs while translation proceeds, before the mature protein is released into the intracisternal space.

The transport process into the chloroplast differs at least in as much as it occurs after translation of the preprotein on free cytoplasmic ribosomes is completed (Chua and Schmidt 1979). In most cases, precursor proteins of higher molecular weight are synthesized which contain a so-called "transit sequence" at the amino terminus. This extension is removed by a processing protease present as a soluble enzyme inside the plastid. Processing is ATP-dependent (Grossman et al. 1980) as far as chloroplasts are concerned, whereas in mitochondria an electrochemical potential, not merely outside ATP, is required to sustain transport in vitro for most proteins (see Ellis and Robinson 1984, for a recent review).

The most extensively studied examples are transit peptide sequences of small subunit proteins of the rubisco enzyme. A characteristic feature in all sequences studied is that the extension is positively charged and that the positions of the charged amino acids occur at relatively fixed positions with a high degree of variability between these sites and in the total length as well (Broglie et al. 1983; Berry-Lowe et al. 1982).

With respect to the transport process whose details are discussed (Chua et al. 1980; Ellis and Robinson 1984), one comment has to be made which is pertinent to the design of fusion protein genes (see below). Cytoplasmic ribosomes do not bind to the chloroplast envelope as soon as the transit peptide is synthesized, but only after the entire preprotein is made. This might mean that the structure of the *whole* protein is the signal and not only the amino-terminal end. Whether this implies that the transit peptide sequence may not be useful in transporting a chimaeric gene remains to be shown.

4.2 Compartmental Communication

The regulation of transcription and translation of genes specifying chloroplast functions which are located in two separating compartments remains the most interesting problem of chloroplast molecular biology (see Bottomley and Bohnert 1982; Bohnert et al. 1982 for review). During plant ontogeny and in a diurnal rhythm, nuclear and chloroplast genes are coordinately expressed. Our knowledge about the underlying control mechanisms, the ways these compartments talk to each other, is still limited and awaits explanation in molecular terms.

5 Function of Plastids

The most obvious and best-studied function of plastids is the involvement of chloroplasts in photosynthetic water splitting and electron transport which leads to the generation of ATP and to $NADP^+$-reduction in the light accompanied by oxygen evolution. The assimilation of carbon dioxide and the synthesis of glucans is another important function which is linked to photosynthesis.

Besides these two main contributions to plant anabolism there are several more functions which – though well studied biochemically – are often not considered as being plastid-located. For example, plastids are involved in the final steps of nitrate- und sulfate-reduction whereby a major amount of the cell's amino acids are synthesized. Plastids are organelles which contain a functional, unique fatty-acid synthase complex of a more "prokaryotic" type than the cytosolic counterpart. Chloroplasts in a complex cooperation with other organelles play a key role in photorespiration, i.e., the light-dependent reversion of CO_2 fixation.

Another major function is the plastid's role in compartmentalization and transport of ions and intermediary metabolites, like compounds connecting the anabolic and catabolic pathways of the carbon and nitrogen metabolisms or its role in the distribution of adenylates and phosphate. Though not complete, as the complex role of plastids in the plant secondary product metabolism is not discussed, this list may highlight the potential of experimentally modifying chloroplast metabolism by introducing chimeric genes in the plant nuclear chromosome.

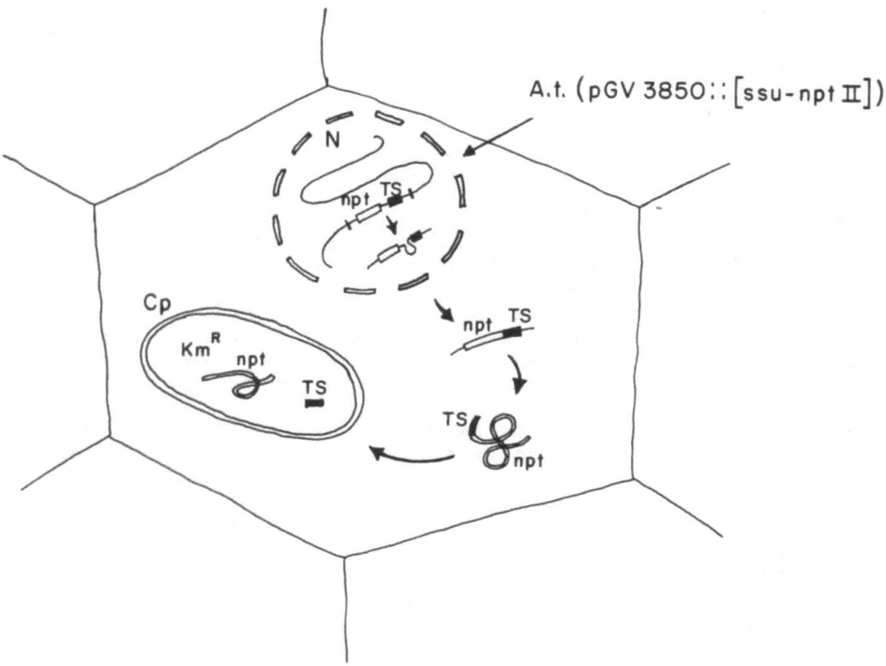

Fig. 1. Conceptual presentation of the chloroplast shuttle gene vector in a plant cell. A chimaeric gene is incorporated into nuclear chromosomes using *Agrobacterium tumefaciens* as vector. Expression of the gene leads to a preprotein whose amino-terminal chloroplast transit sequence accomplishes transport. *N* nucleus; *Cp* chloroplast; *Ts* transit sequence; *npt II* coding region of neomycin phosphotransferase; Km^R kanamycin resistance

6 Engineering of Chloroplast Metabolism

6.1 Transformation of Chloroplasts

The only indication up to now that nucleic acids are exchanged across intracellular membrane bilayers between chloroplasts and the other organelles is indirect. It has been reported that maize mitochondria contain large coherent segments of chloroplast DNA incorporated in their multipartite genome (Stern and Lonsdale 1982) and scrambled parts of mitochondrial genes have been identified in yeast nuclear DNA (Farelly and Butow 1983). There is no evidence yet that intracellular transformation exploiting these observations will be a practical approach.

Experiments to test this can, however, be designed. Recently, several plastid DNA replication origins (Koller and Delius 1982b; Ravel-Chapuis et al. 1982) or putative replication origins (Uchimija et al. 1983; Vallet et al. 1984) have been identified. There is evidence that such DNA segments might act as autonomically replicating sequences in the plant cytoplasm (Rochaix and van Dillewijn 1982). A conceivable experiment is thus to transform large numbers of cells with engineered plasmids which contain a plastid replication origin and selectable marker genes which will have to be present in the chloroplast in order to be effective. As long as the marker gene is, however, a gene which is already a chloroplast gene, or a mutant gene, it will be difficult to see the effect of a single or few transformed genes against the many copies of chloroplast genes in a single chloroplast or cell of a higher plant.

6.2 Gene Shuttle Vectors for Protein Transport

A different concept is outlined in Figure 1. Transcripts from nuclear genes will be translated resulting in chloroplast-located proteins if a transit peptide sequence for import into plastids is present. There are several prerequisites for a successfull modification of chloroplast metabolism following this outline. Most importantly, it has to be demonstrated whether the amino terminal transit sequence is sufficient for transport or whether domains in the mature protein are important in addition. The three-dimensional structure of the chimaeric protein must not mask domains for either recognition, transport and processing. The chimaeric gene to be expressed as an enzmye in the plastid must to a certain degree tolerate modification of its amino terminal end. The test for this concept involved transformation of plants with engineered *Agrobacterium* strains which produce either normal plants (pGV3850) or shoot-producing teratomas (pGV3851) (Zambryski et al. 1984). Included in the T-DNA of these Ti-plasmids is a chimaeric gene consisting of the promoter of a pea rbcS gene (Cashmore A., personal communication), its transit sequence and 24 amino acids from the 5'-end of the mature rbcS sequence. Fused to this truncated gene is the coding region of a modified neomycin-phosphotransferase II gene (from Tn5, Beck et al. 1982; Reiss et al. 1984) followed by a polyadenylation signal from the octopine synthase gene of the *Agrobacterium tumefaciens* Ti-plasmid Ach5 (De Greve et al. 1983). It could be shown that this construction after transfer into tobacco chromosomal DNA is expressed and confers kanamycin resistance to the plantlets (Schreier et al., manuscript in preparation). Expression of the mRNA and protein are light-dependent. Preliminary evidence showing that the protein exists in two molecular weight forms suggests that the protein may be processed and imported into plastids.

6.3 Prospects

Our preliminary results show that a chimaeric gene consisting of the amino terminal part of a rubisco small subunit protein and a kanamycin resistance protein is expressed after transformation of tobacco cells in regenerating plantlets. Expression is light-dependent. This and similar chimaeric genes represent a first generation of constructions which will enable us to study light regulated promoters, to specifically mutate the transit sequence and to study the fate of foreign proteins in plastids.

Introducing mutant genes for genes whose products are normally inside the plastid faces one major problem at present. The wild-type gene product is present in several hundred copies per cell if a gene from chloroplast DNA is concerned, and many nuclear genes coding for chloroplast functions are members of multi-gene families. Research will therefore have to concentrate on the few higher plant chloroplast mutants or on plant systems for which many mutants are already characterized genetically.

A different choice will be to use foreign genes whose products either interfere with chloroplast metabolism by concurring for substrates or to use genes which increase enzyme dosage, thus increasing substrate levels or decreasing concentrations of inhibitors.

Acknowledgements. Our studies were supported by grants from the Deutsche Forschungsgemeinschaft, from USDA (CSRA 174452), and from NSF (PCM 8318166). We wish to thank Dr. Anthony Cashmore (Rockefeller University, New York) for providing a genomic clone for a pea rbcS gene. We thank our colleagues at the MPI für Züchtungsforschung in Köln for many stimulating discussions when the concepts outlined here were developed.

References

Alt J, Westhoff P, Sears BB, Nelson N, Hurt E, Hauska G, Hermann RG (1983) Genes and transcripts for the polypeptides of the cytochrome b6/f complex from spinach thylakoid membranes. EMBO J 2:979-986

Anderson S, Bankier AT, Barrell BG et al. (1981) Sequence and organization of the human mitochondrial genome. Nature (Lond) 290:457-465

Barrell BG, Anderson AT, Baukier AT et al. (1980) Different pattern of codon recognition by mammalian mitochondrial tRNAs. Proc Natl Acad Sci USA 77:3164-3166

Beck E, Ludwig EA, Auerswald B, Reiss B, Schaller H (1982) Nucleotide sequence and exact localization of the neomycin phosphotransferse gene from transposon Tn5. Gene (Amst) 19:327-336

Berry-Lowe SL, McKnight TD, Shan DM, Meagher RB (1982) The nucleotide sequence, expression, and evolution of one member of a multigene family encoding the small subunit of ribulose-1,5-bisphosphate carboxylase in soybean. J Mol Appl Genet 1: 483-498

Blobel G (1980) Intracellular protein topogenesis. Proc Natl Acad Sci USA 77:1496-1500

Bohnert HJ, Loffelhardt W (1982) Cyanelle DNA from Cyanophora paradoxa exists in two forms due to intramolecular recombination. FEBS Lett 150:403-406

Bohnert HJ, Crouse EJ, Schmitt JM (1982) Organization and expression of plastid genomes. In: Pathier B, Boulter D (eds) Encyclopedia Plant Physiol, Vol 14B. Springer, Berlin, Heidelberg, New York, pp 175-530

Bottomley W, Bohnert HJ (1982) The biosynthesis of chloroplast protein. In: Parthier B, Boulter D (eds) Encyclopedia Plant Physiol, Vol 14B. Springer, Berlin, Heidelberg, New York, pp 531-596

Briat J-F, Dron M, Mache R (1983) Is transcription of higher chloroplast ribosomal operons regulated by premature termination. FEBS Lett 163:1-5

Broglie R, Coruzzi G, Lamppa G, Keith B, Chua N-H (1983) Structural analysis of nuclear genes coding for the precursor to the small subunit of wheat ribulose-1,5-bisphosphate carboxylase. Bio Technology 1:55-61

Cashmore AR (1979) Reiteration frequency of the gene coding for the small subunit of ribulose-1,5-bisphosphate carboxylase. Cell 17:383-388

Chua NH, Schmidt GW (1979) Transport of proteins into mitochondria and chloroplasts. J Cell Biol 81:462-483

Chua NH, Grossman AR, Bartlett SG, Schmidt GW (1980) Synthesis, transport, and assembly of chloroplast proteins. In: Bucher C et al. (eds) Biological chemistry of organelle formation. Springer, Berlin, Heidelberg, New York, pp 113-117

Coen DM, Bedbrook JR, Bogorad L, Rich A (1977) Maize chloroplast DNA fragment encoding the large subunit of ribulose bisphosphate carboxylase. Proc Natl Acad Sci USA 74:5487-5491

Corruzi G, Broglie R, Cashmore A, Chua NH (1983) Nucleotide sequences of two pea cDNA clones encoding the small subunit of ribulose-1,5-bisphosphate carboxylase and the major chlorophyll a/b-binding thylakoid polypeptide. J Biol Chem 258: 1399-1402

Crouse EJ, Bohnert HJ, Schmidt JM (to be published 1984) Chloroplast RNA synthesis. In: Ellis RJ (ed) Chloroplast biogenesis. Cambridge University Press

Curtis SE, Haselkorn R (1983) Isolation and sequence of the gene for the cyanobacterium Anabaena 7120. Proc Natl Acad Sci USA 80:1835-1839

Curtis SE, Clegg MT (1984) Molecular evolution of chloroplast DNA sequences. Mol Biol Evol 1:291-301

De Greve H, Dhaese P, Seurinck J, Lemmers M, Van Montagu M, Schell J (1983) Nucleotide sequence and transcript map of the Agrobacterium tumefaciens Ti plasmid-encoded octopine synthase gene. J Mol Appl Genet 6:499-511
Deno H, Shinozaki K, Sugiura M (1983) Nucleotide sequence of tobacco chloroplast gene for the subunit of proton-translocating ATPase. Nucleic Acids Res 11:2185-2191
Dron M, Rahire M, Rochaix JD (1982) Sequence of the chloroplast DNA region of Chlamydomonas reinhardtii containing the gene of the large subunit of ribulose bisphosphate carboxylase and parts of its flanking genes. J Mol Biol 162:715-793
Dunsmuir P, Smith SM, Bedbrook J (1983) The major chlorophyll a/b binding protein of Petunia is composed of seven polypeptides encoded by a number of distinct nuclear genes. J Mol Appl Genet 2:285-300
Edwards K, Bedbrook J, Dyer T, Kossel H (1981) 4.5S rRNA from Zea mays chloroplasts shows structural homology with the 3'-end of prokaryotic 23S rRNA. Biochem Int 2:533-538
Eisenberg D (to be published 1984) J Mol Biol
Ellis RJ (1981) Chloroplast proteins: synthesis, transport and assembly. Annu Rev Plant Physiol 32:11-137
Ellis RJ, Robinson C (to be published 1984) Post-translational transport and processing of cytoplasmically synthesized precursors of organellar proteins. In: Freedman RB, Hawkins HC (eds) The enzymology of the post-translational modifications of proteins. Academic, New York
Farelly F, Butow RA (1983) Rearrangement mitochondrial genes in the yeast nuclear genome. Nature (Lond) 301:296-301
Gray MW, Doolittle WF (1982) Has the endosymbiont hypothesis been proven? Microbiol Rev 46:1-42
Grossman A, Bartlett S, Chua NH (1980) Energy-dependent uptake of cytoplasmically synthesized polypeptides by chloroplasts. Nature (Lond) 285:625-628
Hawley DK, McClure WR (1983) Compilation and analysis of Escherichia coli promoter DNA sequences. Nucleic Acid Res 11:2237-2255
Heinhorst S, Shively JM (1983) Encoding of both subunits of ribulose-1,5-bisphosphate carboxylase by organelle genome of Cyanophora paradoxa. Nature (Lond) 304:373-374
Jolly SO, McIntosh L, Link G, Bogorad L (1981) Differential transcription in vivo and in vitro of two adjacent maize chloroplast genes. Proc Natl Acad Sci USA 78:6821-6825
Kamalay JC, Goldberg RB (1980) Regulation of structural genes expression of tobacco. Cell 19:935-946
Karabin GD, Hallick RB (1983) Euglena gracilis chloroplast transfer RNA transcription units. Nucleotide sequence analysis of a tRNA(Thr)-tRNA(Gly)-tRNA(Met)-tRNA(Ser)-tRNA(Glu) gene cluster. J Biol Chem 258:5512-5518
Koller B, Delius H (1982a) A chloroplast DNA of Euglena gracilis with five complete rRNA operons and two extra 16S rRNA genes. Mol Gen Genet 188:305-308
Koller B, Delius H (1982b) Origin of replication of chloroplast DNA of Euglena gracilis located close to the region of variable size. EMBO J 1:995-998
Koller B, Gingrich JC, Stiegler GL, Farley MA, Delius H, Hallick RB (1984) Nine introns with conserved boundary sequences in the Euglena gracilis chloroplast ribulose-1,5-bisphosphate carboxylase gene. Cell 36:545-553
Krebbers ET, Larrinua IM, McIntosh L, Bogorad L (1982) The maize chloroplast genes for the β and ε subunits of the photosynthetic coupling factor CF_1 are fused. Nucleic Acids Res 10:4985-5002
Link G (1982) Phytochrome control of plastid mRNA in mustard (Sinapsis alba L). Planta (Berl) 154:81-86
Link G, Langridge U (1984) Structure of the chloroplast gene for the precursor of the Mr 32,000 photosystem II protein from mustard. Nucleic Acds Res 12:945-958
Link G, Coen DM, Bogorad L (1978) Differential expression of the gene for the large subunit of ribulose bisphosphate carboxylase in maize cell types. Cell 15:725-731
McIntosh L, Poulsen C, Bogorad L (1980) Chloroplast gene sequence for the large subunit of ribulose-bisphosphate carboxylase of maize. Nature (Lond) 288:556-560
Montandon PE, Stutz E (1984) The genes for the ribosomal proteins S12 and S7 are clustered with the gene for the EF-Tu protein on the chloroplast genome of Euglena gracilis. Nucleic Acids Res 12:2851-2859

Morris J, Herrmann RG (1984) Nucleotide sequence of the gene for the P680 chlorophyll a apoprotein of the photosystem II reaction center from spinach. Nucleic Acids Res 12:2837-2850
Nierzwicki-Bauer SA, Curtis SE, Haselkorn R (to be published 1984) Cotranscription of genes encoding the small and large subunits of ribulose-1,5-bisphosphate carboxylase in the cyanobacterium Anabaena 7120. Proc Natl Acad Sci USA
Palmer JD (1983) Chloroplast DNA exists in two oritentation. Nature (Lond) 301:92-93
Palmer JD, Thompson WF (1982) Chloroplast DNA rearrangements are more frequent when a large inverted repeat sequence is lost. Cell 29:537-550
Ravel-Chapuis P, Heizmann P, Nigon V (1982) Electron microscopic localization of the replication origin of Euglena gracilis chloroplast DNA. Nature (Lond) 300:78-81
Reiss B, Sprengel R, Will H, Schaller H (1984) A new sensitive method for qualitative and quantitative assay of neomycin phosphotransferase in crude cell extracts. Gene (Amst) 30:217-223
Rochaix J-D, van Dillewijn J (1982) Transformation of the green alga Chlamydomonas reinhardii with yeast DNA. Nature (Lond) 296:70-72
Roscoe TJ, Ellios RJ (1982) Two dimensional gel electrophoresis of chloroplast proteins. In: Edelman M et al. (eds) Methods in chloroplast molecular biology. Elsevier/North Holland, Amsterdam, pp 1015-1028
Schatz G, Butow RA (1983) How are proteins imported intc mitochondria. Cell 32:316-318
Shinozaki K, Sugiura M (1982) The nucleotide sequence of the tobacco chloroplast gene for the large subunit of ribulose-1,5-bisphosphate carboxylase oxygenase. Gene (Amst) 20:91-102
Shinozaki K, Deno H, Kata A, Suriura M (1983) Overlap and cotranscription of the genes for the beta and epsilon subunits of tobacco chloroplast ATPase. Gene (Amst) 24:147-155
Spielman A, Stutz E (1983) Nucleotide sequence of soybean chloroplast DNA regions which contain the psbA and trnH genes and cover the ends of the large single copy region and one end of the inverted repeats. Nucleic Acids Res 11:7157-7167
Stern DB, Lonsdale DM (1982) Mitochondrial and chloroplast genomes of maize have a 12-kilobase DNA sequence in common. Nature (Lond) 299:698-702
Subramanian AR, Steinmetz A, Bogorad L (1983) Maize chloroplast DNA encodes a protein sequence homologous to the bacterial ribosomal protein 54. Nucleic Acids Res 11:5277-5286
Sugita M, Sugiura M (1983) A putative gene of tobacco chloroplast coding for ribosomal protein similar to E. coli ribosomal protein S19. Nucleic Acids Res 8:2665-2676
Tobin EM (1978) Light regulation of specific mRNA species in Lemna gibba G3. Proc Natl Acad Sci USA 75:4749-4753
Uchimiya H, Ohtain T, Ohgawara T, Harada H, Sugita M, Sugiura M (1983) Molecular cloning of tobacco chromosomal and chloroplast DNA segments capable of replication in yeast. Mol Gen Genet 193:1-4
Vallet JM, Rahire M, Rochaix J-D (1984) Localization and sequence analysis of chloroplast DNA sequences of Chlamydomonas reinhardii that promote autonomous replication in yeast. EMBO J 3:415-422
Westhoff P, Alt J, Herrmann RG (1983a) Localization of the genes for the two chlorophyll a-conjugated polypeptides (mol. wt. 51 and 44 kd) of the photosystem II reaction center on the spinach plastid chromosome. EMBO J 2:2229-2238
Westhoff P, Alt J, Nelson N, Bottomley W, Bunemann H, Herrmann RG (1983b) Genes and transcripts for the P_{700} chlorophyll a apoprotein and subunit 2 of the photosystem I reaction center complex from spinach thylakoid membranes. Plant Mol Biol 2:95-107
Whitfeld PR, Bottomley W (1983) Organization and structure of chloroplast genes. Annu Rev Plant Physiol 34:279-310
Willey DL, Huttly AK, Phillips AL, Gray JC (1983) Localization of the gene for cytochrome f in pea chloroplast DNA. Mol Gen Genet 189:85-89
Zambryski P, Herrera-Estrella L, de Block M, Van Montagu M, Schell J (to be published 1984) The use of the Ti plasmid of Agrobacterium to study the transfer and expression of foreign DNA in plant cells: new vectors and methods. In: Hollaender A, Setlow J (eds) Genetic engineering, principles and methods, Vol 6. Plenum, New York

Zurawski G, Perot B, Bottomley W, Whitfeld PR (1981) The structure of the gene for the large subunit of ribulose-1,5-bisphosphate carboxylase from spinach chloroplast DNA. Nucleic Acids Res 9:2151-3270

Zurawski G, Bottomley W, Whitfeld PR (1982a) Structure of the genes for the β and ε subunits of spinach chloroplast ATPase indicates a dicistronic mRNA and an overlapping translation stop/start signal. PNAS 79:6260-6264

Zurawski G, Bohnert HJ, Whitfeld PR, Bottomley W (1982b) Nucleotide sequence of the gene for the 32,000-M_r thylakoid membrane protein from Spinach oleracea and Nicotiana debneyi predicts a totally conserved primary translation product of M_r 38,950. PNAS 79:7699-7703

The T-Region of Ti Plasmid Codes for an Enzyme of Auxin Biosynthesis

J. SCHRÖDER[1,4], S. WAFFENSCHMIDT[2], E. W. WEILER[3], and G. SCHRÖDER[1,4]

1 Introduction

Plant cells transformed with Ti plasmids of *Agrobacterium tumefaciens* can be distinguished from normal cells by two important new properties. They produce new substances (opines), and they grow in tissue culture in simple media, independent of the addition of growth hormones like auxins and cytokinins. Both properties are due to expression of genes from the T-DNA, that part of the Ti plasmid which is transferred from *Agrobacterium* into plant cells during induction of tumorous growth (see Schell et al. 1984, for a recent review).

Genetic experiments (Leemans et al. 1982, Joos et al. 1983; Ream et al. 1983; Inzé et al. 1984) indicate that at least two genes of the T-DNA (nos. 1 and 2, see Fig. 1) cooperate in achieving auxin effects, including auxin independence, in crown gall cells. These genes have been sequenced (Barker et al. 1983; Sciaky and Thomashow 1984; Gielen et al. 1984; Klee et al. 1984), and they have been found to express proteins in prokaryotic as well as in eukaryotic cells (Schröder et al. 1983; Schröder and Schröder 1983). Since both crown gall cells and *Agrobacterium* actively synthesize auxins, this suggested a possibility that gene 1 and 2 code in both types of cells for enzymes of auxin

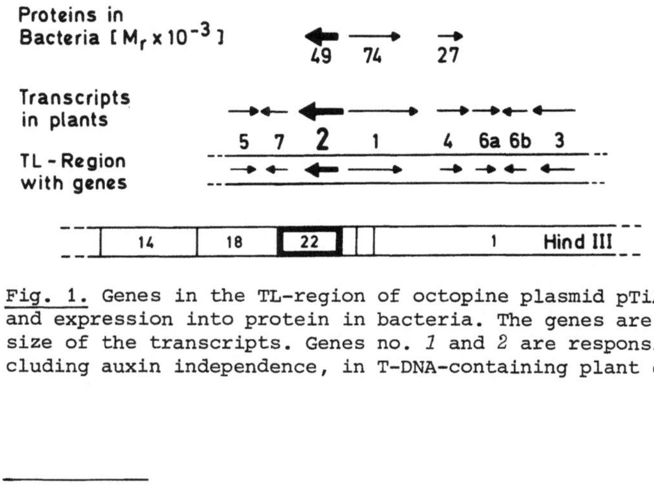

Fig. 1. Genes in the TL-region of octopine plasmid pTiAch5: transcription in plants and expression into protein in bacteria. The genes are numbered according to the size of the transcripts. Genes no. *1* and *2* are responsible for auxin effects, including auxin independence, in T-DNA-containing plant cells

[1]Max-Planck-Institut für Züchtungsforschung, Abt. Schell, 5000 Köln 30, FRG
[2]Institut für Biochemie, Universität Köln, An der Bottmühle 2, 5000 Köln 1, FRG
[3]Lehrstuhl für Pflanzenphysiologie, Ruhr-Universität Bochum, Postfach 102148, 4630 Bochum, FRG
[4]Present address: Institut für Biologie II, Universität Freiburg, Schänzlestr. 1, 7800 Freiburg, FRG

35. Colloquium - Mosbach 1984
The Impact of Gene Transfer Techniques in Eukaryotic Cell Biology

GENE 2 FROM THE T-REGION OF TI-PLASMIDS

-- IS FUNCTIONALLY EXPRESSED IN E. COLI, AGROBACTERIA, AND PLANT CELLS

-- CODES FOR AN AUXIN-SYNTHESIZING ENZYME

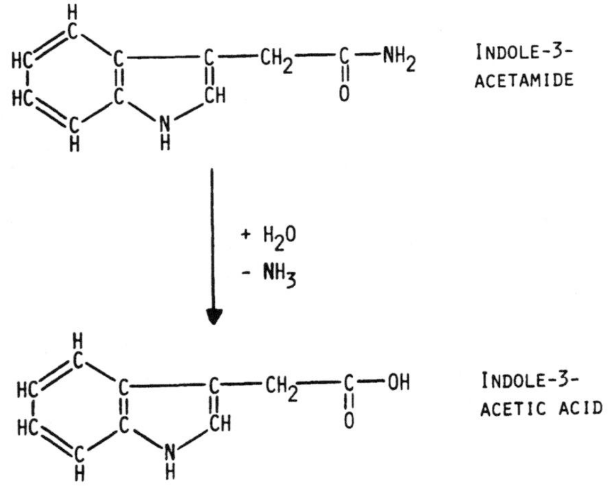

Fig. 2. Amidohydrolase function of the enzyme encoded in Gene 2 (IaaH) with indole-3-acetamide

biosynthesis. In this report we summarize recent findings which indicate that the protein product of gene 2 participates directly in auxin formation, and we also discuss the possible role of gene 1.

2 Results and Discussion

2.1 Role of Gene 2 in Auxin Formation

Gene 2 produces in *E. coli* a protein of M.W. 49,000 (Schröder et al. 1983), but this expression is low. Recloning fragment HindIII 22 (Fig. 1) into vector pINIIA increases expression by raising the rate of transcription reading into the gene (Schröder and Schröder 1983), and *E. coli* cells producing the protein in increased quantity were tested for auxin-synthesizing activity. Among the several precursors of indole-3-acetic acid used as substrates in the first experiments, only indole-3-acetamide showed significant activity. The detailed analysis showed that only extracts from *E. coli* containing and expressing gene 2 hydrolyzed indole-3-acetamide in cell-free extracts into indole-3-acetic acid (Fig. 2), and the product of the reaction was unambiguously identified by HPLC, gas chromatography, and mass spectrometry (Schröder et al. 1984). Since previous experiments suggested that the gene is also expressed into protein in *Agrobacterium*, these enzyme assays were repeated with extracts from a number of *Agrobacterium* strains containing Ti plasmids with or without gene 2. The results indicated a clear correlation between enzyme activity and presence of the gene (Schröder et al., unpublished), indicating that the IaaH gene is functional in *Agrobacterium*. These results established that gene 2 is expressed in prokaryotic cells.

Assays with extracts from gall cells showed that they contained an enzyme activity hydrolyzing indole-3-acetamide, and indole-3-acetic

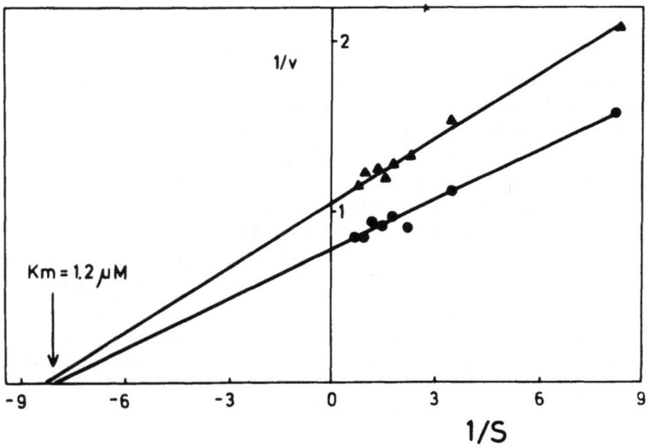

Fig. 3. K_m of the hydrolase with indole-3-acetamide as substrate; (●) in extracts from tobacco crown gall cells; (▲) in extracts from *E. coli* cells expressing the gene

acid as reaction product was unambiguously identified. Extracts from habituated cells (which also grow independently of auxin addition, but contain no T-DNA) were inactive in indole-3-acetamide hydrolysis. This indicated that the T-DNA was responsible for the enzyme activity, and the strict correlation between gene and activity in *E. coli*, *Agrobacterium tumefaciens*, and crown gall cells strongly suggested that the activity in the eukaryotic tumor cells is encoded in gene 2.

Analysis of the properties indicated that the enzyme appears to be a hydrolase with no obvious cofactor requirements; neither $MgCl_2$, EDTA, KCl, nor 2-mercaptoethanol had significant effects. Temperature dependence, pH optimum, and K_m (Fig. 3) for indole-3-acetamide were very similar regardless of whether the enzyme was expressed in *E. coli* or in plant cells, suggesting that eukaryotic cells do not modify the protein post-translationally in such a way that these major properties are changed significantly.

Detailed investigations on the substrate specificity (Kemper et al. 1984) yielded some interesting results. Although indole-3-acetamide seemed to be the best substrate, some other compounds were also hydrolyzed which give rise to indole-3-acetic acid or related substances with auxin activity. Indole-3-acetonitrile which occurs in several plants (Bearder 1980; Sembdner et al. 1980) was hydrolyzed to indole-3-acetic acid. Esters of the auxin with glucose or myo-inositol were also hydrolyzed by the enzyme; this may well be of physiological significance, since they are very widespread in plants and since they are discussed as storage forms for auxins (Sembdner et al. 1980; Cohen and Bandurski 1982). Amide conjugates of the auxin with several different amino acids showed no significant activity. This was somewhat unexpected in view of the amidase properties of the enzyme, but it may be that the assay conditions in vitro did not reflect the in vivo situation in all aspects. Interestingly, phenylacetamide was hydrolyzed to phenylacetic acid, and this substance has been proposed as a major auxin in plant cells (Wightman 1973, Fregeau and Wightman 1983). Taken together, these results suggest a possibility that the enzyme encoded by the IaaH gene has a general function in converting precursor or storage forms into active auxins, and this may be of significance in view of the wide host range of most Ti plasmids, although it is not known at present which of these different physiological substrates contributes to the auxin effects of the gene.

2.2 What is the Role of Gene 1?

Detailed genetic data on T-DNA encoded auxin functions (Inzé et al. 1984) clearly indicate that the IaaH gene alone is not sufficient to produce auxin effects; it needs complementation by T-DNA gene 1. If indole-3-acetamide is the most important physiological substrate, and assuming that this substance is not a natural component in plants (Zenk 1961), it seems reasonable to speculate that the action of gene 1 provides indole-3-acetamide from a precursor common to all plants successfully infected by *Agrobacterium tumefaciens*. An enzyme meeting these requirements has been described in *Pseudomonas savastanoi*. It is a monooxygenase converting tryptophan (and, at a lesser rate, phenylalanin) into indole-3-acetamide (phenylacetamide), and this enzyme cooperates with a hydrolase in producing indole-3-acetic acid in these microorganisms (Kosuge et al. 1966; Comai and Kosuge 1980). In view of this apparent similarity, it is tempting to speculate that the two auxin genes of the T-region are of bacterial origin, and that they are adapted to function in plant cells in auxin biosynthesis, by utilizing two enzyme reactions leading from tryptophan via the intermediate indole-3-acetamide to indole-3-acetic acid. This would satisfactorily explain most of the auxin effects evoked by the T-DNA. It should be noted, however, that it has not been possible so far to demonstrate unequivocally that gene 1 codes for a monooxygenase converting tryptophan into indole-3-acetamide, although recent results indicate that the *Pseudomonas* oxygenase and T-DNA gene 1 share significant homologies on the DNA and protein level (T. Kosuge, S. Hutcheson, personal communication). Thus the similarity of functions is at present based mainly on homology studies for gene 1 and on the fact that indole-3-acetamide is a substrate for the two hydrolases. It is therefore not yet completely excluded that the intriguing similarity with the *Pseudomonas* enzyme system for auxin biosynthesis is misleading, and in this case the other substrates hydrolyzed by the IaaH gene product into auxins may represent more than just side reactions of the enzyme.

2.3 Conclusing Remarks

Regardless of the open question discussed above, all of the available data indicate that at least one, and possibly both of the two T-DNA auxin genes, exert their effects by increasing the concentration of active auxins in plant tumor cells. The hormone independence of crown gall cells has been known for more than four decades, and plant physiologists have always been intrigued by this phenomenon, since it seemed to bear on important questions of growth and differentiation. The present results are a step further in understanding the molecular basis of tumorigenesis and development in plants, and the finding that some of the T-DNA genes appear to be functional in both eukaryotic and prokaryotic cells promises some interesting insights into principles of gene expression.

Acknowledgements. We wish to express our gratitude to J. Schell, the crown gall research groups in Gent and Cologne, and to T. Kosuge and S. Hutcheson, for sharing discussions and results prior to publication.

References

Barker RF, Idler KB, Thompson DV, Kemp JD (1983) Nucleotide sequence of the T-DNA region from the *Agrobacterium tumefaciens* octopine Ti plasmid pTi15955. Plant Mol Biol 2:335-350

Bearder JR (1980) Plant hormones and other growth substances - their background, structures and occurrence. In: MacMillan J (ed) Encyclopedia of plant physiology, New Series: Hormonal regulation of development, Vol 9. Springer, Berlin, Heidelberg, New York, pp 9-112

Cohen JD, Bandurski RS (1982) Chemistry and physiology of the bound auxins. Annu Rev Plant Physiol 33:403-430

Comai L, Kosuge T (1980) Involvement of plasmid deoxyribonucleic acid in indoleacetic acid synthesis in *Pseudomonas savastanoi*. J Bacteriol 143:950-957

Fregeau JA, Wightman F (1983) Natural occurrence and biosynthesis of auxins in chloroplast and mitochondrial fractions from sunflower leaves. Plant Sci Lett 32:23-34

Gielen J, De Beuckeleer M, Seurinck J et al. (1984) The complete nucleotide sequence of the TL-DNA of the *Agrobacterium tumefaciens* plasmid pTiAch5. EMBO J 3:835-846

Inzê D, Follin A, Van Lijsebettens M, Simoens C, Genetello C, Van Montagu M, Schell J (1984) Genetic analysis of the individual T-DNA genes of *Agrobacterium tumefaciens*; further evidence that two genes are involved in indole-3-acetic acid synthesis. Mol Gen Genet 194:265-274

Joos H, Inzê D, Caplan A, Sormann M, Van Montagu M, Schell J (1983) Genetic analysis of T-DNA transcripts in nopaline crown galls. Cell 32:1057-1067

Kemper E, Waffenschmidt S, Weiler EW, Rausch T, Schröder J (1984) T-DNA encoded auxin formation in crown gall cells. Planta (Berl) in press

Klee H, Montoya A, Horodyski F et al. (1984) Nucleotide sequence of the tms genes of the pTiA6NC octopine Ti plasmid: two gene products involved in plant tumorigenesis. Proc Natl Acad Sci USA 81:1728-1733

Kosuge T, Heskett MG, Wilson EE (1966) Microbial synthesis and degradation of indole-3-acetic acid. The conversion of L-tryptophan to indole-3-acetamide by an enzyme system from *Pseudomonas savastonoi*. J Biol Chem 241:3738-3744

Leemans J, Deblaere R, Willmitzer L, De Greve H, Hernalsteens JP, Van Montagu M, Schell J (1982) Genetic identification of functions of TL-DNA transcripts in octopine crown galls. EMBO J 1:147-152

Ream LW, Gordon MP, Nester EW (1983) Multiple mutations in the T region of the *Agrobacterium tumefaciens* tumor-inducing plasmid. Proc Natl Acad Sci USA 80:1660-1664

Schell J, Van Montagu M, Holsters M et al. (1984) Ti plasmids as gene vectors for plants. In: Kumar A (ed) Eucaryotic gene expression. Plenum, New York, pp 141-160

Schröder G, Klipp W, Hillebrand A, Ehring R, Koncz C, Schröder J (1983) The conserved part of the T-region in Ti plasmids expresses four proteins in bacteria. EMBO J 2:403-409

Schröder J, Schröder G (1983) Expression of genes from the T-region of Ti plasmids in bacteria. In: Goldberg RB (ed) UCLA Symposia on molecular and cellular biology, New Series, Vol 12. Plant Molecular Biology, Liss AR, New York, pp 45-53

Schröder G, Waffenschmidt S, Weiler EW, Schröder J (1984) The T-region of Ti plasmids codes for an enzyme synthesizing indole-3-acetic acid. Eur J Biochem 138:387-391

Sciaky D, Thomashow MF (1984) The sequence of the tms 2 transcript locus of the *A. tumefaciens* plasmid pTiA6 and characterization of the mutation in pTiA66 that is responsible for auxin attenuation. Nucleic Acids Res 12:1447-1461

Sembdner G, Gross D, Liebisch HW, Schneider G (1980) Biosynthesis and metabolism of plant hormones. In: MacMillan J (ed) Encyclopedia of plant physiology, New Series: Hormonal regulation of development, Vol 9. Springer, Berlin, Heidelberg, New York, pp 281-444

Wightman F (1973) Biosynthesis of auxins in tomato shoots. Biochem Soc Symp 38:247-275

Zenk MH (1961) 1-(Indole-3-acetyl)-β-glucose, a new compound in the metabolism of indole-3-acetic acid in plants. Nature (Lond) 191:493-494

Cauliflower Mosaic Virus: A Plant Gene Vector

B. GRONENBORN[1]

1 Introduction

The artificial introduction of new traits into living cells using various gene transfer techniques has become a routine procedure applicable for both prokaryotic and some eukaryotic organisms. With respect to higher plants a naturally evolved gene transfer system, the Ti-plasmid of *Agrobacterium tumefaciens*, has been developed into an advanced and sophisticated tool to introduce new genes into the genome of higher plants (Zambryski et al. 1984; see also Schell, Chapter II.8, this Volume).

In addition to a gene vector like the Ti-plasmid, which always causes an insertion of the T-DNA, or derivatives thereof, into the genome of the host plant, an autonomously replicating multicopy vector analogous to bacterial or yeast plasmids is desirable.

A variety of bacterial and animal viruses have been developed into efficient transduction and expression vectors. The vast majority of plant viruses, however, contain RNA as genetic material (Matthews 1981). RNA is not as readily accessible to genetic engineering techniques as DNA. Therefore the only two groups of plant viruses with a DNA genome, the geminiviruses and caulimoviruses, became of immediate interest to molecular biologists.

The intention of this chapter is not to review DNA plant viruses or caulimoviruses. This has been done extensively by Shepherd (1976, 1979), Goodman (1981) and Shepherd and Lawson (1981). Furthermore, there exists an increasing amount of "preview type" literature on the "potential" use of DNA plant viruses, in particular caulimoviruses, as plant gene vectors (Hohn et al. 1982; Hohn and Hohn 1982; Howell 1982; Hull and Davies 1983), to which the reader is referred.

Meanwhile, it has been shown that cauliflower mosaic virus can be used not only to transfer and propagate, but also to express foreign genes in plants (Gronenborn et al. 1981; Brisson et al. 1984). In addition, it has been shown that CaMV-derived promoters are functional in nonhost plants and are independent of a replicating CaMV genome (Paszkowski et al. 1984; Sieg and Gronenborn, in preparation).

Therefore only those features of CaMV pertinent to the topic of its further development as a plant gene vector will be discussed.

2 Cauliflower Mosaic Virus

2.1 The Biology of the Virus

Cauliflower mosaic virus (CaMV) is the group-type member of the caulimoviruses. Members of the same group are for instance dahlia mosaic

[1]Max-Planck-Institut für Züchtungsforschung, 5000 Köln 30, FRG

35. Colloquium - Mosbach 1984
The Impact of Gene Transfer Techniques in Eukaryotic Cell Biology

virus (DaMV), carnation etched ring virus (CERV), strawberry vein banding virus (SVBV), figworth mosaic virus (FMV) and others (Shepherd and Lawson 1981). CaMV infects mainly cruciferous plants. The symptoms of disease range from severe stunting and necrosis to almost invisible mild vein clearing without any gross interference with the growth of the host plant. The type of symptoms is mainly dependent on the virus isolate, but may also be influenced by the growth conditions of the host plant.

The natural vector of CaMV is the aphid (e.g., *Myzus persicae*, Aphidae, Insecta), but the virus is easily mechanically transmitted. A prerequisite for virus transmission by insects is the expression of a functional aphid aquisition factor protein (Lung and Pirone 1973, 1974). Recently, the viral gene encoding the aphid aquisition factor has been identified (Woolston et al. 1983; Armour et al. 1983). Two naturally occurring isolates which are not aphid-transmissible but are transmitted in mixed populations together with wild-type virus have been described (Lung and Pirone 1973; Hull 1980).

2.2 The Virus Particle

CaMV has an isometric capsid of about 50 nm in diameter with a molecular weight of 22.8 million (Hull et al. 1976). The virus particles often appear empty in preparations stained by uranylacetate. Chauvin et al. (1979) propose that the viral DNA is tightly associated with the inner surface of the virion. It is possible that the strict size limitation on the amount of foreign DNA that is stably maintained in the virus genome (Gronenborn et al. 1981) is due to this peculiar entrapment of the DNA inside the virion.

2.3 The DNA of CaMV

The genome of CaMV is a double-stranded DNA molecule of about 8 kb in size. DNA isolated from virus particles appears as a variety of enttangled and twisted circular molecules which are not covalently closed (Shepherd et al. 1970, 1971). Rather, they contain S1 nuclease-sensitive sites in both strands of the DNA (Hull and Shepherd 1977). The S1 nuclease-sensitive sites have been mapped on the CaMV by Hull and Howell (1978) and Volovitch et al. (1978). The first DNA sequence of a CaMV isolate (Cabb-S) by Franck et al. (1980) and later analysis by Richards et al. (1981) revealed that the DNA forms short triple-stranded overlaps at these positions. When complete copies of the CaMV DNA are cloned in plasmid vectors and propagated in *E. coli*, the discontinuities are sealed. Nevertheless, the cloned DNA is infectious on susceptible plants, but after a passage through the plant host, the DNA which becomes encapsidated again contains the discontinuities (Howell et al. 1980; Lebeurier et al. 1980). The introduction and the function of these overlaps remained obscure for some time, but their importance for the replication of the virus became evident when the involvement of a reverse transcription step in the life-cycle of the virus was proposed (see Sect. 2.5).

2.4 The Genetic Organization CaMV

The chromosomes of three different isolates of cauliflower mosaic virus have been sequenced: the Cabb-S isolate (8024 bp) by Franck et al.

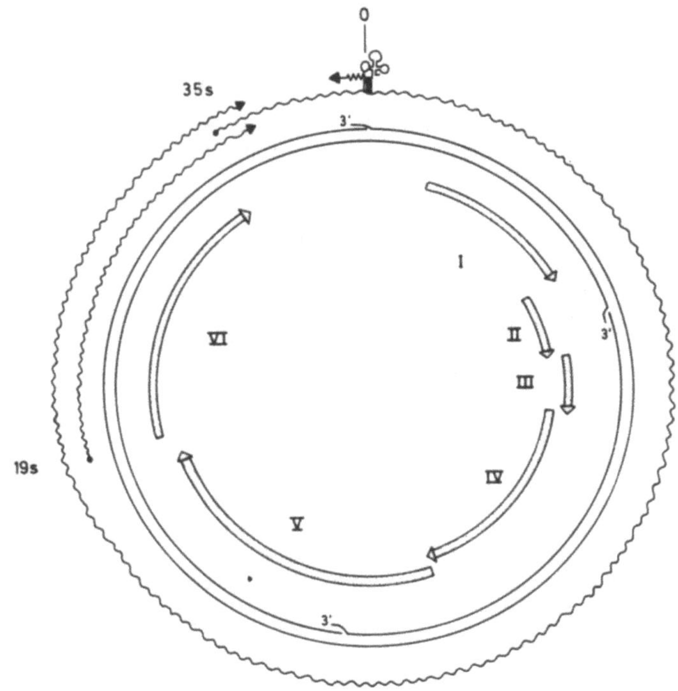

Fig. 1. The genome of organization of CaMV. The CaMV chromosome is represented by the *circular double line*. The three short overlap structures (S1 nuclease-sensitive sites; former "gaps") are indicated. The single discontinuity of the (-) DNA strand is the zero point of the map; the base pair counting is clockwise. The two discontinuities of the (+) DNA strand are located in orf II and orf V; the discontinuity in orf II is dispensable. The met initiator tRNA primer binding site spans map position 1-14. The tRNA primer is symbolized by its *cloverleaf* structure. The two major transcripts are shown as wavy lines, their respective start points are indicated by *19s* and *35s*. The synthesis of the (-) DNA strand by reverse transcriptase is primed by the tRNA hybridizing to the 35s transcript and proceeds anticlockwise as indicated by the *zigzag arrow*. The genes (open reading frames) are shown as *dotted arrows* and numbered *I* through *VI*. Promoters are located in the small intergenic region between orf V and orf VI (19s promoter) and just after orf VI in the large intergenic region (35s promoter)

(1980), the strain CM 1841, a nonaphid-transmissible isolate (8031 bp) by Gardner et al. (1981) and an isolate of lower virulence, the strain D/H (8016 bp) by Balâsz et al. (1982). The three isolates show only 5% sequence divergence, almost exclusively restricted to the coding regions.

2.4.1 The Genes

From the DNA sequence six to eight open reading frames (orfs) have been deduced (Franck et al. 1980; Hohn et al. 1982) (Fig. 1). Five open reading frames are closely linked: orfs I, II, and III are separated by only one base pair, orfs III and IV overlap for a short distance, as do orfs IV and V (assuming the first possible ATG codon deduced from the DNA sequence corresponds to the start codon of the respective gene; see Sect. 3.1.2). A small intergenic region of about 100 base pairs separates orfs V and VI, and a large intergenic region of about 1 kb is located between orf VI and orf I.

To date, functional proteins have been assigned to four open reading frames: Orf II encodes the aphid aquisition factor (Woolston et al. 1983; Armour et al. 1983). Orf IV encodes the virus capsid protein (Franck et al. 1980; Daubert et al. 1982). Orf V most probably encodes a reverse transcriptase (Toh et al. 1983; Volovitch et al. 1984). Orf VI codes for the viroplasm protein which is the main constituent of the inclusion bodies that are formed in virus infected cells (Odell et al. 1981; Xiong et al. 1982). No function has as yet been assigned to the other open reading frames.

2.4.2 Transcripts of CaMV

Only one strand of the CaMV chromosome, the α or (-) strand, is transcribed into RNA (Howell and Hull 1978). Two major polyadenylated transcripts have been isolated from infected leaf material. A small (19s) transcript is derived from orf VI (Odell et al. 1981) and is the message of p66, the matrix protein of the viroplasm (Odell et al. 1980; Ani et al. 1980; Covey and Hull 1981). A large (35s) transcript covers the entire genome and is terminally redundant by 180 bases (Covey et al. 1981; Guilley et al. 1982). Detailed mapping by S1 nuclease and primer extension confined the start of the 19s transcript 11 (Covey et al. 1981), or 12 bases (Guilley et al. 1982) ahead of the first ATG codon of orf VI. Its 3' end was mapped to position 7615, the same position as the 3' end of the major 35s transcript (Guilley et al. 1982; Covey et al. 1981). This means that the 19s and the 35s transcripts coterminate. An AATAAA polyadenylation signal is found 18 bases upstream of the common 3' end of the two transcripts. The 5' end of the major 35s transcript has been mapped to position 7435 on the CaMV genome (Guilley et al. 1982). Both the 19s and the 35s transcript start points are preceded by typical eukaryotic TATAAA sequences and CAAT boxes characteristic for RNA polymerase II initiation sites (Breathnach and Chambon 1981). By these criteria two strong promoters are located in the small intergenic region preceding orf VI and at the 5' extremity of the large intergenic region (Fig. 1).

This is in agreement with the results of Guilfoyle (1980), who has shown that in isolated nuclei from CaMV infected leaf cells RNA polymerase II synthesizes virus specific RNA. In addition, Olszewski et al. (1982) were able to isolate the DNA of CaMV as transcriptionally active covalently closed minichromosomes from the nuclei of infected turnip leaves. This finding resolved the puzzle of how RNA polymerase II was able to transcribe past the discontinuity in the α-strand of the CaMV DNA. Covalently closed forms of CaMV DNA were also found as nonencapsidated molecules by Menissier et al. (1982) and Hull and Covey (1983).

2.5 The Replication of CaMV

Kamei et al. (1969) and later Favali et al. (1973) proposed that the cytoplasmic inclusion bodies are the major site of CaMV replication. Recent findings of several groups can be assembled into a new picture of CaMV replication: Evidence accumulates that CaMV replicates by a reverse transcription mechanism (Guilley et al. 1983; Hull and Covey 1983, Pfeiffer and Hohn 1983; Marco and Howell 1984) (see Fig. 1). According to the current model (Pfeiffer and Hohn 1983) the large 35s transcript is the key replicative intermediate in the life cycle of CaMV. A met initiator tRNA probably serves as a primer for the synthesis of the first (-) DNA strand by the reverse transcriptase. In the region of the terminal redundancy of the 35s RNA a template switch occurs and (-) strand DNA synthesis proceeds until a full length cDNA copy of the 35s RNA is completed. A small covalently linked RNA-DNA molecule of about 700 bases complementary to the 5' end of the 35s transcript has been found (Covey et al. 1983; Turner and Covey 1984). This molecule strongly resembles the "strong stop DNA" of retroviruses (Varmus 1980). The (+) strand DNA synthesis is initiated at a site almost 180 degrees opposite the tRNA primer binding site on the circular genome of CaMV. The mechanism of initiation at that site has not yet been elucidated. This asymmetric mode of replication explains the formation of the overlap structures in either strand of the viral DNA as being the result of strand displacement at the termination sites of either DNA strand synthesis. The formation of "gapped", in the case of CaMV "overlapping" viral DNA molecules resembles some hepatitis B viruses (Summers and Mason 1982).

Taking all these results together, a striking similarity emerges between caulimoviruses, hepatitis B viruses, and retroviruses (Pfeiffer and Hohn 1983; Varmus 1983). Furthermore, Toh et al. (1983) and Volovitch et al. (1984) detected stretches of strong homology at the protein level between retroviral reverse transcriptases, hepatitis B virus polymerases, and the potential protein of CaMV orf V. In addition, Volovich et al. (1984) were able to detect and partially purify an RNA-dependent DNA polymerase activity in infected *Brassica* leaves.

In the light of these recent findings the simple picture of the life-cycle of CaMV had to be revised. Therefore the use of CaMV as gene vector might demand somewhat more sophisticated approaches than initially anticipated.

3 The Development of CaMV into a Plant Gene Vector

3.1 Mutants of CaMV

3.1.1 Mutations in orf II

In order to make use of CaMV for the introduction and expression of foreign genes in plants, sites or regions on the CaMV genome which are dispensable for the virus had to be identified. The existence of the deletion mutant CM 4-184 which had arisen naturally (Hull 1980) already pointed toward a part of the genome which might be unessential. Indeed, small modifications at a unique XhoI site of strain CM 1841, located within the area covered by the deletion in strain CM 4-184, did not interfere with virus multiplication (Gronenborn et al. 1980). These initial experiments were still carried out using uncloned CaMV DNA.

After Howell et al. (1980) demonstrated first that cloned copies of CaMV DNA retain infectivity, the whole plethora of recombinant DNA technology became applicable to CaMV engineering.

Using a cloned copy of CM1841 (Gardner et al. 1981) we found that small sized insertions (65 bp or 256 bp) were tolerated by CaMV when inserted into orf II (Gronenborn et al. 1981). There is, however, a limitation on the size of additional DNA that is stably maintained in the virus genome during systemic spread throughout the plant. An additional 531 bp fragment could not be recovered from encapsidated viral DNA. Similar results have been obtained by Daubert et al. (1983) and Dixon et al. (1983). The largest additional piece of DNA which has been found to be encapsidated has a length of 351 base pairs (Gronenborn, unpublished).

3.1.2 Mutations in Other Regions of the CaMV Genome

The effects of small sized (10 to 30 bp) insertions at various sites scattered around the genome were studied by Howell et al. (1981); Daubert et al. (1983) and Dixon et al. (1983). Only orf II and the large intergenic region tolerated both small sized and larger (up to 250 bp) insertions. In two other open reading frames small insertions were tolerated: 12 bp insertions at two different sites in orf VI leading to the addition of four aminoacids were found to be infectious. This region of orf VI is slightly variable between the CaMV isolates sequenced (Daubert et al. 1983). A 10 bp frameshift linker inserted at the very end of orf IV leads to an alteration of the carboxyterminal sequence of the coat protein and a shortening by three amino acids. This mutation causes a delay in the appearance of disease symptoms (Dixon et al. 1983). It is interesting to note that this particular insertion causes a frameshift between the first and second ATG codon of orf V deduced from the DNA sequence. Therefore the first ATG codon "on paper" probably is not the initiation codon of the orf V protein.

3.2 Translational Polarity

We have carried out a more extended study on the effect of "medium-sized" (351 bp) insertions into orf II of strain CM 1841 and into the truncated short orf II of strain CM 4-184 (Howarth et al. 1981). Based upon sequence analysis of a variety of insertion induced deletions which arose during passage of insertion mutants in orf II, we found polar effects of nonsense mutations in orf II on the expression of orf III (Sieg and Gronenborn, submitted for publication). The same was true for some frameshift mutations in orf II. We therefore hypothesize that genes I, II and III of CaMV are translated from a polycistronic mRNA. These three reading frames are closely linked with only one intercistronic base pair.

A mechanism by which eukaryotic ribosomes are able to translate a polycystronic messenger with closely linked genes is outlined in the "relay-race" model of translation (Sieg and Gronenborn, submitted). It is an intrinsic feature of the proposed mode of translation that any open reading frame at the location of the original orf II of CaMV will be translated into protein in order to ensure expression of the essential orf III.

The model therefore provides us with a clear-cut prediction, on how small-sized foreign genes are to be inserted into the CaMV genome in order to get stable maintainance and translation into protein: Insertions into orf II should not cause any polarity on the expression of the genes downstream of orf II and the intercistronic distance between orf I, the additional new gene and orf III should be as small as possible, preferably a few base pairs only.

3.3 A Transducing Cauliflower Mosaic Virus

The application of the predictions made by the relay-race model of translation on the design of transducing CaMV variants led to the first viable and stable transducing CaMV expressing a foreign gene in plants (Brisson et al. 1984). The gene coding for a methotrexante in-

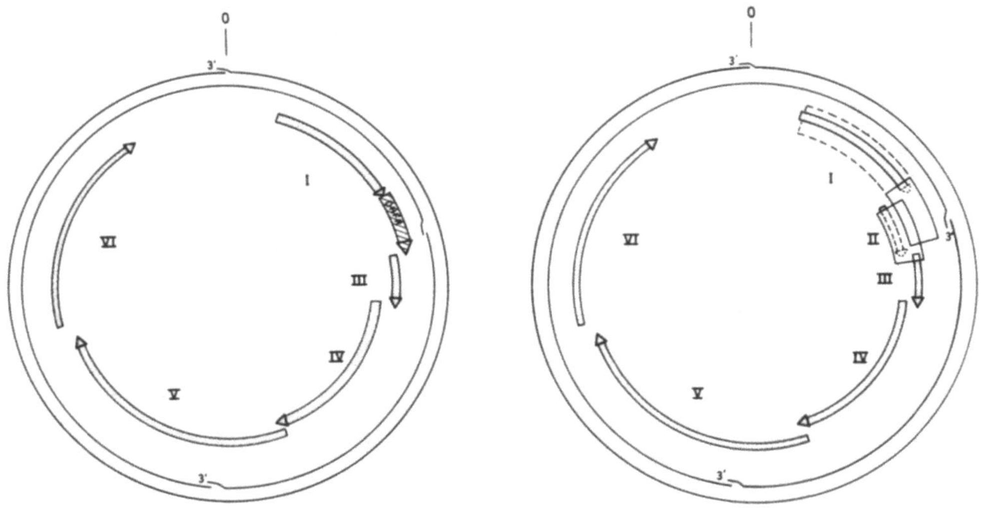

a b

Fig. 2. a Structure of the DHFR transducing CaMV. Orf II has been replaced by the DHFR gene of R67 (not to scale). The discontinuity of the former orf II is no longer present; b Structure of two defective complementing deletion mutants. The extent of the deletions is indicated by the *boxed areas* which overlap in orf II. Note that the figure represents two different individual virus genomes. A potential extension of one deletion in orf I is indicated by the *dashed box*

sensitive dihydrofolate reductase (DHFR), derived from the R67 R-factor of *E. coli* (O'Hare et al. 1981), was tailored in such a way as to precisely fit into a deletion of the entire orf II (Fig. 2A). The new orf II, the DHFR gene, has a length of 240 bp. The 5' intercistronic distance to orf I is 9 bp and the 3' intercistronic distance of orf III is 1 bp. In a, according to the relay-race model, suboptimal constructin which still has a longer intercistronic distance between orf I and the start codon of the DHFR gene (31 bp plus a small part of reading frame II) the genome of the hybrid viurs is less stable and the insert is gradually lost during repeated passages. In the tight construction described initially, the insert is stably maintained in repeated passages. Both transducing viruses render infected plants resistant to methotrexate. The DHFR protein is readily detected in western blots. The DHFR activity was assayed as the methotrexate insensitive incorporation of ^{32}P labeled phosphate into total plant DNA (Brisson et al. 1984).

3.4 Complementation Between Defective CaMV Mutants

Although it has been proven by now that CaMV can be used to transfer and express new genes in plants, the size limitations on the additional DNA that is encapsidated into virus particles still causes an unsatisfactory situation. Furthermore, exact tailoring of foreign genes to replace orf II, to avoid interference with the relay-race translation, makes CaMV laborious to use. Therefore the development of a complementation system between a helper virus and a defective transducing virus or between two defective virus mutants is most desirable.

In the past, numerous attempts to establish a complementation system between defective mutants of CaMV have always failed. In all previous attempts, only wildtype recombinants were recovered (Howell et al. 1981; Lebeurier et al. 1982; Walden and Howell 1982, 1983). This is either due to a very efficient homologous recombination as stated by the authors cited above, or, with the current knowledge of the reverse transcription mode of replication, caused by frequent template switches during the process of reverse transcription.

To overcome these obstacles we have constructed pairs of overlapping deletions which lack homology (Fig. 2B). Deletion mutants spanning orf I plus II and orf II and III failed to cause symptoms of infection on turnip plants when inoculated separately. However, when they were used to co-infect plants, the symptoms of disease were almost normal in speed and severity. Analysis of the viral progeny showed no recombinant virus infectious on its own.

Using this approach we have created CaMV variants with artificial bipartite genomes that are able to fully complement one another and spread systemically throughout the host plant as defective virus (Gronenborn, Schaefer and Sieg, to be published).

Taking advantage of overlapping deletion mutants, it seems feasible to accommodate larger genes into the virus genome and to achieve the systemic spread of defective transducing particles.

3.5 Elements of CaMV in Genetic Engineering

If the CaMV genome, or parts of it, are introduced into plants at the single-cell level, e.g., by transfection or transformation of protoplasts, then there is no longer a need for the virus to function as wild-type or defective transducing virus. In this case only parts of the genome might be necessary to express a foreign gene either transiently or, if the DNA becomes integrated into the genome of the host plant, in a more stable way. In addition, it may be possible to widen the somewhat narrow host range of the virus if the promoters of CaMV are functional in other than cruciferous plants.

In an attempt to follow this approach to use CaMV, we replaced the orf VI, the coding region of the viral inclusion body protein, by the Tn5-derived neomycin phosphotransferase (APH 3'II) gene of *E. coli* (Beck et al. 1982). DNA of this defective virus was used to transform protoplasts of tobacco (*Nicotiana tabacum*, SR1) according to the procedure developed by Krens et al. (1982) for the introduction of Ti-plasmid DNA into plant protoplasts. A number of resistant calli were obtained after selection on a medium containing kanamycin. The presence of the newly introduced aminoglycoside phosphotransferase was verified by an in situ phosphorylation assay of kanamycin in non denaturing polyacrylamide gels, developed by Reiss et al. (1984). Similar results were obtained by Paszkowski et al. (1984), who were able to prove that the DNA becomes integrated into the genome of the plant.

These results prove that regions other than orf II on the CaMV genome can be used to successfully express foreign genes in plants. Furthermore, they provide evidence that not only cruciferous plants may be engineered by CaMV-based hybrid gene constructions.

In conclusion, genetic engineering of plants with cauliflower mosaic virus or with parts of it now seems possible. There remain, however, many interesting questions on the molecular biology of the virus itself.

Acknowledgement. I wish to thank the colleagues at the Institut für Genetik, Universität zu Köln, the Friedrich Miescher-Institut, Basel and the Max-Planck-Institut für Züchtungsforschung, Köln for discussions. Support came from Deutsche Forschungsgemeinschaft (SFB 74), EMBO (long term fellowship) and Bundesministerium für Forschung und Technologie.

References

Al Ani R, Pfeiffer P, Whitechurch O, Lesot A, Lebeurier G, Hirth L (1980) A virus specified protein produced upon infection by cauliflower mosaic virus. Ann Virol (Inst Pasteur) 131E:33-35

Armour SL, Melcher U, Pirone TP, Lyttle DJ, Essenberg RC (1983) Helper component for aphid transmission encoded by region II of cauliflower mosaic virus DNA. Virology 129:25-30

Balázs E, Guilley H, Jonard G, Richards K (1982) Nucleotide sequence of DNA from an altered-virulence isolate D/H of the cauliflower mosaic virus. Gene 19:239-249

Beck E, Ludwig G, Auerswald EA, Reiss B, Schaller H (1982) Nucleotide sequence and exact localization of the neomycin phosphotransferase gene from transposon Tn5. Gene 19:327-336

Breathnach R, Chambon P (1981) Organisation and expression of eukaryotic split genes coding for proteins. Ann Rev Biochem 50:349-383

Brisson N, Paszkowski J, Penswick J, Gronenborn B, Potrykus I, Hohn T (1984) Expression of a bacterial gene in plants using a viral vector. Nature 310:511-514

Chauvin C, Jacrot B, Lebeurier G, Hirth L (1979) The structure of cauliflower mosaic virus. A neutron diffraction study. Virology 96:640-641

Covey SN, Hull R (1981) Transcription of cauliflower mosaic virus DNA. Detection of transcripts, properties, and location of the gene encoding the virus inclusion body protein. Virology 111:463-774

Covey SN, Turner D, Mulder G (1983) A small DNA molecule containing covalently-linked ribonucleotides originates from the large intergenic region of the cauliflower mosaic virus genome. Nucl Acids Res 11:251-264

Daubert S, Richins R, Shepherd RJ, Gardner RC (1982) Mapping of the coat protein gene of cauliflower mosaic virus by its expression in a prokaryotic system. Virology 122:444-449

Daubert S, Shepherd RJ, Gardner R (1983) Insertional mutagenesis of the cauliflower mosaic virus genome. Gene 25:201-208

Dixon LK, Koenig I, Hohn T (1983) Mutagenesis of cauliflower mosaic virus. Gene 25: 189-199

Favali MA, Bassi M, Conti GG (1973) A quantitative autoradiographic study of intracellular sites of replication of cauliflower mosaic virus. Virology 53:115-119

Franck A, Guilley H, Jonard G, Richards K, Hirth L (1980) Nucleotide sequence of cauliflower mosaic virus DNA. Cell 21:285-294

Gardner RC, Howarth AJ, Hahn P, Brown-Luedi M, Shepherd RJ, Messing J (1981) The complete nucleotide sequence of an infectious clone of cauliflower mosaic virus by M13 mp7 shotgun sequencing. Nucleic Acids Res 9:2871-2888

Goodman RM (1981) Geminiviruses. In: Kurstak E (ed) Handbook of plant virus infections and comparative diagnosis. Elsevier/North Holland Biomedical Press, Amsterdam, p 879-910

Gronenborn B, Gardner R, Shepherd R (1980) Introduction of mutations into the genome of cauliflower mosaic virus in vitro. Abstracts of the 6th EMBO Ann Symp Heidelberg, FRG

Gronenborn B, Gardner RC, Schaefer S, Shepherd RJ (1981) Propagation of foreign DNA in plants using cauliflower mosaic virus as vector. Nature 294:773-776

Guilfoyle TJ (1980) Transcription of the cauliflower mosaic virus genome in isolated nuclei from turnip leaves. Virology 107:71-80

Guilley H, Dudley RK, Jonard G, Balázs E, Richards KE (1982) Transcription of cauliflower mosaic virus DNA: detection of promoter sequences, and characterization of transcripts. Cell 30:763-773

Guilley H, Richards KE, Jonard G (1983) Observations concerning the discontinuous DNAs of cauliflower mosaic virus. EMBO J 2:277-282

Hohn B, Hohn T (1982) Cauliflower mosaic virus: a potential vector for plant genetic engineering. In: Kahl G, Schell J (eds) Molecular biology of plant tumors. Academic Press, New York, p 549-560

Hohn T, Richards K, Lebeurier G (1982) Cauliflower mosaic virus on its way to becoming a useful plant vector. Curr Top Microbiol Immunol 96:193-236

Howarth AJ, Gardner RC, Messing J, Shepherd RJ (1981) Nucleotide sequence of naturally occurring deletion in mutants of cauliflower mosaic virus. Virology 112:678-685

Howell SH, Hull R (1978) Replication of cauliflower mosaic virus and transcription of its genome in turnip leaf protoplasts. Virology 86:468-481

Howell SH, Walker LL, Dudley RK (1980) Cloned cauliflower mosaic virus DNA infects turnips (Brassica rapa). Science 208:1265-1267

Howell SH, Walker LL, Walden RM (1981) Rescue of in vitro generated mutants of cloned cauliflower mosaic virus genome in infected plants. Nature 293:483-486

Howell SH (1982) Plant molecular vehicles: Potential vectors for introducing foreign DNA into plants. Ann Rev Plant Physiol 33:609-650

Hull R, Shepherd RJ (1976) Cauliflower mosaic virus: An improved purification procedure and some properties of the virus particles. J Gen Virol 31:93-100

Hull R, Shepherd RJ (1977) The structure of cauliflower mosaic virus genome. Virology 79:216-230

Hull R, Howell SH (1978) Structure of the cauliflower mosaic virus genome. II. Variation in DNA structure and sequence between isolates. Virology 86:482-493

Hull R (1980) Structure of the cauliflower mosaic virus genome. III. Restriction endonuclease mapping of thirty-three isolates. Virology 100:76-90

Hull R, Covey SN (1983) Characterization of cauliflower mosaic virus DNA forms isolated from infected turnip leaves. Nucl Acids Res 11:1881-1895

Hull R, Covey SN (1983) Does cauliflower mosaic virus replicate by reverse transcription? Trends Biochem Sci 8:119-121

Hull R, Davies JW (1983) Genetic engineering with plant viruses, and their potential as vectors. Adv Virus Res 1-33

Kamei T, Rubio-Huertos M, Matsui C (1969) Thymidine-3H-uptake by X-bodies associated with cauliflower mosaic virus infection. Virology 37:506-508

Krens FA, Molendijk L, Wullems GJ, Schilperoort RA (1982) In vitro transformation of plant protoplasts with Ti plasmid DNA. Nature 296:72-74

Lebeurier G, Hirth L, Hohn T, Hohn B (1980) Infectivities on native and cloned DNA of cauliflower mosaic virus. Gene 12:139-146

Lebeurier G, Hirth L, Hohn B, Hohn T (1982) In vivo recombination of cauliflower mosaic virus DNA. Proc Natl Acad Sci USA 79:2932-2936

Lung MCY, Pirone TP (1973) Studies on the reason for differential transmissibility of cauliflower virus isolates by aphids. Phytopathology 63:910-914

Lung MCY, Pirone TP (1974) Acquisition factor required for aphid transmission of purified cauliflower mosaic virus. Virology 60:260-264

Menissier J, Lebeurier G, Hirth L (1982) Free cauliflower mosaic virus supercoiled DNA in infected plants. Virology 117:322-328

Matthews REF (1981) Plant Virology, 2nd ed. Academic Press, New York

Marco Y, Howell SH (1984) Intracellular forms of viral DNA consistent with a model of reverse transcriptional replication of the cauliflower mosaic virus genome. Nucl Acids Res 12:1517-1528

Odell JT, Howell SH (1980) The identification, mapping and characterization of mRNA for p66, a cauliflower mosaic virus-coded protein. Virology 102:349-359

Odell JT, Dudley RK, Howell SH (1981) Structure of the 19S RNA transcript encoded by the cauliflower mosaic virus genome. Virology 111:377-385

O'Hare K, Benoist C, Breathnach R (1981) Transformation of mouse fibroblasts to methotrexate resistance by a recombinant plasmid expressing a prokaryotic dihydrofolate reductase. Proc Natl Acad Sci USA 78:1527-1531

Olszewski N, Hagen G, Guilfoyle TJ (1982) A transcriptionally active, covalently closed minichromosome of cauliflower mosaic virus DNA isolated from infected turnip leaves. Cell 29:395-402

Paszkowski J, Shillito RD, Saul M, Mandák V, Hohn T, Hohn B, Potrykus I (1984) Direct gene transfer to plants. EMBO J: in press
Pfeiffer P, Hohn T (1983) Involvement of reverse transcription in the replication of the plant virus CaMV: A detailed model and test of some aspects. Cell 33:781-789
Reiss B, Sprengel R, Will H, Schaller H (1984) A new sensitive method for qualitative and quantitative assay of neomycin phosphotransferase in crude cell extracts. Gene 30:211-218
Richards KE, Guilley H, Jonard G (1981) Further characterisation of the discontinuities of cauliflower mosaic virus DNA. FEBS Lett 134:67-70
Shepherd RJ, Bruening GE, Wakeman RJ (1970) Double-stranded DNA from cauliflower mosaic virus. Virology 41:339-347
Shepherd RJ, Wakeman RJ (1971) Observation on the size and morphology of cauliflower mosaic virus deoxyribonucleic acid. Phytopathology 61:188-193
Shepherd RJ (1976) DNA viruses of higher plants. Adv Virus Res 20:305-339
Shepherd RJ (1979) DNA plant viruses. Ann Rev Plant Physiol 30:405-423
Shepherd RJ, Lawson RH (1981) Caulimoviruses. In: Kurstak E (ed) Handbook of plant virus infections and comparative diagnosis. Elsevier/North-Holland Biomedical Press, Amsterdam, p 847-878
Summers J, Mason W (1982) Replication of the genome of a hepatitis B-like virus by reverse transcription of an RNA intermediate. Cell 29:403-415
Turner DS, Covey SN (1984) A putative primer for the replication of cauliflower mosaic virus by reverse transcription is virion-associated. FEBS Letters 165: 285-289
Toh H, Hayashida H, Miyata T (1983) Sequence homology between retroviral reverse transcriptase and putative polymerases of hepatitis B virus and cauliflower mosaic virus. Nature 305:827-829
Varmus HE (1982) Form and function of retroviral proviruses. Science 216:812-820
Varmus HE (1983) RNA viruses: Reverse transcription in plants? Nature 304:116-117
Volovitch M, Drugeon G, Yot P (1978) Studies on the single-stranded discontinuities of the cauliflower mosaic virus genome. Nucleic Acids Res 5:2913-2925
Volovitch M, Modjtahedi N, Yot P, Brun G (1984) RNA-dependent DNA polymerase activity in cauliflower mosaic virus-infected plant leaves. EMBO J 3:309-314
Walden RM, Howell SH (1982) Intergenomic recombination events among pairs of defective cauliflower mosaic virus genomes in plants. Mol Appl Genet 1:447-456
Walden RM, Howell SH (1983) Uncut recombinant plasmids bearing nested cauliflower mosaic virus genomes infect plants by intragenomic recombination. Plant Molec Biol 2:27-31
Woolston CJ, Covey SN, Penswick JR, Davis JW (1983) Aphid transmission and a polypeptide are specified by a defined region of the cauliflower mosaic virus genome. Gene 23:15-23
Xiong C, Muller S, Lebeurier G, Hirth L (1982) Identification by immunoprecipitation of cauliflower mosaic virus in vitro major translation product with a specific serum against viroplasm protein. EMBO J 1:971-976
Zambryski P, Herrera-Estrella L, de Block M, Van Montagu M, Schell J (1984) The use of the Ti plasmid of Agrobacterium to study the transfer and expression of foreign DNA in plant cells: new vectors and methods. In: Setlow J, Hollaender A (eds) Genetic engineering, principles and methods, Vol VI. Plenum Press, New York, p 253-278

Introduction of DNA into the Germ Line of Animals

Hybrid Dysgenesis as a Gene Transfer System in *Drosophila melanogaster*

D. J. FINNEGAN[1]

1 Introduction

The ability to introduce cloned DNA into chromosomes of eukaryotic cells is an important aspect of recombinant DNA technology. It is the best way of testing conclusions regarding the function of cloned segments of DNA, or testing the effects of in vitro generated mutations, and will be essential if this technology is to be used to manipulate plant or animal genomes directly. Ideally exogenous DNA should be introduced into the germ-line without rearrangement and, in the case of DNA from the same organism, at its normal chromosome location. At present this can only be achieved in yeast, where transforming DNA can replace equivalent chromosomal sequences by homologous recombination. DNA injected into mouse embryos can integrate into chromosomes of germ cells but does so in a variable and unpredictable fashion and often in a state which is either not expressed or is expressed in an abnormal fashion. The situation in *Drosophila melanogaster* lies somewhere between that of yeast and mouse. Rubin and Spradling (1982) have developed a transformation system for *D. melanogaster* which allows discrete fragments of DNA to integrate into chromosomes of germ cells. The sites of integration are unpredictable, but the DNA is not rearranged, and, in general, is expressed with its normal developmental and tissue specificity. This system uses transposable elements as vectors, taking advantage of their ability to insert themselves at new sites in the genome. The transposable elements used are P elements. These are normally stable in the genome but can be induced to transpose at high frequency under conditions of hybrid dysgenesis.

2 Hybrid Dysgenesis in *D. melanogaster*

Hybrid dysgenesis is the name given to the appearance of a set of unusual characteristics in the progeny of crosses between certain strains of *D. melanogaster*. These characteristics include lowered fertility, recombination in males and increased frequencies of mutations and chromosome aberrations. Excellent reviews of hybrid dysgenesis have been published recently by Bregliano and Kidwell (1983), Kidwell (1983) and Engels (1983) and I shall only summarize the important points here.

Two independent systems of hybrid dysgenesis, P-M and I-R, are known and strains of *D. melanogaster* may be classified as being of one or other of two types in each system. In the P-M system hybrid dysgenesis is seen in the progeny of crosses between *M (maternal)* strain females and *P (Paternal)* strain males. In the I-R system the effect is seen in the progeny of *I (Inducer)* strain males and *R (Reactive)* strain females. In both cases the progeny of reciprocal crosses appear normal.

[1] Department of Molecular Biology, University of Edinburgh, King's Buildings, Mayfield Road, Edinburgh EH9 3JR, Scotland

35. Colloquium - Mosbach 1984
The Impact of Gene Transfer Technique in Eukaryotic Cell Biology

The dysgenic traits produced by the I-R and P-M interactions are similar and in each case seem to be confined to cells of the germ line. There are some differences, the most obvious of which is that the I-R interaction affects only females whereas the P-M interaction affects both sexes. Both interactions result in a degree of infertility, but with different physiological characteristics. The gonads of P-M dysgenic individuals fail to develop, whereas I-R dysgenic females lay eggs which stop developing prior to blastoderm formation. Kidwell (1979) has classified several strains with respect to both P-M and I-R dysgenesis and has found that the two systems are genetically independent.

The characteristics of *P* and *I* strains are determined by transposable genetic determinants known as *P* factors and *I* factors, respectively. Transposition of these elements is controlled in a complex way. *P* factors and *I* factors are stable in *P* and *I* strains but transpose when introduced into the cytoplasmic background of *M* or *R* strains, respectively, in a dysgenic cross. Many P-M induced mutations are unstable in individuals subject to P-M dysgenesis (Engels 1979; Rubin et al. 1982) leading to the suggestion that mutations induced by P-M dysgenesis are due to insertions of *P* factor DNA (Green 1977; Golubovsky et al. 1977; Simmons and Lim 1980). This has been confirmed by Rubin and his colleagues (Rubin et al. 1982). They compared the structure of the *white* gene in DNA from flies carrying P-M induced *white* mutations with that of the parental wild-type allele. The mutations which they studied apparently contained *P* factor sequences since they were unstable in P-M dysgenic individuals. Each mutation had foreign DNA inserted into the *white* gene as expected. These insertions ranged in size from 0.5-1.4kb and were of related sequences. Rubin et al. (1982; O'Hare and Rubin 1983) argued that these insertions were too short to be fully functional *P* factors, which probably code at least for a transposase and a regulatory molecule, but that they might be inactive *P* elements derived from *P* factors by internal deletion. Presumably these *P* elements can transpose if a functional *P* factor is available to provide the necessary transposase.

O'Hare and Rubin (1983) used one of these putative *P* elements as a probe with which to screen a library of recombinant phages containing DNA from a *P* strain. They recovered several clones containing a conserved 2.9kb sequence which they presumed to be the complete *P* factor. The best evidence that these cloned sequences are indeed functional *P* factors would be to demonstrate that they can stimulate P-M dysgenesis when introduced into the cytoplasmic environment of an *M* strain. Spradling and Rubin (1982) have tested this directly by injecting clones of putative *P* factors into *M* strain embryos to see whether or not the resulting flies exhibit dysgenesis. The *M* strain embryos which they injected carried a mutation of the *singed* locus, sn^{w} (Engels 1979). This is a P-M induced mutation which is extremely unstable in dysgenic flies. If the injected DNA were to induce hybrid dysgenesis this would be revealed by mutations of sn^{w} to sn^{+} or more extreme *singed* alleles. This was scored in the progeny of the flies resulting from the injected embryos. Injection of a putative *P* factor did destabilize the sn^{w} mutation, confirming its identity and showing that hybrid dysgenesis can be produced in this way. The instability of sn^{w} was inherited, indicating continued presence of *P* factors. Presumably they were themselves affected by hybrid dysgenesis and had transposed from plasmid DNA onto chromosomes in germ cells of the injected embryos. They had integrated by transposition rather than recombination since the complete *P* factors had inserted into the chromosomes without any flanking DNA.

O'Hare and Rubin (1983) have determined the base sequences of one complete *P* factor and of several *P* elements. The *P* factor is 2907 bp long and is bounded by perfect inverted repeats of 31 bp. One strand of the *P* factor contains three long open reading frames which could code for polypeptides of 238, 264 and 218 amino acids, respectively. Restriction mapping and base sequence data indicate that *P* factors are highly conserved and that *P* elements can be derived from *P* factors by simple deletions. Each *P* factor and *P* element is flanked by 8 bp duplications of target site DNA. *P* elements do not insert at random since three of the five P-M-induced *white* gene mutations studied by Rubin et al. (1982) were due to insertions at exactly the same position. O'Hare and Rubin (1983) have been able to derive a consensus sequence from the 8 bp duplications generated by *P* elements inserted at eight different sites but they could find no obvious homology between the sequences flanking these sites.

The *I* factor which controls I-R hybrid dysgenesis has also been identified by examination of the molecular lesions associated with I-R induced *white* mutations. Five white mutations induced by I-R dysgenesis are due to insertions of apparently identical 5.4 kb transposable elements (Finnegan et al. 1983; Bucheton et al. 1984). One of these mutations is very tightly linked to a genetically active *I* factor (Pelisson 1981) indicating that the 5.4 kb element is the *I* factor. The DNA's of two *I* factors have been cloned. No homology can be detected by hybridization between the ends of an *I* factor, indicating that it does not contain long direct or inverted repeats. The *I* factor is thus not a member of the *copia*-like or fold-back classes of transposable elements. There is no detectable homology between *I* and *P* factors, as expected from the observation that the I-R and P-M systems of hybrid dysgenesis are genetically independent. We do not yet know whether or not *I* factors can induce I-R dysgenesis when injected into *R* strain embryoes, and if they do, whether they can transpose onto the host chromosomes. Experiments are underway to test this.

3 Drosophila Transformation Using P Element Vectors

Having found that P factors can transpose when injected into *M* strain embryos, Rubin and Spradling (1982) wondered whether this could be exploited to introduce any DNA into *D. melanogaster* chromosomes. In order to test this they inserted within a *P* element, a fragment of DNA carrying the *rosy* gene. This gene was chosen because it had been cloned and because its effects are not cell autonomous, and very low levels of enzyme activity have detectable effects on eye colour. A plasmid carrying the *P-rosy* element was injected into embryos of an *M* strain, together with a second plasmid containing a complete *P* factor to provide any functions necessary for transposition. The DNA was injected at the posterior pole of the embryos in order to maximise the chance of incorporation in the germ-line.

Up to 50% of fertile adults derived from injected embryos produced Rosy$^+$ progeny, indicating that the *rosy* gene had been incorporated into chromosomes in at least some germ cells. The *P-rosy* element inserted at many sites in the genome without altering the structure of the *rosy* gene. Again this was the result of transposition and not recombination, since the complete *P-rosy* element had inserted into the chromosomes without any flanking plasmid DNA. *D. melanogaster* embryos have now been successfully transformed with *P* elements carrying the *rosy* (Rubin and Spradling 1982; Spradling and Rubin 1983), chorion (Rubin and Spradling 1982), dopa decarboxylase (Scholnick et al. 1983), alcohol dehydrogenase (Goldberg et al. 1983) and *white* (Hazelrigg et al. 1984) genes, and there is probably no restriction on sequences which can be used in these experiments.

The developmental and tissue-specific expression of transformed ADH, DDC, and *rosy* genes has been examined in some detail. In each case the transformed genes are expressed in approximately the same fashion as the corresponding genes in their normal chromosome locations. Spradling and Rubin (1983) have isolated about 70 Rosy$^+$ transformants and have studied xanthine dehydrogenase expression in 36. Xanthine dehydrogenase is the product of the *rosy* gene. The level of activity was at least half that of the normal gene in 25 of them. The only consistant change in gene expression was shown by the autosomal genes, *rosy* and DDC, when inserted on the X chromosome. The DDC and *rosy* genes are expressed at a level about 1.6-fold higher in males than in females under these circumstances. This presumably indicates that they are being affected by the dosage compensation mechanism which ensures that genes on the X chromosome in males are twice as active as those on each of the two X chromosomes in females.

Very similar results have been obtained with the *white*-eye gene (Hazelring et al. 1984). This is normally present on the X chromosome and its expression is higher in males than females. The same is true of transformed *white* genes on autosomes, again indicating a normal pattern of expression. In these experiments each gene was present on a fragment less than 12 kb long. This indicates that any cis-acting sequences essential for the expression of these genes must be close to them, and that their activity is little affected by sequences more than a few kilobases away.

4 Why is P Factor Transformation so Successful?

The results obtained with *P* factor transformation in *D. melanogaster* contrast with those found in mouse after DNA is introduced into embryos by injection, and subsequently incorporated into the germ line. In experiments of this sort, the number of copies, and structure of integrated sequences is variable and their expression unpredictable (Constantini and Lacy 1981; Brinster et al. 1981; Palmiter et al. 1982). The structure of fragments of DNA introduced into into *D. melanogaster* chromosomes by P factor transformation is almost certainly maintained because these fragments are carried by transposable elements. The sequences to be integrated are flanked by the ends of a *P* element and are integrated by a mechanism which recognizes these ends specifically. This suggests that a search for transposable elements which could be used as transformation vectors in other eukaryotes would be worthwhile. Retroviral vectors are already being developed for this purpose (Mulligan 1983; Mann et al. 1983; Lewis et al. 1984).

It is not so easy to understand why transformed genes should be expressed so readily in *D. melanogaster* but not in mammals. The answer could be trivial. Expression of the few genes which have been tested in *D. melanogaster* might be unusually insensitive to chromosome location. Alternatively, the result could be an experimental artifact, since transformants have been selected for study because they showed expression of the transforming genes. Both explanations seem unlikely. Firstly, each of the genes used has a different pattern of expression and unpublished data from several laboratories indicates that other transforming genes are also expressed normally. Secondly, Spradling and Rubin (1983) have found several Rosy$^+$ transformants with more than one copy of the transforming *rosy* gene. When each of these is tested separately, it is found to function normally.

The expression of transforming genes in the mouse may often be abnormal because they are affected by the structure of the chromatin at sites into which they integrate. This may not happen in *D. melanogaster* because there are relatively few nonexpressing sites in its small genome or because *P* elements, which are known to insert nonrandomly (Green 1977; Simmons and Lim 1980; Rubin et al. 1982; O'Hare and Rubin 1983), avoid these regions. The *P* element sequences which flank *Drosophila* transforming genes in these experiments may also protect them from the influence of surrounding sequences. This may be a property which transposable elements evolve to allow them to move about the genome with impunity (Spradling and Rubin 1983; Flavell 1983). Two insertions in *D. melanogaster* have been mapped near centromeric heterochromatin, and so might be expected to show position effect (Spradling and Rubin 1983; Hazelrigg et al. 1984). Their expression is somewhat abnormal, but no more so than that of some insertions elsewhere in the genome.

The ease with which *D. melanogaster* can be transformed should encourage those working with other organisms. Only a short while ago *Drosophila* seemed to be totally untransformable. The advances which have been made have come from increased knowledge of the mechanism of DNA rearrangements in *Drosophila*, and should stimulate a search for similar systems in other organisms.

References

Bregliano JC, Kidwell MG (1983) Hybrid dysgenesis determinants. In: Shapiro J (ed) Mobile genetic elements. Academic, New York, pp 363-410

Brinster RL, Chen HY, Trumbauer M, Senear AW, Warren R, Palmiter RD (1981) Somatic expression of herpes thymidine kinase in mice following injection of a fusion gene into eggs. Cell 27:223-231

Bucheton A, Sang HM, Paro R, Pelisson A, Finnegan DJ (1984) The molecular basis of I-R hybrid dysgenesis in *Drosophila melanogaster:* identification, cloning and properties of the I factor. Cell 38:153-163)

Constantini F, Lacy E (1981) Introduction of a rabbit β globin gene into mouse germ line. Nature (Lond) 294:92-94

Engels WR (1979) Extrachromosomal control of mutability in *Drosophila melanogaster*. Proc Natl Acad Sci USA 76:4011-4015

Engels WR (1983) The P family of transposable elements in *Drosophila*. Annu Rev Genet 17:315-344

Finnegan DJ, Bucheton A, Sang HM (1983) The molecular basis of hybrid dysgenesis in *Drosophila melanogaster*. In: Chater KF, Cullis CA, Hopwood DA, Johnston AWB, Woolhouse HW (eds) Genetic rearrangements. Croom Helm, London, pp 75-92

Flavell AJ (1983) *Drosophila* takes off. Nature (Lond) 305:96-97

Goldberg DA, Posakony JW, Maniatis T (1983) Correct developmental expression of a cloned alcohol dehydrogenase gene transduced into the *Drosophila* germ line. Cell 34:59-73

Golubovsky MD, Ivanov YN, Green MM (1977) Genetic instability in *Drosphila melanogaster:* putative multiple insertion mutants at the singed bristle locus. Proc Natl Acad Sci USA 74:2973-2975

Green MM (1977) Genetic instability in *Drosophila melanogaster*. De novo induction of putative insertion mutants. Proc Natl Acad Sci USA 74:3490-3493

Hazelrigg T, Levis R, Rubin GM (1984) Transformation of *white* locus DNA in Drosophila: dosage compensation, *zeste* interaction and position effects. Cell 36:469-481

Kidwell MG (1979) Hybrid dysgenesis in *Drosophila melanogaster:* the relationship between the P-M and I-R interactive systems. Genet Res 33:205-217

Kidwell MG (1983) Intraspecific hybrid sterility. In: Ashburner M, Carson HL, Thompson JW (eds) Genetics and Biology of *Drosophila*, Vol 3c. Academic, London, pp 125-153

Lewis S, Gifford A, Baltimore D (1984) Joining of V_K to J_K gene segments in a retroviral vector introduced into lymphoid cells. Nature (Lond) 308:425-428

Mann R, Mulligan R, Baltimore D (1983) Construction of a retroviral packaging mutant and its use to produce helper-free defective retroviruses. Cell 33:153-159

Mulligan R (1983) Construction of highly transmissible mammalian cloning vehicles derived from murine retroviruses. In: Inoye M (ed) Experimental manipulation of gene expression. Academic, New York

O'Hare K, Rubin GM (1983) Structure of P transposable elements and their sites of insertion and excision in the *Drosophila melanogaster* genome. Cell 34:25-35

Palmiter RD, Chen HY, Brinster RL (1982) Differential regulation of metallothionin-thymidine kinase fusion genes in transgenic mice and their offspring. Cell 29: 701-710

Pelisson A (1981) The I-R system of hybrid dysgenesis in *Drosophila melanogaster:* are I factor insertions responsible for the mutator effect of the I-R interaction? Mol Gen Genet 183:123-129

Rubin GM, Spradling AC (1982) Genetic transformation of *Drosophila* with transposable element vectors. Science (Wash DC) 218:348-353

Rubin GM, Kidwell MG, Bingham PM (1982) The molecular basis of P-M hybrid dysgenesis: the nature of induced mutations. Cell 29:987-994

Scholnick SB, Morgan BA, Hirsh J (1983) The cloned dopa decarboxylase gene is developmentally regulated when reintegrated into the Drosophila genome. Cell 34:37-45

Simmons MJ, Lim JK (1980) Site-specificity of mutations arising in dysgenic hybrids of *Drosophila melanogaster*. Proc Natl Acad Sci USA 77:6042-6046

Spradling AC, Rubin GM (1982) Transposition of cloned P elements into *Drosophila* germ line chromosomes. Science (Wash DC) 218:341-347

Spradling AC, Rubin GM (1983) The effect of chromosomal position on the expression of the *Drosophila* xanthine dehydrogenase gene. Cell 34:47-57

Introducing Genes into Mice and into Embryonal Carcinoma Stem Cells

E. F. WAGNER, U. RÜTHER, and C. L. STEWART[1]

1 Introduction

How genes are regulated during the development of a multicellular organism is still unclear. The formulation of new concepts for cellular differentiation, determination, and regeneration, as well as for normal and neoplastic cell growth should be facilitated by an understanding of developmentally regulated gene expression. The experimental approach must be sought within the context of a developing organism. Recent results obtained in two experimental systems will be reviewed: the introduction, expression, and fate of recombinant DNA following microinjection into the fertilized mouse egg and gene transfer studies using embryonal carcinoma (EC) cells.

2 Gene Transfer into Mice by Egg Injection

Simultaneously with the development of various gene transfer systems for cultured cells (Pellicer et al. 1980), similar techniques were being utilized for microinjection of cloned DNA into fertilized mouse eggs. The aim of this experimental approach is to study developmentally regulated and tissue-specific gene expression at the level of the whole organism.

By microinjection into one of the two pronuclei of the fertilized egg, various cloned genes have been introduced into mice (see Table 1). In most cases the injected genes are stably integrated into the host cell DNA and are transmitted through the germ line to offspring. New strains of mice homozygous for the acquired genotype can be obtained by selective breeding. The drawback of this approach has been the failure to find regulated tissue or cell specific expression of the foreign gene. In Table 1, some of the characteristics concerning the mode of expression of various genes together with the references are listed: expression of a newly introduced gene was often found in inappropriate tissues with great variations in the level of expression. If the gene was linked prior to injection to an inducable promoter such as the mouse metallothionein (MT), expression has usually, but not always ocurred. However, first indications of a possible cell type-specific expression have come from mice carrying the chicken transferrin gene or mouse immunoglobulin genes. In the former example, a strong preference for expression of the transferrin gene was found in mouse liver, where transferrin is normally synthesized. The introduced immunoglobulin kappa gene is expressed in the spleen, but not in the liver of mice carrying the injected gene.

There are probably many reasons why most genes introduced into mice are not regulated, whereas correct tissue specific expression after

[1]European Molecular Biology Laboratory, Postfach 10.2209, 6900 Heidelberg, FRG

35. Colloquium - Mosbach 1984
The Impact of Gene Transfer Technique in Eukaryotic Cell Biology

Table 1. Modes of expression of genes microinjected into fertilized mouse eggs

Gene injected	Expression Specific[a]	Unspecific	No	Reference
HSV-TK		+		Wagner et al. 1981
MT-TK		+		Brinster et al. 1981
		+		Palmiter et al. 1982
β-Globin:				
Human			+	T.A. Stewart et al. 1982
Rabbit		+	+	Costantini and Lacy 1981
		+	+	Lacy et al. 1983
M-MuLV		+		Harbers et al. 1981
		+		C. Stewart et al. 1983
Growth hormone:				
MT-GH rat		+		Palmiter et al. 1982
MT-GH human		+		Palmiter et al. 1983
GH human			+	Wagner et al. 1983
Transferrin chicken	+	+		McKnight et al. 1983
Immunoglobulin kappa	+[b]			Brinster et al., 1983

[a]Specific expression is defined as expression in the appropriate tissue type

[b]Expression only examined for spleen and liver

P- element mediated gene transfer in Drosophila seems to be the rule (Goldberg et al. 1983; Hazelrigg et al. 1984; Scholnick et al. 1983; Spradling and Rubin 1983). The regulation of stage- and tissue-specific gene expression in mammals might be quite different from *Drosophilia* (e.g., the presence of methylated sequences in mammals; Jaenisch and Jähner 1984) and various classes of genes might be regulated in very different ways. Some genes may be regulated exclusively by structural sequences within or around the gene and be expressed independently of the position on a particular chromosome. Others might require the correct position, the right chromatin structure and additional cis- or trans-acting elements which we are just beginning to understand. However, the latest findings on the preferential expression of newly introduced genes in mice are encouraging, and should lead to the dissection of events governing tissue-specific gene expression during differentiation and development.

What are the other possible consequences when changing the genetic make-up of a mammal? The integration of newly DNA may disrupt important chromosomal regions, which could lead to developmental defects, sterility, neoplasia, or to defects in metabolism. There have been numerous reports on insertional mutagenesis in nonmammalian systems (Mc Clintock 1956; Roeder and Fink 1980; Kidwell et al. 1977) and more recently in mice (Jenkins et al. 1981; Copeland et al. 1983; Jaenisch et al. 1983). In one case a recessive embryonic lethal mutation was experimentally induced in mice by insertion of a Moloney murine leukemia virus copy into the α1(I) collagen gene (Schnieke et al. 1983). We have tested the mutagenic potential of recombinant DNA in the germ line of mice as a possible alternative tool for a molecular identification of gene functions essential for normal development (Wagner et al. 1983).

Six individual strains of mice (HUGH strains) were obtained from eggs injected with the cloned human growth-hormone gene in a pBR322 vector. All were heterozygotes containing 1-20 intact copies of the foreign sequences in unique and often complex integration patterns. The six HUGH strains stably transmitted the newly inserted sequences through the germ line to their progeny and no indication of human growth hormone expression was found (see Table 1). When the individual heterozygotes were mated with each other, only four of the HUGH strains yielded viable homozygotes. In HUGH/3 and HUGH/4, no postnatal homozygotes were found, when a total of 31 and 52 offspring, respectively, were analyzed. In addition, litter sizes at birth in these two strains were markedly reduced compared to the other HUGH strains, suggesting that homozygous animals die prenatally. In Mintz' laboratory experiments are now in progress to establish the developmental stage at which these animals die. The molecular cloning of the integration site in HUGH/3 and HUGH/4 may reveal genomic coding sequences which are disrupted. Identification of these sequences may indicate whether a normal cellular housekeeping function is affected or if, more optimistically speaking, a novel gene could be identified, which is essential for a given developmental process. Additional experiments are needed to establish if this high incidence (33%) of obtaining homozygous recessive prenatal lethals reflects a mutagenic potential of recombinant DNA introduced into the germ line of mice by egg injection.

3 Gene Transfer into Mice Using Mouse Embryonal Carcinoma Stem Cells

Almost 10 years ago it was found that EC cells can lose their malignant phenotype and participate normally in development (to form all somatic as well as germ cells) after reintroduction into early embryos (for review see Mintz and Fleischman 1981; Stewart 1984). Therefore, these cells can be potentially used as vectors for introducing new genetic information into the germ line of mice. As an experimental system to study regulation of gene expression, the major advantage of using EC cells vs. direct DNA injection into fertilized eggs is that cells carrying the desired genetic change can first be selected and characterized prior to the reintroduction into the animal. In addition, these cells serve as an excellent tool for in vitro differentiation studies of early mouse development. The current state of the field is summarized in *Embryonic and Germ Cell Tumours in Man and Animals*, 1983 and in *Teratocarcinoma Stem Cells*, 1983.

A large number of established EC cell lines, mostly derived from tumors of grafted embryos are available, and recently stem cell lines derived directly from embryos grown in vitro have been isolated (Evans and Kaufman 1981; Martin 1981; Axelrod 1984). Using the calcium phosphate method, selectable genes, along with unselectable ones, were introduced into established EC cells (Pellicer et al. 1980; Linnenbach et al. 1980; Wagner and Mintz 1982; Bucchini et al. 1983; Nicolas and Berg 1983). Problems were encountered with respect to the relative low transformation efficiency (about 100-fold less compared to L-cell transformations), the integration of multiple and rearranged gene copies, the low or nonexpression of co-transferred genes and the instability of expression under nonselectable conditions.

Recently the infectious properties of retroviruses have been exploited so that they can function as vectors for introducing new genes into cells and animals (Gilboa 1983; Mann et al. 1983). These retroviral based vectors have a number of advantages over the calcium phosphate method: (a) Infection and incorporation of the virus into a cell's chromosomal DNA is extremely efficient; every single cell can be found to carry the vector sequences. (b) Generally, a single copy of the viral genome integrates into host cell DNA. (c) The integrated retroviral copy is usually stable in remaining within the cell's chromosome and with respect to its expression. The disadvantages in using retroviruses as vectors for introducing genes into mouse embryos or EC cells is that early embryonic cells of the mouse, including EC cells, suppress the expression of the retroviral genomes (C.L. Stewart et al. 1982; Gautsch and Wilson 1983). In addition, the insertion of large genomic DNA fragments into a retroviral vector is presently rather difficult and by no means a standard procedure.

Two approaches are currently used to fully utilize the advantages of retroviruses. The first approach has been to use several vectors, containing Moloney murine leukemia virus regulatory regions as well as selectable (neo or gpt) and nonselectable (human β-interferon cDNA) genes for infection of various established EC cell lines. We have been able to obtain a high frequency of clones under selectable conditions (efficiency 10^{-1}-10^{-3}). The clones contain a single intact copy of the recombinant gene and seem to stably express the gene in the absence of selection (unpublished results, partly in collaboration with E. Gilboa and B. Vennström). We are also infecting preimplantation embryos with these vectors to see if these genes can also be expressed in vivo.

As a second approach some of these viruses containing the neomycin resistance gene were used to infect feeder dependent pluri- and totipotent EC cells. We have obtained EC cell clones from a cell line known to form chimeras (unpublished results). The developmental capability of these cells was tested by introducing them back into normal mouse embryos by the aggregation technique (Stewart 1982; Fuji and Martin 1983). At present 34 mice have been born following manipulation, and 11 of them show overt phenotypic chimerism in the coat. These chimaeras are currently being analyzed to establish the extent of chimerism in various tissues and whether expression of the introduced gene(s) has been maintained during differentiation. It thus appears that it is possible to select first for the desired, genetically altered EC cell clones in vitro, that are still able to undergo subsequent differentiation and development in the embryo. Since the microinjection of DNA into pronuclei does not ensure the expression of the introduced gene, this system offers new possibilities for examining the action of specific gene products in mice by preselecting for the expression of these sequences in tissue culture.

4 Transfer of c-*onc* Genes into F9 Teratocarcinoma Stem Cells

In a different set of experiments we asked whether a cellular oncogene (c-*onc*) might exert a biological effect upon transfection into F9 cells (Müller and Wagner 1984). For this approach we chose the c-*fos* gene, which has been suggested to have a function during prenatal development, since it is expressed at high levels in the placenta and late-gestation fetal membranes (for review see Müller and Verma 1984). It was found that introduction of either the normal mouse or human c-*fos* gene along with a selectable marker gene (neo) in undifferentiated F9 cells results in colonies of cells that are morphologically different from the stem cells. The transformed cells are greatly enlarged, very flat, grow in an epitheloid fashion, and stop proliferating after reaching a maximum clone size of about 400 cells. Presently, the phenotype of these cells is being analyzed with respect to expression of various differentiation markers. To test whether the c-*fos* mediated induction of transformed cells requires a continuous c-*fos* expression or if only a temporary expression is necessary, various vectors have been constructed, where *fos* genes are under the control of a number of inducable promoters (Rüther et al. in preparation). The aim here is to control the level of c-*fos* expression and to correlate this to the induction of differentiation. In one series, the c-*fos* gene was combined with the inducable human metallothionein promoter (Karin et al. 1984). Transformed cells were obtained with these vectors and we are currently analyzing their potential to induce differentiation by activating the c-*fos* gene. These data suggest that the normal c-*fos* gene can induce differentiation in vitro and lends support to the hypothesis that c-*onc* genes have a vital function in cell differentiation and embryonic development.

Acknowledgments. The work on the introduction of human growth hormone genes into the germ line of mice was carried out in Beatrice Mintz' laboratory at the Institute for Cancer Research in Philadelphia. We wish to thank Becky Adkins for stimulating discussions, Mirka Vanek for technical assistance and Ines Benner for typing the manuscript. U.R. is a recipient of an EMBO post-doctoral fellowship.

References

Axelrod HR (1984) Embryonic stem cell lines derived from blastocysts by a simplified technique. Dev Biol 101:225-228

Brinster RL, Chen HY, Trumbauer M, Senear AW, Warren R, Palmiter RD (1981) Somatic expression of herpes thymidine kinase in mice following injection of a fusion gene into eggs. Cell 27:223-231

Brinster RL, Ritchie KA, Hammer RE, O'Brien RL, Arp B, Storb U (1983) Expression of a microinjected immunoglobulin gene in the spleen of transgenic mice. Nature (Lond) 306:332-336

Bucchini D, Lasserre C, Kunst F, Lovell-Badge R, Pictet R, Jami J (1983) Stable transformation of mouse teratocarcinoma stem cells with the dominant selective marker Eco. gpt and retention of their developmental potentialities. EMBO J 2: 229-232

Copeland NG, Jenkins NA, Lee BK (1983) Association of the lethal yellow (A^y) coat color mutation with an ecotropic murine leukemia virus genome. Proc Natl Acad Sci USA 80:247-249

Costantini F, Lacy E (1981) Introduction of a rabbit β-globin gene into the mouse germ line. Nature (Lond) 294:92-94

Evans MJ, Kaufman MH (1981) Establishment in culture of pluripotential cells from mouse embryos. Nature (Lond) 292:154-156

Fuji JT, Martin GR (1983) Developmental potential of teratocarcinoma stem cells in utero following aggregation with cleavage stage mouse embryos. J Embryol Exp Morphol 74:79-96

Gautsch JW, Wilson MC (1983) Delayed de novo methylation in teratocarcinoma suggests additional tissue-specific mechanisms for controlling gene expression. Nature (Lond) 301:32-37

Gardner RL (ed) Embryonic and germ cell tumours in man and animals. Cancer Survey 2(1)

Gilboa E (1983) Use of retrovirus derived vectors to introduce and express genes in mammalian cells. In: Papas TS, Rosenberg U, Chirikjian J (eds) Expression of cloned genes in procaryotic and eucaryotic vectors. Elsevier/North-Holland, New York

Goldberg DA, Posakony JW, Maniatis T (1983) Correct development expression of a cloned alcohol dehydrogenase gene transduced into the drosophila germ line. Cell 34: 59-73

Harbers K, Jähner D, Jaenisch R (1981) Microinjection of cloned retroviral genomes into mouse zygotes: integration and expression in the animal. Nature (Lond) 293: 540-542

Hazelrigg T, Levis R, Rubin GM (1984) Transformation of *white* locus DNA in Drosophila: dosage compensation, *zeste* interaction and position effects. Cell 36:469-481

Jaenisch R, Harbers K, Schnieke A et al. (1983) Germline integration of Moloney murine leukemia virus at the *Mov*13 locus leads to recessive lethal mutation and early embryonic death. Cell 32:209-216

Jaenisch R, Jähner D (to be published 1984) Methylation, expression and chromosomal position of genes in mammals. Biochim Biophys Acta

Jenkins NA, Copeland NG, Taylor BA, Lee BK (1981) Dilute (*d*) coat colour mutation of DBA/2J mice is associated with the site of integration of an ecotropic MuLV genome. Nature (Lond) 293:370-374

Karin M, Haslinger A, Holtgreve H, Richards RI, Krauter P, Westphal HM, Beato M (1984) Characterization of DNA sequences through which cadmium and glucocorticoid hormones induce human metallothionein-II_A gene. Nature (Lond) 308:513-519

Kidwell MG, Kidwell JF, Sred JA (1977) Hybrid dysgenesis in Drosophila melanogaster: a syndrome of aberrant traits including mutation, sterility and male recombination. Genetics 86:813-833

Lacy E, Roberts S, Evans EP, Burtenshaw MD, Costantini F (1983) A foreign β-globin gene in transgenic mice: integration at abnormal chromosomal positions and expression in inappropriate tissues. Cell 34:343-358

Linnenbach A, Huebner K, Croce CM (1980) DNA-transformed murine teratocarcinoma cells: regulation of expression of simian virus 40 tumor antigen in stem versus differentiated cells. Proc Natl Acad Sci USA 77 (8) 4875-4879

Mann R, Mulligan RC, Baltimore D (1983) Construction of a retrovirus packaging mutant and its use to produce helper-free defective retrovirus. Cell 33:153-159

Martin G (1981) Isolation of a pluripotent cell line from early mouse embryos cultured in medium conditioned by tetratocarcinoma stem cells. Proc Natl Acad Sci USA 78 (12) 7634-7638

McClintok B (1956) Controlling elements and the gene. Cold Spring Harbor Symp Quant Biol 21:197-216

McKnight GS, Hammer RE, Kuenzel EA, Brinster RL (1983) Expression of the chicken transferrin gene in transgenic mice. Cell 34:335-341

Mintz B, Fleischman RA (1981) Teratocarcinomas and other neoplasms as developmental defects in gene expression. Adv Cancer Res 34:211-278

Müller R, Verma IM (1984) Expression of cellular oncogenes. Curr Top Microbiol Immunol 112:73-115

Müller R, Wagner EF (1984) Differentiation of F9 teratocarcinoma stem cells after transfer of C-fos proto-oncogenes. Nature (Lond) 311:438-442

Nicolas JF, Berg P (1983) Regulation of expression of genes transduced into embryonal carcinoma cells. In: Silver LM, Martin GR, Strickland S (eds) Teratocarcinoma stem cells..Cold Spring Harbor Conf Cell Proliferation, vol 10, pp 469-485

Palmiter RD, Chen HY, Brinster RL (1982a) Differential regulation of metallotioneinthymidine kinase fusion genes in transgenic mice and their offspring. Cell 29: 701-710

Palmiter RD, Brinster RL, Hammer RE, Trumbauer ME, Rosenfeld MG, Birnberg NC, Evans RM (1982b) Dramatic growth of mice that develop from eggs microinjected with metallotionein-growth hormone fusion genes. Nature (Lond) 300:611-615

Palmiter RD, Norstedt G, Gelinas RE, Hammer RE, Brinster RL (1983) Metallothioneinhuman GH fusion genes stimulate growth of mice. Science (Wash DC) 222:809-814

Pellicer A, Robins D, Wold B et al. (1980a) Altering genotype and phenotype by DNA-mediated gene transfer. Science (Wash DC) 209:1414-1422

Pellicer A, Wagner EF, El Kareh A et al. (1980b) Introduction of a viral thymidine kinase gene and the human β-globin gene into developmentally multipotential mouse teratocarcinoma cells. Proc Natl Acad Sci USA 77(4):2098-2102

Roeder GS, Fink GR (1980) DNA rearrangements associated with a transposable element in yeast. Cell 21:239-249

Schnieke A, Harbers K, Jaenisch R (1983) Embryonic lethal mutation in mice induced by retrovirus insertion into the α1(I) collagen gene. Nature (Lond) 304:315-307

Scholnick SB, Morgan BA, Hirsh J (1983) The cloned dopa decarboxylase gene is developmentally regulated when reintegrated into the Drosophila genome. Cell 34:37-45

Silver LM, Martin GR, Strickland S (eds) (1983) Teratocarcinoma Stem cells. CHS Conf Cell Proliferation 10

Spradling AC, Rubin GM (1983) The effect of chromosomal position on the expression of the Drosophila xanthine dehydrogenase gene. Cell 34:47-57

Stewart CL (1982) Formation of viable chimaeras by aggregation between teratocarcinoma and preimplantation mouse embryos. J Embryol Exp Morphol 67:167-179

Stewart CL, Stuhlmann H, Jähner D, Jaenisch R (1982) De novo methylation, expression and infectivity of retroviral genomes introduced into embryonal carcinoma cells. Proc Natl Acad Sci USA 79:4098-4102

Stewart CL, Harbers K, Jähner D, Jaenisch R (1983) X-chromosome-linked transmission and expression of retroviral genomes microinjected into mouse zygotes. Science (Wash DC) 221:760-762

Stewart CL (to be published 1984) Teratocarcinoma chimaeras and gene expression. In: Douarin L, McLaren A (eds) Chimaeras in developmental biology. Academic, New York

Stewart TA, Wagner EF, Mintz B (1982) Human β-globin gene sequences injected into mouse eggs, retained in adults, and transmitted to progeny. Science (Wash DC) 217:1046-1048

Wagner EF, Stewart TA, Mintz B (1981) The human β-globin gene and a functional viral thymidine kinase gene in developing mice. Proc Natl Acad Sci USA 78 (8):5016-5020

Wagner EF, Mintz B (1982) Transfer of nonselectable genes into mouse teratocarcinoma cells and transcription of the transferred human β-globin gene. Mol Cell Biol 2 (2):190-198

Wagner EF, Covarrubias L, Stewart TA, Mintz B (1983) Prenatal lethalities in mice homozygous for human growth hormone gene sequences integrated in the germ line. Cell 35:647-655

Gene Transfer in Living Organisms

S. Rusconi[1]

1 Cloned Genes Injected Into *Xenopus* Zygotes Give Rise to Transgenically Mosaic Frogs

1.1 Introduction

Oocytes and eggs of *Xenopus laevis* are valuable tools for testing the activity of biological macromolecules. Oocytes have been injected with protein (reviewed by Lane 1981), RNA (Gurdon et al. 1974; Woodland and Wilt 1980a), and DNA (reviewed by Gurdon and Melton 1981) molecules. The unfertilized egg has been particularly useful in studying the replication of injected DNA (Harland and Laskey 1980; Hines and Benbow 1982). mRNA (Woodland and Wilt 1980b) and genomic DNA (Gurdon and Brown 1977) injected into fertilized eggs is maintained in the active state for a while, but seems to disappear in later developmental stages. Only with the advent of DNA cloning techniques has it become possible to inject defined protein-coding genes into the fertilized frog egg and to study their fate throughout development. To this end, cloned genes coding for the rabbit β-globin gene (Rusconi and Schaffner 1981) and for sea urchin histone genes (Bending 1981) were used in our laboratory. Both kinds of gene were replicated as extrachromosomal circles within developing *Xenopus* embryos, and were correctly transcribed around blastula-gastrula stage. Whereas most injected DNA was degraded after gastrula stage, a low number of rabbit β-globin genes became stably associated with the *Xenopus* genome and persisted throughout metamorphosis. In the present chapter we take a closer look at the state of such injected DNA sequences.

1.2 Materials and Methods

Microinjection of fertilized eggs of *Xenopus laevis* and isolation of nucleic acids at different developmental stages have been previously described (Rusconi and Schaffner 1981). For the analysis shown in Fig. 1, undigested material was used whereas, in the other cases, the DNA (usually 3 µg) was digested with various restriction enzymes as indicated in the figure legends. Incubation with HpaII and MspI was performed under conditions of 20-fold overdigestion (5 x enzyme excess and 4 x incubation time). Other enzymes were used as recommended by the manufacturers (Boehringer Mannheim and New England Biolabs). If two successive digestions were required on the same sample, they were separated by phenol-chloroform extraction and ethanol precipitation. Agarose gel electrophoresis and Southern blot analysis were carried

[1] c/o Prof. Keith Yamamoto, Department of Biochemistry and Biophysics, School of Medicine, University of California, San Francisco, CA 94143 USA

35. Colloquium - Mosbach 1984
The Impact of Gene Transfer Technique in Eukaryotic Cell Biology

out as described (Rusconi and Schaffner 1981). Dot blot analysis was performed essentially according to the protocol of Kafatos et al. (1979).

1.3 Results and Discussion

1.3.1 Extrachromosomal Replication of Exogenous DNA in Early Developmental Stages

In a typical experiment, about 10^7 DNA molecules were injected into each fertilized egg. One might have expected that the majority of the transferred molecules would not integrate into the host genome and would progressively disappear. However, in very early stages of development (from 0 to about 24 h) every DNA tested so far (including pure prokaryotic vectors such as pBR322 plasmid DNA) is able to undergo autonomous replication. In many cases this replication is so efficient that the newly synthesized DNA forms a visible band in agarose gel electrophoresis of undigested material from single embryos. Figure 1a shows an example of such DNA bands which appear in undigested material extracted from blastula or gastrula (see lanes 2 and 3) but which cannot be detected in material extracted immediately after injection of 250 pg of recombinant DNA (lane 1). As mentioned above, this ability to replicate is a general property of all the DNA molecules tested so far. Harland and Laskey (1980) suggested that during early developmental stages of *X. laevis* there may be no requirement for specific origins of replication. For example, they showed that each one of the six circularized HindIII fragments of simian virus 40 DNA is replicated after injection into activated *X. laevis* eggs. However, this still does not exclude a preference for sequences related to specific origins of replication (Harland and Laskey 1982). During early embryogenesis frog DNA replicates at a very high rate (two to three times every hour). Use of sequences related to specific origins of replication might facilitate recruitment of a large number of sites for initiation of replication. At the beginning of gastrulation, when the frequency of cell divisions decreases, a more stringent mode of replication would again prevail. Consistent with this hypothesis, we observe that all the extrachromosomal recombinant DNA which may have accumulated because of

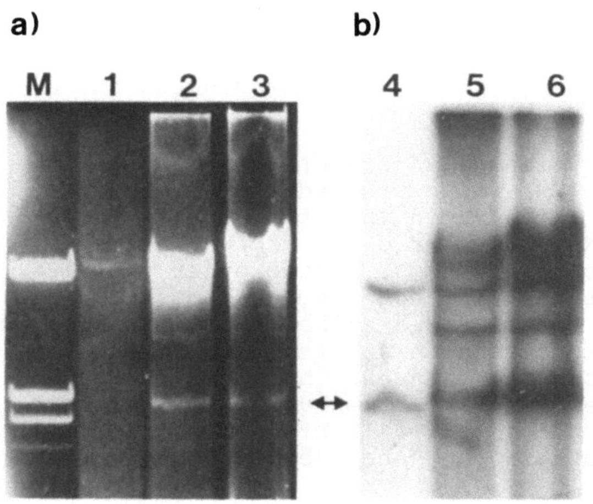

Fig. 1a,b.

c)

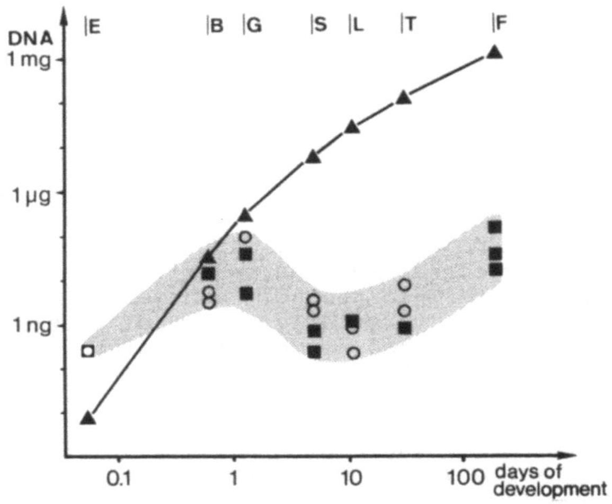

Fig. 1c

Fig. 1a-c. Fate of the injected DNA during development. a Undigested material (1 embryo equivalent per slot) was separated by gel electrophoresis in 0.8% agarose and 1 mgl^{-1} ethidium bromide. *Lane 1* nucleic acids extracted immediately after injection; *lanes 2* and *3* material from blastula and gastrula stage respectively. *Arrow* indicates the position of the supercoiled form of the injected β-globin gene recombinant plasmid (pJKd-; see Rusconi and Schaffner 1981); b The samples shown in a after Southern transfer and hybridization with a ^{32}P-labeled probe homologous to the injected DNA. *Lanes 4,5* and *6* of the resulting autoradiograph correspond to *lanes 1,2* and *3* of a respectively; c Double logarithmic plot representing the amount of foreign DNA (*symbols* in the *shaded area*) and host genomic DNA (▲) For various stages of development (*E* egg; *B* blastula; *G* gastrula; *S* swimming larva; *L* feeding larva; *T* 3-week-old tadpole; *F* 5-month-old frog). The amount of foreign DNA molecules was estimated by Southern blots (■) or dot blots (O). Data have been collected from several independent experimental batches, where different DNA clones were injected

a transient high permissivity for replication gradually disappears after gastrulation. After about 30 days of development, the amount of foreign DNA increases in parallel with genomic DNA indicating that the two DNAs replicate at the same rate (see Fig. 1c).

1.3.2 Mosaic Distribution of the Persisting Foreign Sequences in the Tissues of Transgenic Frogs

We also studied the tissue distribution of injected DNA at later developmental stages. For this purpose DNA was extracted from different tissues of 5-6 month-old frogs which had been injected at the zygote stage with rabbit β-globin gene DNA. Only three out of ten analyzed individuals lacked detectable rabbit sequences. In two other individuals the foreign genes were present at low copy number (less than one copy per cell) in at least one of the organs examined. The remaining individuals (50%) contained higher copy numbers (from 2 to 50 copies per cell) unevenly distributed among the tissues. These observations are in good agreement with the previously reported values of three to ten copies per cell (Rusconi and Schaffner 1981) obtained with pools of frogs. Figure 2a shows four Southern blot analyses of PstI-digested

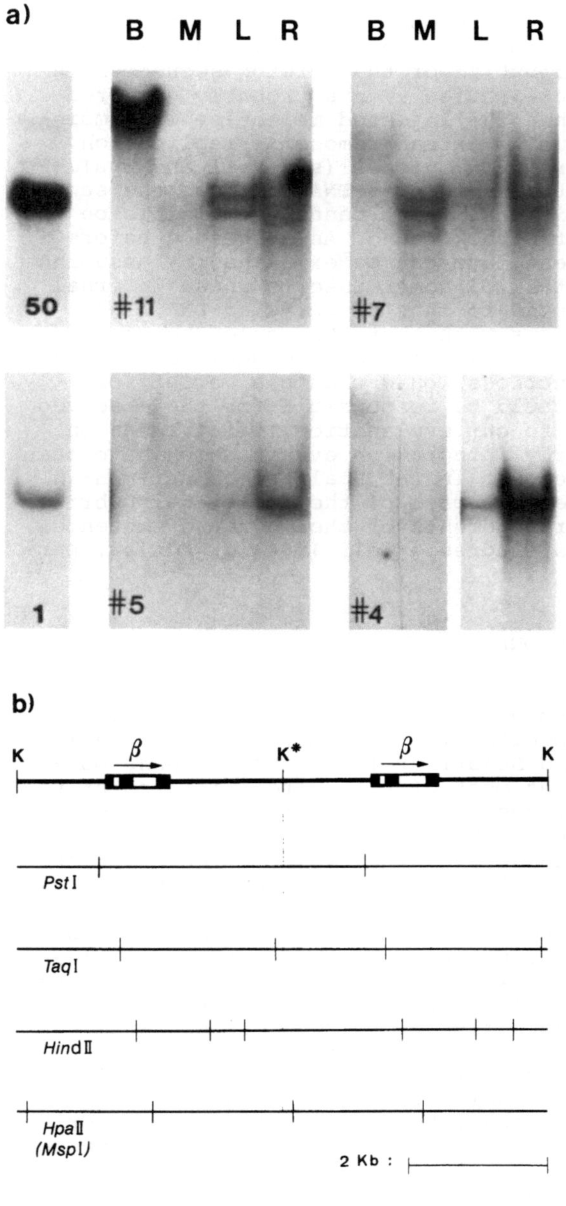

Fig. 2a,b. Mosaic distribution of the foreign sequences in young frogs. DNA was extracted from blood (*B*), leg muscle (*M*), liver (*L*), and the rest of the body (*R*), using the same procedure as for entire embryos. For the Southern blots shown here, 3 µg of PstI-digested DNA were loaded in each lane. a Southern blot of DNA extracted from tissues of different individuals (*11*, *7*, *5* and *4*). The two separate lanes at the left side show a reconstruction experiment (cloned rabbit β-globin DNA mixed with excess frog DNA) representing the intensities of hybridization expected when 1, respectively 50 gene copies per cell are present. The digestions of blood DNA are incomplete; b Restriction map of the DNA used for most of the injections described in this article. The DNA is a dimer of the genomic 4.7 kb KpnI(K) fragment whose middle KpnI site (indicated as *K**) has been destroyed. The circularized product obtained after incubation with T4 ligase was purified on agarose gel prior to injection. The PstI digest of such circularized dimers yields two globin gene fragments of different size, due to a short deletion which occurred during the elimination of the intervening KpnI site by S1 nuclease. Black bars and open bars represent exons and intervening sequences of the rabbit globin gene, respectively. *Arrow* direction of transcription

DNA from blood (B), leg muscle (M), liver (L) and the remaining parts of the body (R). Every lane contains the same amount of DNA so that the number of persisting gene copies can be directly estimated by comparison with known standards (shown in Fig. 2: 1 and 50 copies per cell). The intensity of hybridization is subject to extensive variation within the same organism. The analyzed tissues or organs are also equally permissive for the presence of the heterologous sequences. For example, the rabbit globin DNA is highly represented in the blood of frog No. 11 but is absent in the blood of other frogs. We therefore conclude that the transformed frogs were mosaics for the heterologous gene. Integration into the host genome at the one cell stage would have resulted in homologous distribution of the foreign sequences in the frogs. Therefore, multiple integration events probably occurred throughout the cleavage stages when the injected molecules were maintained at a high concentration due to extrachromosomal replication. Upon digestion with various restriction enzymes (see PstI and HpaII in Fig. 2 and 3, respectively), the integrated DNA releases the same bands as input circular DNA. Additional faint bands are visible on top of a widely spread smear in the background. As suggested before (Rusconi and Schaffner 1981), these data can be explained by assuming the existence of tandem head-to-tail oligomer inserts whose internal restriction pattern would correspond to that of circular DNA. The fragments generated at the junctions with host DNA would therefore have a different molecular weight and would generate additional submolar bands. A number of such junctions would create a variety of fragments of different size and could be responsible for the observed "smear" hybridization. According to our explanation, faint bands on top of the smear reflect some early integration events which have been clonally propagated to a larger extent in the analyzed compartment.

Finally, we cannot yet exclude that part of the additional hybridizing bands may result from rearrangements of the injected sequences, as it has been observed by others (Andres et al. 1984; G. Köhler, personal communication).

1.3.3 Expression of the Exogenous DNA is Restricted to Early Developmental Stages

In our previous work on frog transformation, we noted that the injected rabbit β-globin genes were specifically transcribed during early phases of development (Rusconi and Schaffner 1981). S1-nuclease mapping of the transcripts showed that most of them were initiated at the same point as in rabbit reticulocytes. The amount of steady-state transcripts was approximately equimolar to the number of gene copies present in the early embryo, reaching a maximal value at gastrula stage and decreasing as degradation of the heterologous episomes progressed.

Analysis at later stages of development failed to reveal any rabbit β-globin specific transcripts in tadpoles older than three weeks (data not shown). Even in frog tissues with relatively high copy numbers of foreign genes (for example the blood of frog No. 11 with about 50 copies per cell), the amount of specific heterologous transcripts was below the detection level (i.e., less than 1 transcript per cell). These observations suggest that expression of the heterologous globin gene is restricted to the period in which the injected genes persist in episomal state. Our results are consistent with the data of Andres et al. (1984), who also showed that an exogenous vitellogenin-gene derivative is inactive in post-metamorphic frogs, while being transcribed during early developmental stages. It seems as if integration into the host genome reduces or abolishes further transcription of the foreign gene.

1.3.4 Lack of de Novo Methylation of the Exogenous DNA

The conversion of the persisting rabbit genes into such a "silent" state could have been due to de novo DNA methylation, since methylation has been repeatedly correlated with the inactive state of several vertebrate genes (reviewed by Felsenfeld and McGhee 1982). The degree of methylation of some DNA sequences can be inferred from the digestion

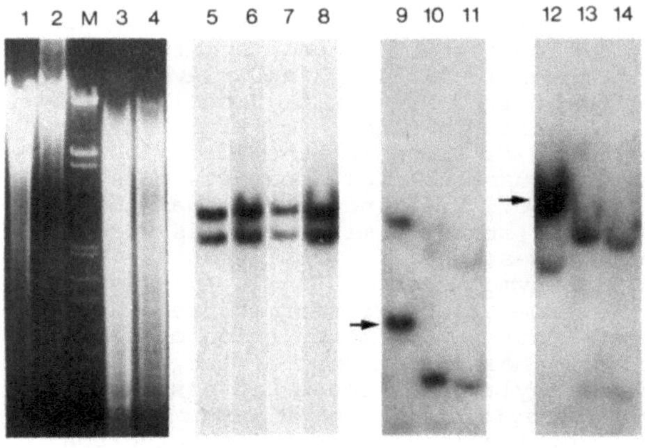
1 2 M 3 4
5 6 7 8
9 10 11
12 13 14

bined digestion experiments were also done by using HpaII and MspI together with enzymes insensitive to CpG methylation, like HindII and Taq I. We were particularly interested in the HpaII site located in the large intron of the rabbit β-globin gene, since its methylation pattern in the rabbit is tissue specific (Waalwijk and Flavell 1978). As expected, digestion of the DNA of frog Nr. 4 with Hind II release a hybridizing fragment of 1.3 kb (arrow in panel c of Fig. 3), which contains the HpaII site of interest. As shown in lane 10 of the same figure, subsequent HpaII digestion quantitatively converts this fragment to the smaller products (0.7 + 0.6 kb), indicating that this HpaII site persists in an unmodified state. Similar results were obtained with combined Taq I/HpaII digestion (Fig. 3c, lanes 12-14), with other methylation-sensitive enzymes such as Sal I, Ava I and Xho I (not shown), and also with DNA from other transformed frogs (data not shown). Therefore we conclude that *de novo* methylation of these sites is negligible, which probably means that most of the persisting heterologous sequences remain unmethylated at these CpG positions. It is not so surprising that *de novo* methylation does not occur in activated eggs during the first two replication cycles of injected DNA (Harland 1982), but it is remarkable that there is still no methylation of the injected sequences after several months as it has also been recently confirmed by others (Bendig and Williams 1983; Andres et al. 1984). Genes integrated into the host genomes of mouse eggs or early embryos are extensively *de novo* methylated (Palmiter et al. . 1982a; S.R. unpublished) whereas integration into post-implantation embryos does not result in *de novo* methylation (Jähner et al. 1982). As mentioned, in our frog transformations integration events seem to occur late in embryonic development and this may be related to the absence of *de novo* methylation. Therefore, it remains to be seen whether or not the correlation between DNA methylation and gene activation also applies to vertebrates other than mammals and birds. Whatever the role of DNA methylation may be for *Xenopus*, it is obvious that *de novo* methylation cannot explain the inactivity of the rabbit β-globin gene in transformed frogs. Our results are also compatible with the observation of MacLeod and Bird (1982) who reported that hypomethylation of certain sites on *X. borealis* rRNA genes is not sufficient to ensure their activation in *X. laevis-borealis* hybrid tadpoles.

1.3.5 Conclusion and Perspectives

There is a further possible explanation for conversion of the exogeneous gene into a quiescent state. For example, boundaries between euchromatin and heterochromatic regions of eukaryotic chromosomes seem to be preferred sites for sister chromatid exchange (Latt 1981) and mitotic crossover (Tartof 1975). One could argue that in most cases the foreign sequences are trapped by the host genome during these crossover/exchange events and therefore end up near to, or within highly packaged heterochromatin. These flanking regions could exert an unpredictable cis-effect, in a fashion analogous to the phenomena of position variegation (Baker 1968). Therefore it may be useful to develop techniques whereby the exogenous gene is directed into its homologous chromosomal locus, or maintained in an episome-like state. In our case it would be interesting to see whether passage through the germ line to the offspring may allow the clonal propagation of simpler integration patterns, and whether the newly acquired genes, previously devoid of CpG methylation, would then have a methylation pattern more similar to their natural counterparts. The handicap of the relatively long generation time of *X. laevis* (about one year under optimal conditions) would be partly compensated by the very high fertility of the frogs. Furthermore, one could clone single nuclei from transformed early embryos by transplanting them into enucleated eggs (Etkin and Roberts 1983) as an alternative to the germ line passage.

2 Transformation of Mice with an in Vitro Mutated Mouse Alpha Globin Gene

2.1 Introduction

In recent years recombinant DNA technology has dramatically improved our knowledge of eukaryotic gene structure and function. Cultured cells and frog oocytes have become indispensable tools for testing the engineered molecules, which can be transferred into the novel eukaryotic environment by a variety of techniques. In number of laboratories, including our own, similar approaches have been extended to the transformation of entire organisms, in order to study gene regulation during development. The initial work in organism transformation was done using cloned genes derived from heterologous organisms, in part to identify them easily in the host organism. Experiments carried out in our laboratory showed that rabbit globin, SV40, and sea urchin histone genes can be transferred into *Xenopus laevis* fertilized eggs and maintained during development (Rusconi and Schaffner 1981; Bendig 1981; preceding sections of this volume). Transformation with heterologous genes, as in the above-mentioned experiments, has undoubtedly provided many general insights into the behavior of the newly acquired sequences in the transgenic organisms. However, we soon became aware that experiments of greater biological significance can be done by using homologous genes for transformation, and by transferring them into an organism which allows a pedigree analysis within a reasonable amount of time (not the 2 years required for *Xenopus*!). Among the vertebrates, the mouse (in addition to its relatively short 60-day generation time) offers a number of advantages, such as well-defined genetics (and pathogenetics) and the existence of many different inbred strains. Therefore, it is not surprising that a steadily increasing number of molecular biologists are dedicating more and more attention to this system. Successful transformation of mice via microinjection of various cloned eukaryotic genes into zygote pronuclei has already been reported by several groups (Gordon et al. 1980; Costantini and Lacy 1981; Wagner et al. 1981, 1983; Buerki and Ullrich 1982; Brinster et al. 1981, 1983; Palmiter et al. 1982a, 1982b, 1983; McKnight et al. 1983). We developed similar transformation protocols independently in order to follow the fate and expression of a mutated homologous α-globin gene in developing mice. As described in the following sections, the gene was mutated in vitro in its 3'untranslated mRNA coding region in a way that allows the detection of both the exogenous gene and its transcripts in the transgenic mouse.

2.2 Materials and Methods

2.2.1 In Vitro Mutagenesis of the Mouse α-Globin Gene

The phage containing the mouse α1-globin gene, gtWESMα1 (Leder et al. 1978), was a gift of P. Leder. The 10.4 kb EcoRI insert was recloned in pBR327 (Soberon et al. 1980) and this recombinant was called pαwt. A 0.56 kb BamHI-XhoI fragment containing the 3'portion of the gene

(Nishioka and Leder 1979) was excized with the corresponding enzymes and the ends were dephosphorylated. Then, the fragment was cut which HaeIII and the products were religated in the presence of an equimolar amount of plasmid vector (BamHI-XhoI ends) and a 100-fold molar excess of double-stranded XbaI linkers (BRL). The recombinant obtained was called sub α(x). An analogous procedure was adopted for the introduction of an additional SacI linker into the AluI site two nucleotides downstream of the translation-termination codon, thus creating the recombinant sub α(xs). The entire 10.4 kb mouse DNA segment was then reconstituted by substitution of the native 0.56 kb BamHI-XhoI fragment with its mutated analogs. Technical details on the further construction of the $\alpha\beta$-derivative (shown in Fig. 4) will be given elsewhere (S. Rusconi, in preparation). Plasmid DNA preparation and manipulation were performed according to standard cloning procedures (Telford et al. 1979; Birnboim and Doly 1979).

2.2.2 Mice, Eggs, and Micromanipulation

Mice of the outbred strain ICR/JZ (obtained from Tierspital, Zuerich) were kept on a 12 h-dark schedule (1700-0500 h). Females selected in proestrus were caged in groups of two to three with either fertile or vasectomized males to obtain egg donors and recipients respectively. The following morning, females showing a vaginal plug were separated and eggs were collected from 1400 to 1600 h from the extirpated oviducts of killed donor females by puncturing the ampullae. Removal of the cumulus cells was done at room temperature in presence of 0.2 mg ml^{-1} of bovine testes Hyaluronidase (Sigma) dissolved in Hank's balanced salt solution (Gibco) supplemented with 4 mg ml^{-1} of bovine serum albumin (Sigma) and 35 mgl^{-1} sodium pyruvate (Gibco). Most of the further manipulations (including the transfer into foster mothers) were carried out as described by others (see references in the Introduction). The microinjection was performed at room temperature (in absence of CO2) and suspending the eggs in the same medium used for Hyaluronidase treatment. Embryos were allowed to develop to term and the DNA was extracted from the tail tip as described by Palmiter et al. (1982a).

2.2.3 DNA and RNA Analysis, Transient Expression Assays

0.5 to 1 microgram of DNA was digested with the appropriate restriction enzyme, run on small horizontal agarose gels (8 x 7 x 0.3 cm, 2.5 mm-wide slots) and the blots were hybridized to the probes indicated in Figs. 4 and 6. Transient expression assays and RNA extraction have been essentially carried out as described (Banerji et al. 1981). The 3'end-labeled BstEII S1-probe was obtained by filling up the the protruding BstEII (see Fig. 4) 5'ends in a DNA polymerase (Klenow enzyme) reaction in the presence of 32P α-dCTP (Amersham). S1-mapping was performed according to the protocol described by Weaver and Weissmann (1979) with minor modifications.

2.3 Results and Discussion

2.3.1 Transcripts from Genes Modified by Linker Insertions in the 3'Trailer Are Not Detected in Transgenic Mice

Modifications were performed on the 10.4 EcoRI segment of mouse genomic DNA harboring the α1-globin gene (Leder et al. 1978; Nishioka and Leder 1979) illustrated in Fig. 4 (top). In the first series of transformation experiments we used constructs of the type indicated as αX-DNA or a derivative of it, αXS-DNA, which contained an additional SacI linker insertion adjacent to the termination codon (see Materials and Methods section). It was established that in transient expression assays (results not shown) both mutants (or artificial alleles) gave rise to transcripts distinguishable from the wild type RNA by an S1-mapping strategy analogous to the one presented in Fig. 4 (bottom). 16 mice which contained various amounts of the exogenous DNA were identified among the first 85 which developed from zygotes microinjected with the α-XSDNA recombinant. Their blood cells RNA was analyzed by S1-mapping by using a probe analogous to the one described in Fig. 4D. Transcripts originating from the exogenous DNA could not be identified in any of the animals. Several hypotheses have been proposed to explain this apparent lack of expression. The following sections are devoted to a description of the three most probable hypotheses and the experimental approaches undertaken to test them.

2.3.2 Construction of a Chimeric αβ-Gene

The strategy of inserting small linker sequences, while being very suitable for DNA analysis, turned out to be much less advantageous for the S1-mapping of the RNA in the presence of an extremely large excess of wild type transcripts (hybridization in RNA excess). Conditions under which the α-X or the α-XS RNA's could have been detected in a mixture containing 95% or more wild-type α-globin RNA could not be reproducibely obtained. It was frequently observed that in a significant percentage of the hybrid molecules formed between the probe and wild-type RNA, the discontinuities provided by the linkers were not efficiently recognized and digested by the S1 enzyme. The labeled DNA molecules which in this way escaped S1 hydrolysis gave rise to a high background which could have masked the presence of specifically protected fragments derived from hybrid molecules formed with genuine αXS transcripts. Therefore we cannot exclude the possibility that there might well have been a low level transcription of the newly acquired sequences in the transgenic mice. Consequently, the mutation scheme was slightly revised and the difference between the endogenous and the exogenous DNA was increased in order to improve the sensitivity of the S1-mapping. As shown in Fig. 4D, a 400 bp DNA segment containing the 3'untranslated region of the rabbit β-globin mRNA (Van Ooyen et al. 1979), including the poly-A addition signal and some downstream sequences, was inserted into the artificial XbaI site of the α-X DNA. This chimaeric version, called αβ-DNA, could be easily distinguished from the endogenous couterpart, as is shown later. The S1-mapping strategy adopted for DNA analysis is shown in Fig. 4E. As illustrated, wild-type RNA can only protect a 135 nucleotide fragment on the BstEII 3'-endlabelled αβ-probe, whereas the transcripts from the chimaeric gene will protect a 237 nucleotide fragment, provided that the rabbit poly-A addition site is correctly used.

2.3.3 Construction of a "Driver" Gene, One Case of Ectopic Expression

The lack of expression of the exogenous α-globin DNA observed in the first 16 transgenic mice (see Sect. 2.3.1), and in general when using globin genes (Lacy et al. 1983; Wagner et al. 1983; E. Wagner, personal communication), is a rather disappointing observation, especially in light of the successful use of the metallothionein (=MT)-fusion genes (Brinster et al. 1981; Palmiter et al. 1982a,b). A possible explanation for this striking difference (globin genes giving very reduced expression versus MT fusion genes being expressed in a high percent

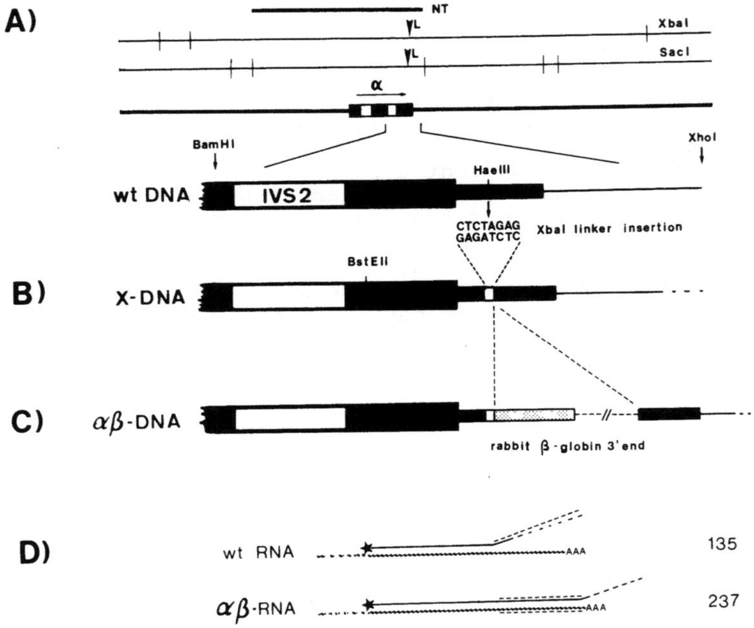

Fig. 4A-D. Mutagenesis of the alpha-globin gene. A The XbaI and the SacI restriction maps of the 10 kb genomic DNA segment harboring the α1-globin gene (Leder et al. 1978) are shown. The position at which the corresponding linkers were introduced are indicated with *L*. Other symbols: *open bars* intervening sequences (IVS); *arrow* direction of transcription; *NT* SacI fragment subcloned in pBR and used as a nick-translated probe for the Southern hybridization shown in Fig. 6. *WT-DNA* an expanded version of the 0.56 kb BamHI-XhoI fragment showing the position of the HaeIII restriction site used for in vitro mutagenesis: *thick bar* translated region; *thick line* 3'untranslated region; *thin line* flanking sequences downstream of the poly-A addition site; B The insertion of a XbaI linker is shown (*open thick line*). A second modified version, the XS-DNA was obtained upon insertion of an additional SacI linker in the AluI site located 2 nucleotides downstream of the termination codon (not shown). C The αβ-modification was obtained by splicing the rabbit β-globin 3'trailer (*shaded thick line*) to the 3'β-globin trailer (see text and Methods section for details); D The S1-mapping strategy using the BsteII 3'end-labeled αβ-globin probe is illustrated. The hybrid molecules formed with a wild type and a chimeric mRNA are shown (*wavy line* mRNA; *straight line* DNA probe; *black star* 32P label; sequences emphasized with a *broken line* represent rabbit β-globin moiety. The numbers at the right (135 and 237) refer to the sizes (in nucleotides) of the corresponding S1-resistant fragments

of the transgenic animals) could reside in the events which immediately follow DNA microinjection into the pronucleus. It has been shown that MT-fusion genes are transcribed almost immediately after introduction into the zygotes (Brinster et al. 1982). On the basis of these results it is reasonable to assume that a MT-fusion gene is readily compartmentalized in an actively transcribing region of the nucleus and is therefore preferentially taken up by "permissive" domains of the host genome. On the other hand, a globin gene might be expected to remain silent after injection. Such a DNA segment might not be correctly compartmentalized and therefore might end up in heterochromatic regions of the genome (see also Sect. 1.3.5). To circumvent this possible drawback, we decided to link the αβ-gene to be tested to a "driver gene",

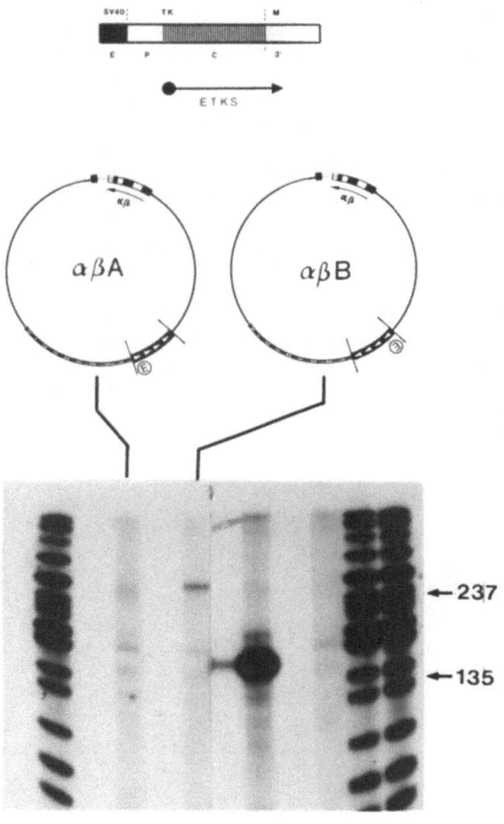

Fig. 5. The composite unit ETKS, the recombinants αβA and αβB in a transient expression system. *Top* the driver gene ETKS: SV40, the SV40 enhancer segment (*E*); *TK* the thymidine kinase gene (herpes simplex) with its promoter (*P*) and the coding region (*hatched area, C*) interrupted at the SacI (*broken line*); *M* the 3'portion of the mouse α-globin gene. *Arrow* indicates the direction of TK transcription. *Middle* the two αβ-DNA recombinants which differ in the orientation of the ETKS unit (borders delimited with *slashes*). Circled letter *E* indicates the position of SV40 enhancer in the two recombinants. *Thick line with white arrows* ETKS unit; *second thickest line* pBR sequences. Other symbols as in Fig. 4. *Lower panel* autoradiogramm showing S1-nuclease analysis of transcripts obtained by transfection of αβA and αβB in HeLa cells. The two lanes with a band at position *135* nucleotides result from the S1 analysis of different amounts of wild type RNA with the same probe. Lanes at the extremities were loaded with molecular weight marker (pBR322, HpaII digested, end-labeled)

e.g., a DNA segment capable of expression immediately after injection. For this purpose the recombinant shown on Fig. 5 (top) was assembled. It consists of an SV40 enhancer segment (72 bp region, Banerji et al. 1981) linked to a thymidine kinase (TK) gene (McKnight and Kinsburg 1982) which was truncated to prevent toxicity problems. Similar combinations (SV40 enhancer and TK promoter) are apparently actively transcribed in embryocarcinoma cells (F. Cuzin, personal communication) and in pre-implantation mouse embryos (R.L. Badge, personal communication). This unit, called ETKS, was linked to the αβ-10.4 kb segment in two different drientations (see Fig. 5, middle), yielding the recombinants αβ-A and αβ-B, respectively. In the type A construct the SV40 enhancer cannot directly influence the αβ-promoter because of the presence of intervening inhibitory sequences (de Villiers et al. 1982) on both sides (TK promoter and pBR, see Fig. 5, middle). However, in constructs of the B-type we expect the SV40 enhancer to stimulate transcription from the αβ-promoter. The difference between these two recombinant molecules is easily demonstrated in a transient expression assay, as shown by the S1-mapping analysis of the transcripts isolated from HeLa cells transfected with the corresponding constructs (see Fig. 5, lower panel). As expected, when recombinants of the B-type are transfected, one can detect the 237 nucleotide protected fragment. The lane representing the analysis of cells transfected with the A-type is clearly lacking a radioactive band at the same position.

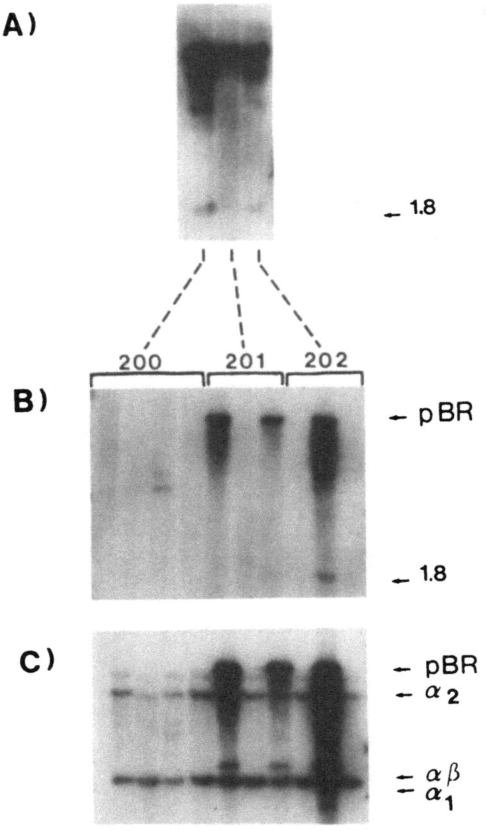

Fig. 6A-C. Southern blot analysis of three transgenic mice and their progeny. DNA was digested with SacI and separated on minigels (2.5 mm wide slots). Blots A and B were hybridized to nicktranslated pBRTK DNA (a gift of CH. Weissmann, Zuerich); blot C was hybridized with the α-globin probe described in Fig. 4. Blot A shows the three line founders (Nr. 200, 201, 202; testes DNA). Bot B and C are equivalent and show part of the corresponding progeny (tail tip DNA from male offspring). Symbols: *1.8* the 1.8 kb fragment hybridizing to the TK probe and diagnostic for the B-type constructs; α-1, αβ and α2 indicate the fragments (3.0, 3.5 and 8kb) hybridizing to the α-globin probe and corresponding to the endogenous (α-1 and α-2) and to the newly acquired (αβ) genes respectively

Both types (αβ-A and αβ-B) were microinjected into zygote pronuclei and a number of transgenic mice was obtained (four containing the A- and eight containing the B-type). A few selected examples of these transgenic animals and their progeny are shown in Fig. 6. Southern blot hybridization using either pBRTK (panels A and B) or the α-globin probe (panel C) revealed the presence of the foreign sequences in these mice and their offspring. A case in which the exogenous genes are not transmitted to the progeny is also shown (see mouse Nr. 200). The blood cell RNA was analyzed by S1-mapping using the probe described in Fig. 4D. Again, to our disappointment, none of the samples yielded a detectable band at position 237 (not shown). This indicates that in all the analyzed mice transcription of the exogenous genes was very low or nonexistent. Tissues other than blood were also analyzed in a few cases. As shown in Fig.7, some transcripts from the chimaeric gene could be detected in the brain (Br) of mouse Nr. 200. This result is very reminiscent of the data of Lacy et al. (1983) who also report evidence for the ectopic expression of the exogenous rabbit β-globin genes in some mice. Unfortunately, mouse Nr. 200 did not transmit the exogenous αβ-genes to its progeny (13 analyzed, 4 of them shown on Fig. 6 B and C) and we could not analyze this interesting case in more detail.

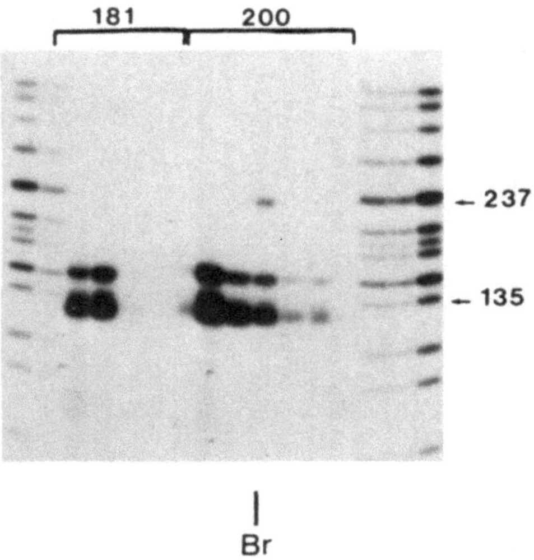

Fig. 7. Ectopic expression of the αβ-genes in mouse No. 200. 15 to 40 ug of total RNA from various tissues was analyzed by S1-mapping as illustrated in Figs. 4 and 5. Samples analyzed (starting from the *left*): (No. *181*) kidney, blood, muscle, testes, liver; (No. *200*) blood, kidney, brain, testes, liver. The lane with the analysis of brain RNA of mouse 200 is indicated by *Br*. The occurrence of intermediate size fragments (about 160 nucleotides) are due to incomplete S1 digestion

2.3.4 Summary of Results

The results obtained by transforming mice with the modified versions of the α1-globin gene in its 10.4 kb genomic context can be summarized as follows. (a) Almost 30 transgenic animals containing the various α-globin recombinants were generated. (b) By using the 3'end-labeled BstEII probe one could not detect transcripts generated from the newly acquired genes in the peripheral blood cell RNA. This analysis does not allow the detection of rare transcripts (less than 2-5%). More sensitive approaches are under way to test this possibility. (c) One case of ectopic expression (brain of mouse Mr. 200) was discovered within six analyzed individuals. (d) Neither the linkage of a "driver" unit such as ETKS, nor the exposure of the αβ-promoter to the action of the SV40 enhancer (eight cases) seemed to positively influence the transcription of the exogenous genes in the transgenic animals. Therefore, one might tentatively conclude that within the 10 kb of genomic α-globin DNA used for transformation, there is not sufficient information for the efficient and developmentally regulated transcription of the transferred sequences. We therefore decided to isolate more sequences surrounding the 1-globin gene, as described in the last section.

2.3.5 Linkage of More Flanking Sequences to the Modified α-Globin Gene

The 37 kb genomic segment illustrated in Fig. 8 (top) was isolated from a cosmid library (see Materials and Methods). The restriction map was found to be in good agreement with the published data concerning the α-globin locus (Leder et al. 1981). We could also detect the second nonallelic α-globin gene within these sequences (α2 of Fig. 8) by cross hybridization with the α1-probe (results not shown). Since an embryonic gene (α-emX) has been mapped 6-8 kb upstream of the α1-gene (Leder et al. 1981, one expect it to also be present in our genomic segment, as it spans more than 11 kb of the α1-upstream region (see Fig. 8, α-e). The replacement of the α1-gene by its modified αβ-counterpart was obtained after a complex series of sub- and recloning steps. The modified version of the 37 kb genomic insert was recloned into an ETKS (see Fig. 5) containing vector, giving rise to cos αβ-a6,

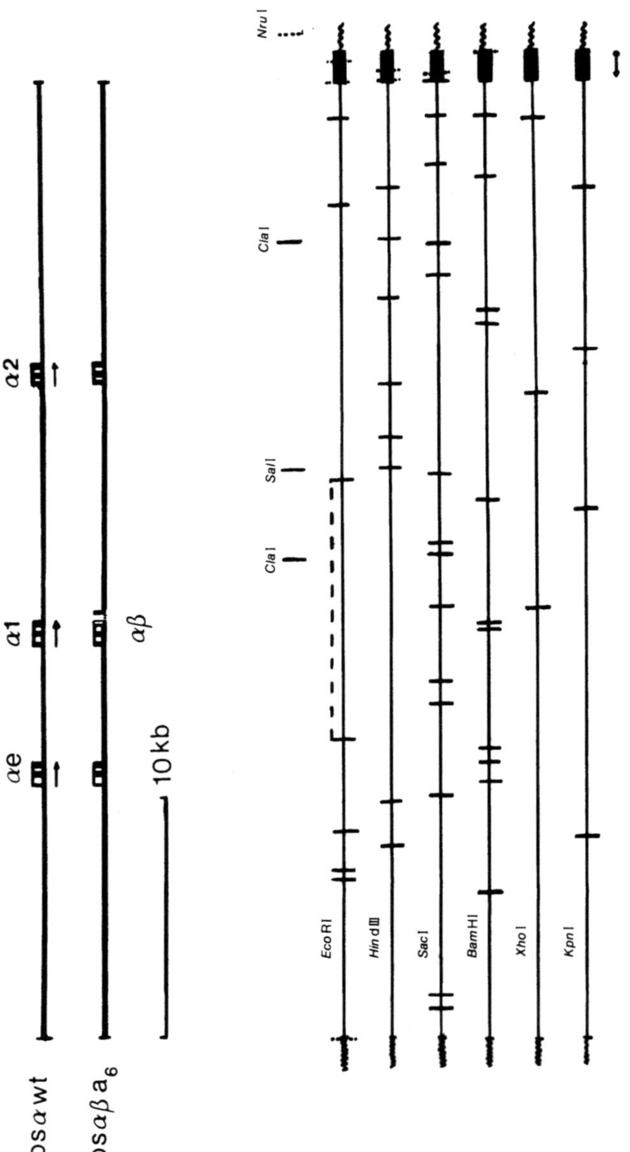

Fig. 8. Native and modified version of a 37 kb genomic DNA segment harboring three α-globin genes. A mouse genomic cosmid library (Grosveld et al. 1982) was screened with a nicktranslated probe consisting of the two SacI-EcoRI extremities of the 10.4 kb EcoRI segment (see Fig. 4 for orientation). A cosmid clone containing a 37 kb insert was thereby isolated and characterized (cos wt shown at the top). The replacement of the α-1 gene by its αβ-modified version yielded the clone cos αβa6 depicted below cos αwt. In the nomenclature of this paper α-1, α-2 and α-e correspond to α-ad1, α-ad2 and α-emX as described by Leder et al. (1981). Graphic symbols: *wavy line* plasmid sequences; *thickest bars* ETKS unit; *arrows* direction of transcription; *broken vertical segments* restriction sites outside of the genomic DNA insert; *broken horizontal line* delimitation of the 10.4 kb EcoRI fragment used for all the transformation experiments described in this work

which seemed very suitable for gene transfer studies in mice. Since microinjections of this recombinant DNA have been started only recently no mice have yet been obtained which are transgenic for cos αβa6.

Nevertheless, there are several reasons which lead us to believe that additional flanking sequences may render the exogenous globin genes capable of a more efficient expression in the host organism. In that context, the high transcriptional activity of promoters linked to a few hundred base pairs regulatory sequence of the metallothionein gene (Brinster et al. 1981; Palmiter et al. 1983) may be somewhat misleading. Short controlling regions might expected to be typical for constitutive genes, or for genes which have to exert a quick response to sudden environmental and metabolic changes such as the heat shock genes, interferon genes, etc. It should also be noted that the exogenous metallothionein-fusion genes, while being inducible by heavy metals, do not respond correctly to hormonal stimulation (Brinster et al. 1981). Therefore, one might argue that developmentally regulated genes may require a larger number of (possible scattered) controlling elements. This could explain why, for example, a 14.2 kb segment bearing a rearranged immunoglobulin (=Ig) K light chain gene is efficiently expressed in transgenic mice (Brinster et al. 1983), whereas shorter versions of Ig K genes (e.g., 7 kb depleted at the 3'flanking region) apparently remain silent in the same system (our unpublished data). The fact that a recombinant molecule containing large portions of Ig-DNA such as pRHL-TNP (Ochi et al. 1983) was found to be highly expressed in transgenic animals (S.R. and G. Koehler, manuscript in preparation) seems to reaffirm this point of view. In contrast to these positive results, a number of works describing "non-expressing" exogenous DNA can be listed, for example: the rabbit β-globin (20 kb, Lacy et al. 1983), the mouse α-globin (10 kb, this chapter), the human β-globin (10 kb, Stewart et al. 1982), the human insulin (12 kb, Buerki and Ullrich 1982) and the human growth hormone gene (2.6 kb, Wagner et al. 1983). It may be that in these cases some important cis-controlling elements are still missing in the transferred molecules although some of them reach a quite respectable size. Transcriptional regulation may be the end-effect of a complex interaction of several regulatory sequences, whose positive (or negative) effects may propagate over a long range, as has been demonstrated in the case of transcriptional enhancers (Banerji et al. 1981, 1983; de Villiers et al. 1983). For some genes the number of controlling elements may be very reduced (see the metallothionein promoter), whereas other developmentally controlled transcriptional units, may reach conspicuous sizes, due to an elevated number of interspersed cis-controlling elements. The minimal size may range from 10-15 kb, as tentatively defined by the experiments with the immunoglobulin genes (see above) or the chicken transferrin gene (McKnight et al. 1983) to the yet unknown size (more than 20 kb ?) required for the globin genes. Studies on several thalassemia mutants led to the proposal that globin genes may be organized in domain-like structures consisting of "gene doublets" and substantial amounts of flanking sequences (Bernards and Flavell 1980; Vanin et al. 1983). Therefore, it may be that even by using almost 40 kb of globin DNA (see Fig. 8) we will not obtain efficient expression of the transferred genes in the transgenic animals. Fortunately, our current experiments are driven by a more optimistic point of view.

Acknowledgments. I am grateful to Drs. P. Leder, Ch. Weissmann and F. Grosveld for the gift of α1-globin DNA, pBRTK DNA and the use of the mouse genomic cosmid library respectively. I also want to mention several people who generously contributed to the final draft, and namely: S. Oberholzer who typed part of it, F. Ochsenbein who took care of the photographic documentation and my wife Augusta, who ran a patient search through the bibliographic references. I am also indebted to Dr. E. Serfling for critical reading and helpful suggestions. The experiments have been carried out during my doctoral and post-doctoral period in the group of Prof. W. Schaffner whom I am grateful for encouragements and competent advice. This work has been supported by the Schweizerische Nationalfonds, grant 3.687.80, and the Kanton Zürich.

References

Andres AC, Muellener DB, Ryffel GU (1984) Persistence, methylation and expression of vitellogenin gene derivatives after injection into fertilized eggs of *Xenopus laevis*. Nucleic Acids Res 12:2283-2302

Baker WK (1968) Position-effect variegation. Adv Genet 14:133-169

Banerji J, Olson L, Schaffner W (1983) A lymphocyte-specific cellular enhancer is located downstream of the joining region in immunoglobulin heavy chain genes. Cell 33:729-740

Banerji J, Rusconi S, Schaffner W (1981) Expression of a β-globin gene is enhanced by remote SV40 DNA sequences. Cell 27:299-308

Bendig MM (1981) Persistence and expression of histone genes injected into *Xenopus* eggs in early development. Nature (Lond) 292:65-67

Bendig MM, Williams JG (1983) Replication and expression of *Xenopus laevis* globin genes injected into fertilized *Xenopus* eggs. Proc Natl Acad Sci USA 80:6197-6201

Bernards R, Flavell RA (1980) Physical mapping of the globin gene deletion in hereditary persistence of foetal haemoglobin (HPFH). Nucleic Acids Res 8:1521-1534

Birnboim HC, Doly J (1979) A rapid alkaline extraction procedure for screening recombinant plasmid DNA. Nucleic Acids Res 7:1513-1523

Brinster RL, Chen HY, Trumbauer M, Senear AW, Warren R, Palmiter RD (1981) Somatic expression of herpes thymidine kinase in mice following injection of a fusion gene into eggs. Cell 27:223-231

Brinster RL, Chen HY, Warren R, Sarthy A, Palmiter RD (1982) Regulation of metallothionein-thymidine kinase fusion plasmids injected into mouse eggs. Nature (Lond) 296:39-42

Brinster RL, Ritchie KA, Hammer RE, O'Brien RL, Arp B, Storb U (1983) Expression of a microinjected immunoglobulin gene in the spleen of transgenic mice. Nature (Lond) 306:332-336

Bürki K, Ullrich A (1982) Transplantation of the human insulin gene into fertilized mouse eggs. EMBO J 1:127-131

Busslinger M, Hurst J, Flavell RA (1983) DNA methylation and the regulation of globin gene expression. Cell 34:197-206

Costantini F, Lacy E (1981) Introduction of a rabbit β-globin gene into the mouse germ line. Nature (Lond) 294:92-94

De Villiers J, Olson L, Banjeri J, Schaffner W (1982) Analysis of the transcriptional "enhancer" effect. Cold Spring Harbor Symp Quant Biol 47:911-919

Etkin LD, Roberts M (1983) Transmission of integrated sea urchin histone genes by nuclear transplatation in *Xenopus laevis*. Science (Wash DC) 221:67-69

Felsenfeld G, McGhee J (1982) Methylation and gene control. Nature (Lond) 296:602-603

Gordon JW, Scangos GA, Plotkin DJ, Barbosa A, Ruddle FH (1980) Genetic transformation of mouse embryos by microinjection of purified DNA. Proc Natl Acad Sci USA 77:7380-7384

Grosveld FG, Lund T, Murray EJ, Mellor AL, Dahl HHM, Flavell RA (1982) The construction of cosmid libraries which can be used to transform eukaryotic cells. Nucleic Acids Res 10:6715-6732

Gruenbaum Y, Cedar H, Razin A (1981) Restriction enzyme digestion of hemimethylated DNA. Nucleic Acids Res 9:2509-2415

Gurdon GB, Brown DD (1977) Towards an in vivo assay for the function of nuclear macromolecules. In: T'so P (ed) The molecular biology of the mammalian genetic apparatus. North-Holland, Amsterdam, pp 111-123

Gurdon GB, Melton DA (1981) Gene transfer in amphibian eggs and oocytes. Annu Rev Genet 15:180-218

Gurdon GB, Woodland HR, Lingrel JB (1974) The translation of mammalian globin mRNA injected into fertilized eggs of *Xenopus laevis*. Dev Biol 39:125-133

Harland RM (1982) Inheritance of DNA methylation in microinjected eggs of *Xenopus laevis*. Proc Natl Acad Sci USA 79:2323-2327

Harland RM, Laskey RA (1982) Replication origins in the *Xenopus* eggs. Cell 31:503

Harland RM, Laskey RA (1980) Regulated replication of DNA microinjected into eggs of *Xenopus laevis*. Cell 21:761-771

Hines PJ, Benbow RM (1982) Initiation of replication at specific origins in DNA molecules injected into unfertilized eggs of the frog *Xenopus laevis*. Cell 30:459-468

Jaehner D, Stuhlmann H, Steward CL, Harbers K, Loehler J, Simon I, Jaenisch.R (1982) De novo methylation and expression of retroviral genomes during mouse embryogenesis. Nature (Lond) 298:623-628

Kafatos FD, Jones CW, Efstratiadis A (1979) Determination of nucleic acid sequence homologies and relative concentration by a dot blot hybridization procedure. Nucleic Acids Res 7:1541-1552

Lacy E, Roberts S, Evans EP, Burtenshaw MD, Costantini F (1983) A foreign β-globin gene in transgenic mice: integration at abnormal chromosomal positions and expression in inappropriate tissues. Cell 34:343-358

Lane CD (1981) The fate of foreign proteins introduced into Xenopus oocytes. Cell 24: 281-282

Latt SA (1981) Sister chromatid exchange formation. Annu Rev Genet 15:11-15

Leder A, Miller HI, Hamer DH, Seidman JG, Norman B, Sullivan M, Leder P (1978) Comparison of cloned mouse α and β globin genes: conservation of intervening sequence locations and extragenic homology. Proc Natl Acad Sci USA 75:6187-6191

Leder A, Swan D, Ruddle F, D'Eustachio P, Leder P (1981) Dispersion of α-like globin genes of the mouse to three different chromosomes. Nature (Lond) 293:196-200

Macleod D, Bird A (1982) DNaseI sensitivity and methylation of active versus inactive rRNA genes in *Xenopus* species hybrids. Cell 29:211-218

Mc Knight GS, Kinsburg R (1982) Transcriptional control signals of a eukaryotic protein-coding gene. Science (Wash DC) 217:316-324

Mc Knight GS, Hammer RE, Künzel EA, Brinster RL (1983) Expression of the chicken transferrin gene in transgenic mice. Cell 34:335-341

Nishioka Y, Leder P (1979) The complete sequence of a chromosomal mouse α-globin gene reveals elements conserved throughout vertebrate evolution. Cell 18:875-882

Ochi A, Hawley RG, Hawley T, Shulman MJ, Traunecker A, Köhler G, Hozumi N (1983) Functional immunoglobulin M production after transfection of cloned immunoglobulin heavy and light chain genes into lymphoid cells. Proc Natl Acad Sci USA 80: 6351-6355

Palmiter RD, Chen HY, Brinster RL (1982a) Differential regulation of metallothionein-thymidine kinase fusion genes in transgenic mice and their offspring. Cell 29: 701-71o

Palmiter RD, Brinster RL, Hammer RE, Trumbauer ME, Rosenfeld MG, Birnberg NC, Evans RM (1982b) Dramatic growth of mice that develop from eggs microinjected with metallothionein-growth hormone fusion genes. Nature (Lond) 300:611-615

Palmiter RL, Norstedt G, Gelinas RE, Hammer RE, Brinster RL (1983) Metallothionein-human GH fusion genes stimulate growth of mice. Science (Wash DC) 222:809-814

Razin A, Riggs A (1980) DNA methylation and gene function. Science (Wash DC) 210: 604-610

Rusconi S, Schaffner W (1981) Transformation of frog embryos with a rabbit β-globin gene. Proc Natl Acad Sci USA 78:5051-5055

Soberon X, Covarrubias L, Bolivar F (1980) Construction and characterization of new cloning vehicles IV. Deletion derivatives of p-BR 322 and p-BR 325. Gene (Amst) 9:287-305

Stewart TA, Wagner EF, Mintz B (1982) Human β-globin gene sequences injected into mouse eggs, retained in adults, and transmitted to progeny. Science (Wash DC) 217:1046-1048

Tartof KD (1975) Redundant genes. Annu Rev Genet 9:355-385

Telford JL, Kressmann A, Koski RA, Grosschedl R, Müller F, Clarkson SG, Birnstiel ML (1979) Delimitation of a promoter for RNA polymerase III by means of a functional test. Proc Natl Acad Sci USA 76:2590-2594

Van Ooyen A, Van Den Berg J, Mantel N, Weissmann C (1979) Comparison of total sequence of a cloned rabbit β-globin gene and its flanking regions with a homologous mouse sequence. Science (Wash DC) 206:337-244

Vanin EF, Henthorn PS, Kioussis D, Grosveld F, Smithies O (1983) Unexpected relationships between four large deletions in the human β-globin gene cluster. Cell 35:701-709

Waalwijk C, Flavell RA (1978) DNA methylation at a CCGG sequence in the large intron of the rabbit β-globin gene: tissue specific variations. Nucleic Acids Res 5: 4631-4641

Wagner EF, Stewart TA, Mintz B (1981) The human β-globin gene and a functional viral thymidine kinase gene in developing mice. Proc Natl Acad Sci USA 78:5016-5020

Wagner EF, Covarrubias L, Stewart TA, Mintz B (1983) Prenatal lethalities in mice homozygous for human growth hormone gene sequences integrated in the germ lime. Cell 35:647-655

Weaver T, Weissmann C (1979) Mapping of RNA by a modification of the Berk-Sharp procedure: the 5'termini of 15 S β-globin mRNA precursor and mature 10 S β-globin mRNA have identical map coordinates. Nucleic Acids Res 7:1175-1193

Woodland HR, Wilt RH (1980a) The functional stability of sea urchin histone mRNA injected into oocytes of *Xenopus laevis*. Dev Biol 75:199-213

Woodland HR, Wilt RH (1980b) The stability and translation of sea urchin histone mRNA molecules injected into *Xenopus laevis* eggs and developing embryos. Dev Biol 75:214-221

Applications of Genetic Engineering

Genes Involved in Resistance Reactions in Higher Plants: Possible Candidates for Gene Transfer?

K. HAHLBROCK[1], J. CHAPPELL[2], and D. SCHEEL[1]

1 Introduction

The preceding chapters have dealt with the methods and have presented some first examples of introducing foreign, or reintroducing altered, genetic material into eukaryotic organisms. One prerequisite for a successful and desirable genetic manipulation is the availability of defined genes which would endow a transformed organism with new, potentially useful traits.

In higher plants, useful traits are commonly associated with nutritional value, crop yield, storage properties, and, with increasing importance, properties allowing a drastic reduction of soil, water and air pollution with pesticides and other potentially harmful chemicals. Many of these considerations are more or less directly linked to questions concerning the resistance of crop plants to a variety of natural or anthropogenic environmental hazards (Table 1). It is to a large part for this reason that greatly increasing research efforts are now being made toward a better understanding of the biochemistry and physiology of the various resistance mechanisms in higher plants. The obvious aim is to exploit the naturally occurring defense reactions optimally for the benefit of a modern agriculture, which must cope with its often conflicting responsibilities for both the nutrition of a still-growing human population and the survival of the large variety of living organisms, including man, under acceptable conditions. Future developments will show to what extent the new techniques of genetic manipulation can contribute to this goal.

Table 1. List of some important environmental factors to which certain plants might be resistant

RESISTANCE (TOLERANCE) OF PLANTS TO	
Herbivores	natural
Pathogens	
High-energy irradiation	↑
Extreme temperatures	
Extreme ionic strengths	
Drought	
Flooding, Anaerobiosis	
Heavy metal ions	
Air, water, soil pollutants	↓ anthro-
Herbicides	pogenic
etc.	

[1]Max-Planck-Institut für Züchtungsforschung, 5000 Köln 30, FRG

[2]Present address: Department of Biochemistry C-016, University of California, San Diego, La Jolla, CA 92093, USA

35. Colloquium - Mosbach 1984
The Impact of Gene Transfer Technique in Eukaryotic Cell Biology

During several hundred million years of evolution, plants have acquired resistance mechanisms to many of the adverse environmental situations listed in Table 1. With increasing growth and technological development of the human population, more and more of these situations are either aggravated or newly created by man on a time-scale far exceeding that of most previous rates of changes in the natural environment. It seems therefore appropriate to express deep concern in this connection about the possible dangers associated with the breeding, whether by conventional methods or gene transfer in vitro, of plants which are highly tolerant or even totally resistant to man-made toxic chemicals. Strictly controlled confinement to restricted areas of exposure, as well as satisfactory results concerning toxicology and persistence, would be absolute requirements, especially for agricultural crop plants.

Although it would be very interesting from a biochemical point of view to discuss in more detail some striking similarities between the various resistance mechanisms, the following discussion will be restricted to resistance to pathogens, to UV light, and to one selected herbicide. More detailed information about herbicide resistance will be given in the following chapter.

Most of our own work on resistance mechanism has been carried out with cultured plant cells, and in the case of disease resistance, live fungal pathogens have frequently been substituted for by "elicitor" preparations from mycelial cell walls. The valuable advantages and inevitable limitations of such greatly reduced systems are important to realize and have recently been discussed elsewhere (Hahlbrock et al. 1984).

2 Induced Resistance to Pathogens and UV Light

Both disease (pathogen) resistance and resistance to UV light involve specific, induced defence reactions which are triggered ("induced resistance") either by infection with a potentially pathogenic microorganism or by UV irradiation. We have used cultured parsley cells (*Petroselinum hortense*) as a model system in which the responses are large and rapid and can be studied at the level of gene transcription. One particular advantage of this system is that the two responses are partly overlapping. UV light induces the formation of UV-absorbing flavonoid glycosides, whereas an elicitor from a fungal pathogen elicits the formation of furanocoumarins (antimicrobial substances, or "phytoalexins"). Both pathways utilize 4-coumaroyl-CoA, a central intermediate of phenylpropanoid metabolism in higher plants, as a substrate (Fig. 1). It appears therefore that the genes coding for the three enzymes generating this precursor molecule from the primary metabolite, phenylalanine, are useful candidates for studying both the mechanisms of induced resistance and the structures of resistance-related genes.

3 Model Systems: Cell Cultures and Elicitors Versus Natural Complexity

A true picture of the entire complexity of a plant's resistance response to UV light or a pathogen can of course only be obtained with the intact system. On the other hand, the sensitivity levels of many of our present analytical methods are too low to enable the rapid detection of newly induced molecules (such as mRNAs, enzymes, and their products) in, for example, one or a few cells of a whole leaf. A typical situation, where only one leaf cell is challenged with a pathogen in the initial phase of interaction, is schematically depicted in Fig. 2. The scheme

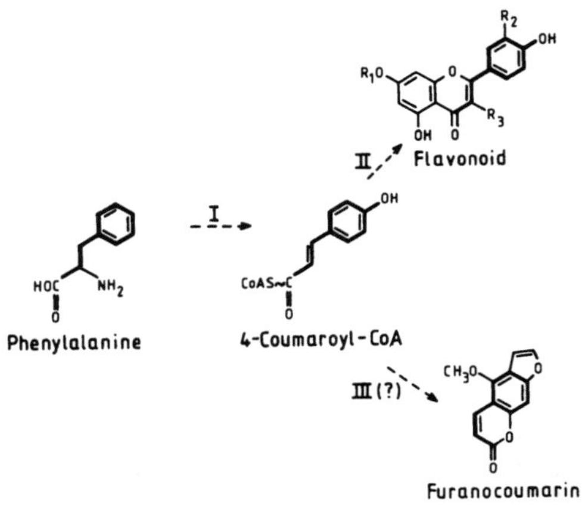

Fig. 1. Central role of 4-coumaroyl-CoA in the biosynthesis of flavonoid glycosides (UV-protective compounds) and furanocoumarins (phytoalexins) in parsley. For details, see text and Hahlbrock et al. 1982

Fig. 2. Schematic illustration of the difference in number of leaf cells responding to exposure of the organ to UV light (most or all cells of the upper epidermis reacting) or a pathogen (only one or a few cells reacting as primary targets)

also illustrates the principal difference between treatments with UV light or with a pathogen. It is very likely that most or all cells of the epidermal cell layer react upon irradiation of a leaf with UV light, whereas usually only one or a few cells are simultaneously challenged with a pathogen. But even if UV-irradiated leaves might therefore appear more favorable for biochemical studies with fully differentiated tissue, the presence of large amounts of hydrolytic enzymes in such plant material often prevents mRNA's and other readily degraded molecules from being isolated intact and in sufficiently high yield. This is not the case with cultured parsley cells, which can be used as a convenient source of UV-light- or elicitor-induced enzymes or mRNA's.

Although suspension cultures are far from being completely uniform populations of cells, the physiological states of single cells and cells inside or on the surface of aggregates in a "habituated" culture differ much less in various respects than the physiological states of fully differentiated cells, e.g., within a leaf. It is therefore perhaps not surprising that most, if not all, parsley cells in suspension culture accumulate flavonoid glycosides upon irradiation with UV light (U. Matern, personal communication), whereas in several plants their occurrence has been demonstrated to be restricted to certain types of cells (Wellmann 1974). A similar large response is likely to occur in elicitor-treated cell cultures.

The full complexity of all defence reactions of a given plant cell challenged with a given pathogen is unknown and is probably impossible to define precisely, due to the constant changes in environmental and physiological conditions of the interacting organisms. A scheme illustrating a few of the typical mechanisms which might be involved in plant-pathogen interactions and induced resistance responses of plant cells is given in Fig. 3. Apart from the direct chemical interactions, designated here as recognition and binding (for which specific molecules have not been unequivocally identified), both organisms possess a variety of means to seriously impair or kill the respective counterpart. For the plant cell, whose defense reactions are of particular

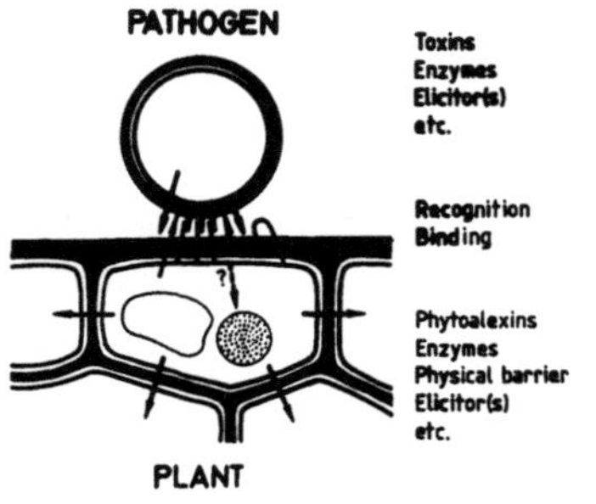

Fig. 3. Some of the known mechanisms involved in plant-pathogen interactions. See text for further explanations

interest in this context, the responses include the rapid synthesis of phytoalexins, that is, of induced low-molecular weight antimicrobial substances, the formation of a rigid physical barrier by reinforcing the cell wall with newly synthesized material, the synthesis and/or release of hydrolytic enzymes capable or degrading the pathogen's cell wall, and probably many other more or less immediate reactions (Horsfall and Cowling 1980; Wood 1982; Callow 1983).

Despite many unsolved questions (Hahlbrock et al. 1984), elicitors of microbial origin have proved to be extremely valuable tools, especially for biochemical work with cultured plant cells. One major advantage of using elicitors in combination with cell cultures is that sterile, undamaged cells are treated with nonliving material, thus ensuring that the measured responses are solely due to reactions of the plant cells. For many purposes, the principle limitations of work with cell cultures versus intact plant tissue, and with elicitors as compared with living pathogens, are compensated by far by the many advantages associated with the use of such greatly simplified systems.

4 Genes Involved in Resistance to Pathogens and UV Light

When cell suspension cultures of parsley are irradiated with UV light, the three enzymes of general phenylpropanoid metabolism (Group I, see Fig. 1) are rapidly induced together with the ca. 13 enzymes comprising the flavone and flavonol glycoside pathway (Group II) in this system (Ebel and Hahlbrock 1982). We have measured, both in vivo and in vitro, the time courses of induction of two representative mRNA activities from each group. Within experimental error, the timing was the same within each group, but differed between the groups. Following a standard UV irradiation period of 2.5 h, the mRNA activities for phenylalanine ammonia-lyase (PAL) and 4-coumarate: CoA ligase (4CL), the first and the last of the three enzymes of Group I, increased to a sharp peak around 7 h after the onset of induction, whereas maximal mRNA activities for chalcone synthase (CHS) and UDP-apiose synthase (UAS), two enzymes of Group II, were reached about 4 h later. Subsequently, all light-induced mRNA activities returned rapidly to the original, low or undetectable, levels with apparent half-lives of approximately 3.5 h (PAL and 4CL), 6 h (CHS) and 8-10 h (UAS) (Schröder et al. 1979; Gardiner et al. 1980; Ragg et al. 1981).

The rates of enzyme synthesis, or mRNA activities, were measured using specific antisera for immunoprecipitation and quantitation of newly synthesized proteins. Similarly, specific cDNA probes were required for the determination of amounts and rates of synthesis of the mRNA's. Therefore, cDNA libraries were generated for both light-induced and elicitor-induced mRNAs, and clones containing plasmids with cDNA inserts complementary to PAL, 4CL, CHS, and several other light- or elicitor-induced mRNAs whose translation products have not yet been identified, were characterized by hybrid-arrest and hybrid-select translations (Kuhn et al. 1984). Several plasmids with cDNA copies of CHS mRNA, which is by far the most abundant of the light-induced mRNA's, were isolated. Three of these cDNA's were used to determine the nucleotide sequence of the coding region of CHS mRNA and to compare the deduced amino acid sequence with some related properties of the native enzyme protein. Agreement of the respective results provided further evidence for the identity of the cloned CHS cDNAs (Reimold et al. 1983).

For comparison with the translational activities of the mRNA's, the time courses of changes in the amounts of PAL, 4CL and CHS mRNA's in UV-irradiated and elicitor-treated cells were measured by RNA blot hybridizations with the respective cDNAs (Kreuzaler et al. 1983; Kuhn et al. 1984). The changes in mRNA amount and activity coincided in all cases, with one exception. The exception was an apparent continued increase in the amount of PAL mRNA far beyond the peak in PAL mRNA activity in elicitor-treated cells (Kuhn et al. 1984). At present, we cannot exclude the possibility that this latter, unexplained result is an artifact due to cross-hybridization of PAL cDNA with some mRNA(s) of unknown function, whose timing of induction with elicitor differs considerably from that expected for PAL mRNA.

Interestingly, multiple spots were obtained after two-dimensional gel electrophoresis for immunoprecipitates of both PAL and 4CL, and the patterns were similar whether UV light or elicitor was used for induction (Kuhn et al. 1984). It is therefore possible, but by no means certain, that the same genes are involved in the resistance responses to UV light and fungal pathogens.

In recent experiments (J. Chappell and K. Hahlbrock, 1984) we have determined the rates of transcription of the PAL, 4CL and CHS genes in nuclei isolated from UV-irradiated and elicitor-treated cells, again using the respecitve cDNA's as hybridizations probes. The results were in agreement with those predicted from the kinetics of induced changes in the mRNA amounts.

Representative data for the timing of UV-light- and elicitor-induced changes in enzyme activities, mRNA activities and amounts, and rates of transcription are shown in Fig. 4.

All individual steps in the biosynthesis of flavonoid glycosides in UV-irradiated parsley cells can be related to light-induced, transient increases in the rates of transcription of the genes coding for the enzymes of this pathway (Fig. 5). The coincidence of the calculated curves (Hahlbrock et al. 1982) with the corresponding experimental data is taken as evidence that regulatory control is exerted in this system mainly at the level of gene expression. Although such detailed calculations have not been possible for the elicitor-induced synthesis of phytoalexins (due to the lack of sufficient detailed measurements), the involvement of gene transcription has been demonstrated in this case as well.

5 Mechanisms of Herbicide Resistance

In contrast to phytoalexins and UV-protective compounds, herbicides are man-made chemicals designed primarily to control weeds in crop plant cultivations. Resistance, or tolerance (LeBaron and Gressel 1982), of plants for herbicides is not a trait acquired on the same evolutionary time-scale as disease or UV resistance. While induced defense re-

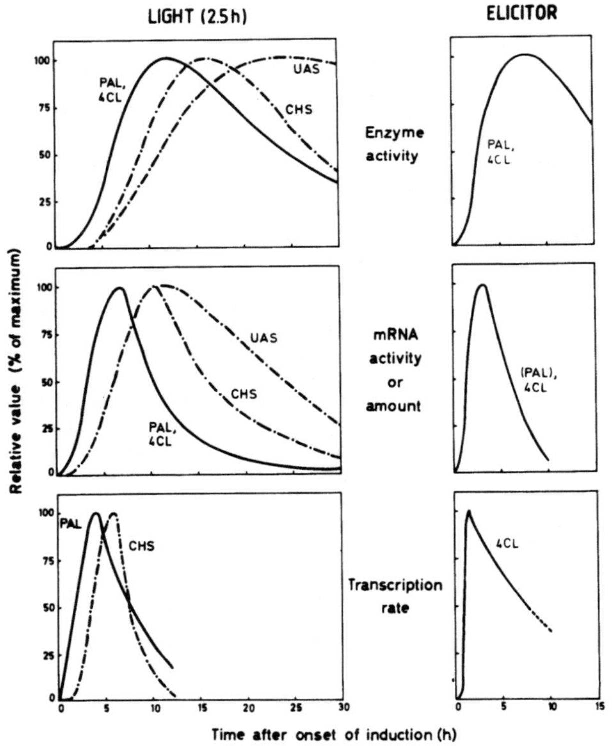

Fig. 4. Summary of induction kinetics for enzymes involved in flavonoid and furanocoumarin biosynthesis in UV-irradiated or elicitor-treated parsley cells respectively. References are given in the text

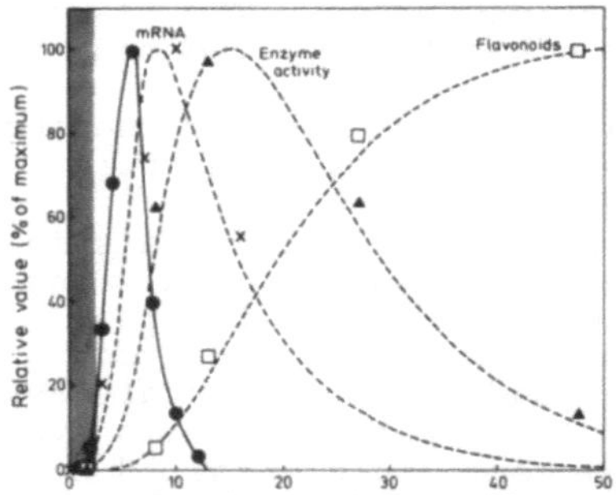

Fig. 5. Sequence of events leading from changes in the rates of transcription, mRNA activity (or amount) and enzyme activity for CHS, the key enzyme of flavonoid biosynthesis, to product accumulation in UV-irradiated parsley cells. *Shaded area* 2.5-h period of irradiation: *broken curves* calculated from solid curve (Hahlbrock et al. 1984). Previously unpublished data from Chappell and Hahlbrock (1984)

actions play an important role in the resistance to natural hazards, plants usually tolerate herbicides by preformed mechanisms. The observation that some weeds have developed tolerance for herbicides was the basis for studies of the underlying mechanisms and for experiments aimed at the introduction of herbicide tolerance to susceptible crop plants.

Tolerance or resistance of plants to herbicides can be caused by effects on uptake, translocation or binding to the target site, by detoxification, or by general physiological differences between resistant and susceptible plants. In the following, we will focus on one new class of herbicides which might at first view appear particularly promising with regard to the potential application of new breeding techniques, including gene transfer in vitro. However, we will also point out some problems possibly arising from the use of herbicide resistant plants in agriculture.

6 Resistance to Chlorsulfuron: An Example for Possible Gene Transfer?

Sulfonylurea derivatives, applied at concentrations below 100 g ha^{-1}, are very potent, selective herbicides, as compared with many other herbicides which have to be applied at concentrations around 1 kg ha^{-1} or more. Most broadleaf plants are highly sensitive to chlorsulfuron, the first sulfonylurea derivative with known herbicidal activity, whereas cereals show a broad spectrum of sensitivity, from corn, being susceptible, to wheat, being resistant. The work described in the following was carried out with cultured soybean cells (*Glycine max*) and chlorsulfuron, whose herbicidal activity involves the inhibition of growth and of thymidine incorporation into DNA under conditions where photosynthesis, respiration, cell elongation, as well as RNA and protein synthesis are not affected (Ray 1982a,b; Hatzios and Howe 1982).

Chlorsulfuron inhibited growth in cell suspension cultures of soybean very quickly. Fifty percent growth inhibition were achieved by 170 parts per billion, based on initial cell dry weight, or 0.6 nmol l^{-1}, based on the concentration of chlorsulfuron in the culture medium (Scheel and Casida 1985). This inhibition was overcome by 50 µg ml^{-1} of valine or leucine, whereas isoleucine was without any effect. Most effective in the reversal of the inhibition was a mixture of valine, leucine and isoleucine, and the common precursor for valine and leucine, 2-ketoisovalerate, also reversed the inhibition, whereas pyruvate showed no alleviating effect (Scheel and Casida 1985). Among the free amino acids, valine and leucine, but not isoleucine, decreased rapidly in soybean cells treated with chlorsulfuron, indicating a specific block at one of the steps involved in valine and leucine biosynthesis (Fig. 6). The high chlorsulfuron sensitivity of acetolactate synthase (ALS), the first common enzyme in this pathway, further supports this assumption (Ray 1984). A comparison of ALS preparations from resistant and susceptible plants (Ray 1984) suggests that differential sensitivity of this putative primary target is not the cause of differential plant responses to the herbicide. In this case, resistance to chlorsulfuron is more likely to be due to differences in its metabolism in susceptible and resistant plants (Sweetster et al. 1982). As shown in Fig. 7, plants hydroxylate chlorsulfuron in the para position to the chlorine residue, followed by glucosylation of the newly generated hydroxyl group. The resulting metabolites have no herbicidal activity. In plants which are sensitive to chlorsulfuron, the conversion to these metabolites is slow, whereas resistant plants detoxify the herbicide very rapidly.

Thus, the primary site of action of chlorsulfuron is assumed to be ALS. Apparently, inhibition of this enzyme leads to depletion of the plant of branched-chain amino acids, and this might cause, by unknown mechanisms, inhibition of DNA synthesis and growth. Similar effects can be observed as a result of feed-back inhibition of ALS by valine (Bourgin 1983).

Increasing difficulties opposing the introduction, and increasing costs counteracting the development, of new agrochemicals have added to an interest in transferring herbicide resistance from already existing tolerant to susceptible plants. One method (Chaleff and Ray 1984), which is only applicable in the case of herbicides not interfering with photosynthesis, is to select for herbicide resistance at the level of protoplats or cells in suspension culture. For a number of plant

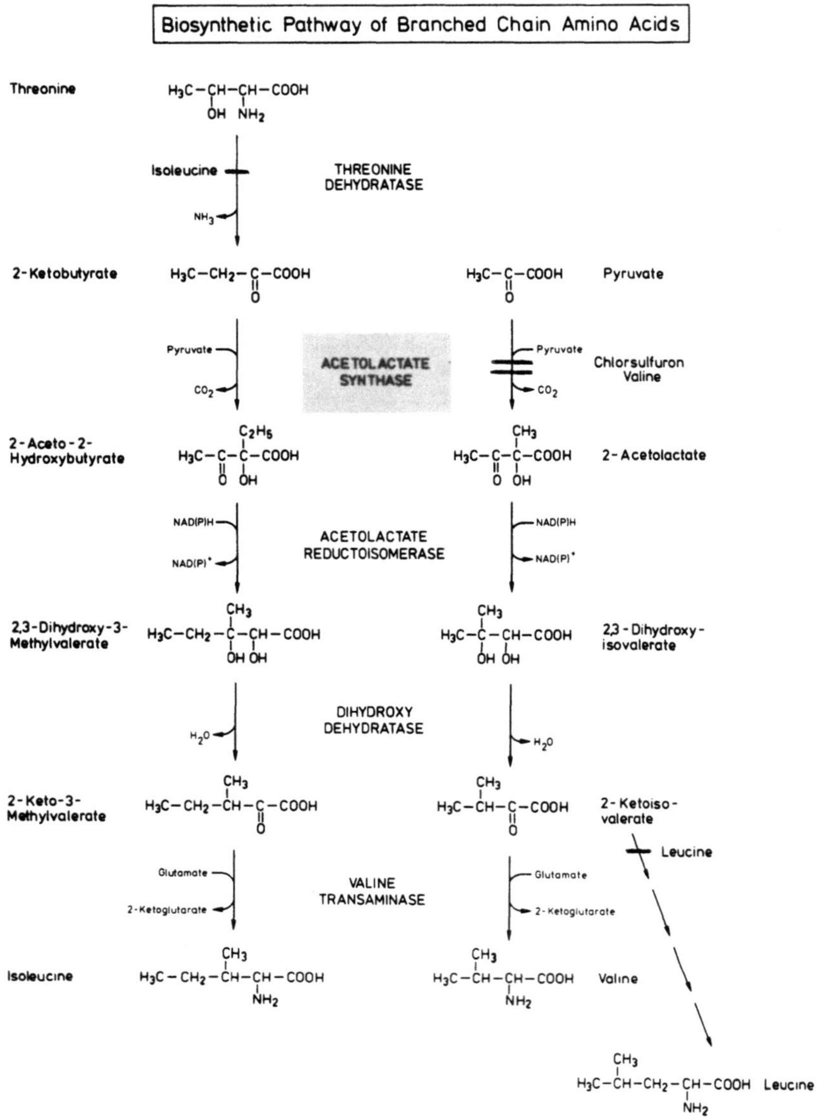

Fig. 6. Biosynthetic pathway of branched-chain amino acids. Sites of inhibition by chlorsulfuron and feed-back inhibition by end products are indicated by *horizontal bars*

species, resistant clones can be isolated and plants regenerated via callus cultures. Plants with stably inherited resistance to sulfonylurea derivatives, sulfometuron methyl, several other herbicides, or various amino acids have been obtained by this method (Bourgin 1983; Chaleff and Ray 1984; Chaleff and Parsons 1978; Singer and McDaniel 1984). However, since growth in the presence of the selecting agent is the only criterion for resistance, the actual mechanism may be different in many or all of the resulting plants. It has been shown, for

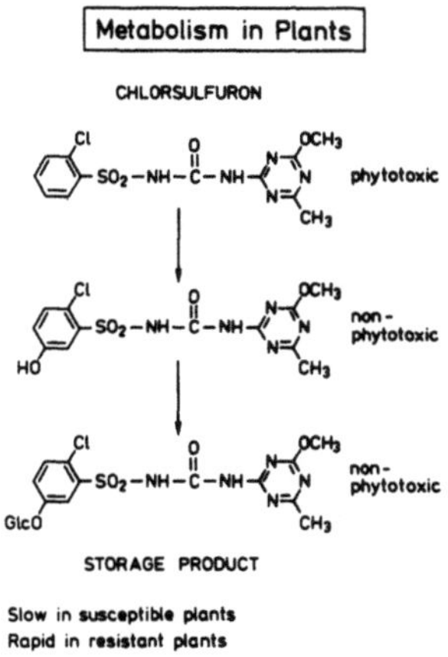

Fig. 7. Mechanism of chlorsulfuron detoxification

example, that valine resistance of plants which were obtained according to this selection scheme can be caused either by reduced valine uptake or by the loss of competence of ALS for feed-back inhibition by valine (Bourgin 1983). If a specific resistance mechanism is desired, the gene(s) related to, e.g., uptake, translocation, detoxification or interaction of the herbicide with the target site would preferably have to be identified, isolated, cloned, and used for transfer to the plant of interest by the methods discussed in the previous chapters. Resistance to chlorsulfuron might be an example where such an approach is feasible.

However, it seems worth repeating the above-mentioned concern about toxicology and persistence problems associated with the use of herbicides and herbicide-resistant plants, or with the large-scale application of man-made chemicals in general. A brief summary of a few major effects and sites of action of most, or all, herbicides is given in Fig. 8. Obviously, it is not sufficient to elucidate the biochemical basis of resistance as a prerequisite for the isolation and transfer of genes. Metabolites accumulating in the plant and thus entering the food chain have to be identified and their toxicology has to be evaluated, especially when herbicide-detoxifying mutants are used. Moreover, herbicides with high soil persistence are strongly adsorbed to the soil and might persist for several years. Only resistant plants can be grownown for one or more growing seasons following the application of chlorsulfuron (O'Sullivan 1982). This problem might even be aggravated when uptake or translocation mutants are used.

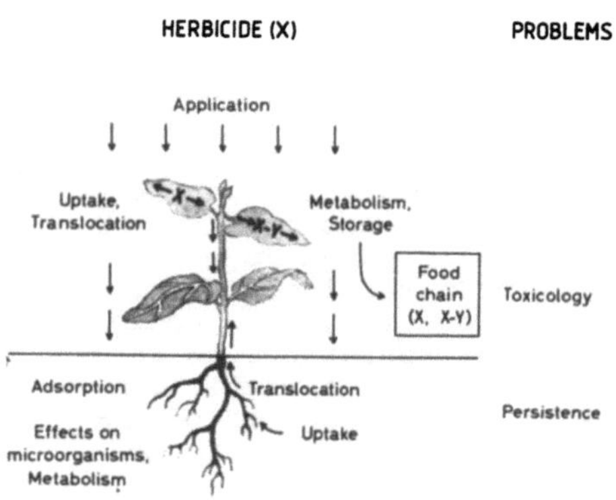

Fig. 8. Some reactions and problems associated with the application, metabolism, and effects of herbicides

7 Conclusions

Using UV-irradiated and elicitor-treated cell suspension cultures of parsley as a model system, we have shown that each of the two treatments triggered the formation of different types of phenylpropanoid derivatives, whose functions can be related to the resistance response of the plant to the respective inducing agent. Transient changes in the rates of gene transcription are involved in both cases, indicating that resistance to UV light or to potential pathogens is dependent on the induced expression of "resistance-related genes". The present state of knowledge concerning the mechanisms of induction of UV-protective flavonoids by UV light and of furanocoumarins, compounds with antimicrobial activity, by fungal elicitor(s) is summarized in Fig. 9.

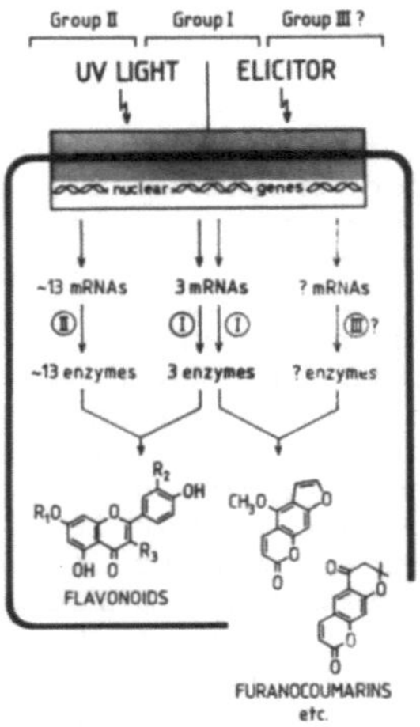

Fig. 9. Scheme summarizing the known steps in the two signal-reaction chains leading either to the UV light-induced formation of flavonoids or to the elicitor-induced formation of furanocoumarins and related compounds with antimicrobial activity in parsley cells. (After Hahlbrock et al. 1982). *Shaded area* present lack of knowledge in the areas of signal perception and transmission

The number of enzymatic steps, and hence probably the minimal number of genes, involved in the induction of UV light resistance appears to be about 16 (3 of Group I and probably 13 of Group II). No such estimations can yet be made for disease resistance. Although even fewer enzymes might be involved in furanocoumarin than in flavonoid synthesis, there is a wealth of information indicating that the induction of additional pathways (among others those involved in the formation of new cell-wall material) is at least of equal importance in the entire, complex disease resistance response. Further work on the biochemistry of induced resistance of plants to potential pathogens will therefore be required before the most important resistance-related genes can be defined. The existing data support the view that this is a difficult, but certainly not an impossible task.

The case study of the sulfonylurea herbicide, chlorsulfuron, has shown that resistant (or tolerant) mutants can be selected without prior characterization of "resistance genes". In cultured soybean cells, the primary site of action of chlorsulfuron was localized in the biosynthetic pathways of branched-chain amino acids, with acetolactate synthase being the target enzyme. Lateral resistance to chlorsulfuron is due to detoxification of the herbicide by enzymatic conversion. In such a case, detailed knowledge of the mechanisms of action and metabolism of the herbicide might enable the definition of resistance genes which might then be used to transform susceptible crop plants without changes in the genetic background. Such transfer of herbicide resistance may be beneficial in terms of food production, but associated with environmental hazards if necessary precautions are not taken.

References

Bourgin JP (1983) Selection of tobacco protoplast-derived cells for resistance to amino acids and regeneration of resistant plants. In: Lurquin PF, Kleinhofs A (eds) Genetic engineering in eukaryotes. Plenum, New York, pp 195

Callow JA (ed) (1983) Biochemical plant pathology. Wiley, Chichester

Chaleff RS, Parsons MF (1978) Direct Selection in vitro for herbicide-resistant mutants of *Nicotiana tabacum*. Proc Natl Acad Sci USA 75:5104-5107

Chaleff RS, Ray TB (1984) Herbicide-resistant mutants from tobacco cell cultures. Science (Wash DC) 223:1148-1151

Chappell J, Hahlbrock K (1984) Transcription of plant defence genes in response to UV light or fungal elicitor. Nature 311:76-78

Ebel J, Hahlbrock K (1982) Biosynthesis of flavonoids. In: Harborne JB, Mabry TJ (eds) Advances in flavonoid research 1975-1980. Chapman and Hall, London, pp 641-679

Gardiner SE, Schröder J, Matern U. Hammer D, Hahlbrock K (1980) mRNA-dependent regulation of UDP-apiose synthase activity in irradiated plant cells. J Biol Chem 255:10752-10757

Hahlbrock K, Chappell J, Kuhn DN (to be published 1984) Rapid induction of mRNAs involved in defense reactions in plants. In: Lea PJ, Steward GR (eds) The genetic manipulation of plants and its application to agriculture, vol 23. Oxford University Press

Hahlbrock K, Kreuzaler F, Ragg H, Fautz E, Kuhn DN (1982) Regulation of flavonoid and phytoalexin accumulation through mRNA and enzyme induction in cultured plant cells. In: Jaenicke L (ed) Biochemistry of differentiation and morphogenesis. Springer, Berlin Heidelberg New York, pp 34-43

Hatzios KK, Howe CM (1982) Influence of the herbicides hexazinone and chlorsulfuron on the metabolism of isolated soybean leaf cells. Pestic Biochem Physiol 17:207-214

Horsfall JG, Cowling EB (eds) (1980) Plant disease, vol 5. Academic, New York

Kreuzaler F, Ragg H, Fautz E, Kuhn DN, Hahlbrock K (1983) UV induction of chalcone synthase mRNA in cell suspension cultures of *Petroselinum hortense*. Proc Natl Acad Sci USA 80:2591-2593

Kuhn DN, Chappell J, Boudet A, Hahlbrock K (1984) Induction of phenylalanine ammonia-lyase and 4-coumarate: CoA ligase mRNAs in cultured plant cells by UV light or fungal elicitor. Proc Natl Acad Sci USA 81:1102-1106

LeBaron HM, Gressel J (eds) (1982) Herbicide resistance in plants. Wiley, New York

O'Sullivan PA (1982) Response of various broad-leaved weeds, and tolerance of cereals, to soil and foliar applications of DPX-4189. Can J Plant Sci 62:715-724

Ragg H, Kuhn DN, Hahlbrock K (1981) Coordinated regulation of 4-coumarate: CoA ligase and phenylalanine ammonia-lyase mRNAs in cultured plant cells. J Biol Chem 256:10061-10065

Ray TB (1982a) The mode of action of chlorsulfuron: a new herbicide for cereals. Pestic Biochem Physiol 17:10-17

Ray TB (1982b) The mode of action of chlorsulfuron: the lack of direct inhibition of plant DNA synthesis. Pestic Biochem Physiol 18:262-266

Ray TB (1984) The primary site of action of the herbicide chlorsulfuron. Annual meeting of the weed Science Society of America. Book of Abstracts, p 87

Reimold U, Kröger M, Kreuzaler F, Hahlbrock K (1983) Coding and 3' non-coding nucleotide sequence of chalcone synthase mRNA and assignment of amino acid sequence of the enzyme. EMBO J 2:1801-1805

Scheel D, Casida JE (1985) Sulfonylurea herbicides: growth inhibition in soybean cell suspension cultures and in bacteria correlated with block in biosynthesis of valine, leucine or isoleucine. Pestic Biochem Physiol, in press

Schröder J, Kreuzaler F, Schäfer E, Hahlbrock K (1979) Concomitant induction of phenylalanine ammonia-lyase and flavanone synthase mRNAs in irradiated plant cells. J Biol Chem 254:57-65

Singer SR, McDaniel CN (1984) Selection of amitrole tolerant tobacco calli and the expression of this tolerance in regenerated plants and progeny. Theor Appl Genet 67:427-432

Sweetster PB, Show GS, Hutchinson JM (1982) Metabolism of chlorsulfuron by plants: biological basis for selectivity of a new herbicide for cereals. Pestic Biochem Physiol 17:18-23

Wellmann E (1974) Gewebespezifische Kontrolle von Enzymen des Flavonoidstoffwechsels durch Phytochrom in Kotyledonen des Senfkeimlings (*Sinapis alba* L.). Ber Dtsch Bot Ges 87:275-279

Wood RKS (1982) Action Defense Mechanisms in Plants. Plenum, New York

Herbicide Resistance Through Gene Transfer? Biochemical and Toxicological Aspects

H. SANDERMANN, JR.[1]

1 Introduction[2]

1.1 Natural Pesticides in Plants

Recent years have seen a greatly increased interest in plant biochemistry, stimulated to a large part by the possible role of plant secondary products as natural pesticides (Grisebach and Ebel 1983; Harborne 1982; Hedin 1982, 1983; Schildknecht 1981; Schlösser 1983). For example, pyrethroids, precocenes and certain proteinase inhibitors may function as natural insecticides, phytoalexins and certain saponins as natural fungicides. Plant compounds with additional defensive roles have been described, including allelopathic chemicals which are excreted by certain plants and have herbicidal activity against other plants. No allelopathic chemicals directed by crop plants against weeds seem to be known.

Natural pesticides isolated from plant sources usually cannot be sprayed in the field because of their marked instability. The disadvantage can be overcome by 'biorational' chemical synthesis of more stable analogues, for example of pyrethroids and of plant compounds with juvenile hormone activity (Menn 1980; Hedin 1982). An alternative method consists in generating a desired natural pesticide within the crop plant by use of genetic engineering techniques. Some of the plant enzymes for the biosynthesis of defensive substances have already been isolated. The induction of such enzymes has been studied at the levels of translation and transcription, and cDNA's for some biosynthetic enzymes have been cloned as a prerequisite for the isolation of the respective genes (Hahlbrock et al. 1984). Attempts are underway in some laboratories to introduce the genes coding for the biosynthesis of natural pesticides into crop plants. This appears to be a long-term goal because, as in the case of nitrogen fixation (Barton and Brill 1983), several genes may have to be simultaneously transferred. In view of this experimental difficulty, much attention has recently been placed on the agricultural trait, herbicide resistance, which may require transfer of only a single gene (Barton and Brill 1983; Drummond 1983; Kearney 1983; Schell 1983).

[1] Institut für Biologie II, Universität Freiburg, 7800 Freiburg i. Br., FRG

[2] The trivial names of herbicides are used throughout the text. The corresponding systematic designations have been listed (Ashton and Crafts 1981; Hatzios and Penner 1982). For the sake of simplicity, herbicide resistance and herbicide tolerance are not distinguished. Detailed discussions of the difference between these terms may be found in a recent authoritative volume on herbicide resistance in plants (LeBaron and Gressel 1982)

35. Colloquium - Mosbach 1984
The Impact of Gene Transfer Technique in Eukaryotic Cell Biology

Table 1. Transfer of chimeric antibiotic resistance genes into plant cells

Bacterial gene product	*Transferred into*	*Test for gene expression*	*Reference*
Chloramphenicol acetyltransferase	Tobacco seedlings	Transferase assay	Herrera-Estrella et al. 1983b
Aminoglycoside phosphotrans-ferase II	Regenerating to-bacco protoplasts	Selection, with kanamycin, Enzyme assay.	Herrera-Estrella et al. 1983a
Aminoglycoside phosphotrans-ferase II	Tobacco stem explant	Selection with G418.	Bevan et al. 1983
Aminoglycoside phosphotrans-ferase I and II	Regenerating to-bacco and petunia protoplasts	Selection with kanamycin, Enzyme assay.	Fraley et al. 1983
Dihydrofolate reductase	Regenerating to-bacco protoplasts	Selection with methotrexate	Herrera-Estrella et al. 1983a

1.2 Herbicide Resistance

A transfer of single defined resistance genes into plant cells has in fact already been achieved, as summarized in Table 1. Known bacterial resistance genes for antibiotics with phytotoxic activity were employed in these studies. Furthermore, the regeneration of whole plants with retention of the transferred marker has been successful in the case of tobacco (Barton et al. 1983; Caplan et al. 1983; Horsch et al. 1984). In view of these advances in vector construction and gene transfer, it is now the task of the plant biochemist to define and isolate genes whose transfer could result in herbicide resistance. Unfortunately, knowledge on the plant biochemistry of xenobiotics (Lamoureux and Frear 1979; Shimabukuro et al. 1982; Sandermann 1982) is much less developed than knowledge on the plant biochemistry of secondary plant products (Stumpf and Conn 1981). Generation of herbicide resistance is of immediate scientific interest, since it provides a direct selection marker in studies on plant gene regulation. Furthermore, the introduction of herbicide resistance genes into crop plants could improve the agricultural use of herbicides. Some of the presently employed herbicides or derived metabolites suffer from serious disadvantages, such as toxicity (e.g., mutagenicity), insufficient specificity, persistence in the plant or in soil, interference with crop rotation or mixed cropping. Use of another herbicide without such undesirable properties is not feasable in cases where the crop plant is susceptible to the alternative herbicide. This applies in particular to broad-spectrum herbicides. Once the required techniques have been developed, it may be easier and less costly to transfer herbicide resistance into crop plants than to develop a new herbicide.

Table 2. One-step detoxification of herbicides by bacterial enzymes. The sites of enzymatic modification are indicated by arrows

Herbicide	*Enzyme*	*Reference*
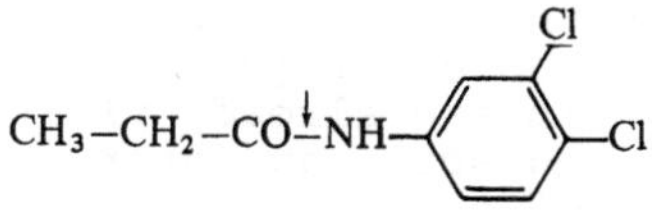 *Propanil*	Amidase	Johnson and Talbot 1983
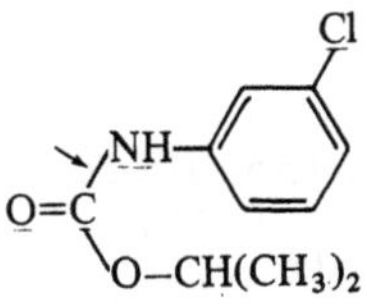 *Chlorpropham*	Amidase	Johnson and Talbot 1983
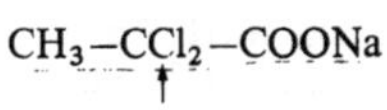 *Dalapon*	Dehalogenase	Motosugi and Soda 1983
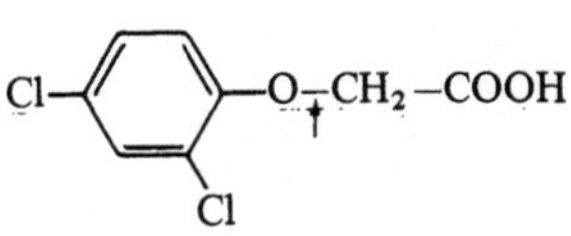 *2,4-D*	Cleavage enzyme	Johnson and Talbot 1983
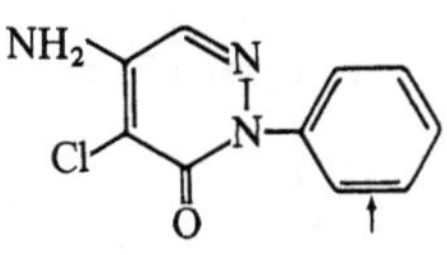 *Chloridazon*	Oxygenase	Eberspächer and Lingens 1981

2 Sources for Herbicide Resistance Genes

2.1 Microorganisms

It is known for some time that microorganisms of soil and water can degrade a wide variety of herbicides and other industrial chemicals (Bollag 1974; Cripps and Roberts 1978; Kaufman and Kearney 1976; Kobayashi and Rittmann 1982). A number of herbicides are completely mineralized by defined bacterial strains, the degradative enzymes often being plasmid-encoded. These herbicides include 2,4-D and MCPA (*Alcaligenes;* Don and Pemberton 1981), 2,4,5-T (*Pseudomonas;* Kilbane et al. 1982; Karns et al. 1983), chlorpropham (*Pseudomonas;* Cripps and Roberts

1978), pentachlorophenol (*Pseudomonas, Arthrobacter;* Karns et al. 1983; Stanlake and Finn 1982), chloridazon (*Phenylobacter;* Eberspächer and Lingens 1981), dalapon (*Pseudomonas;* Cripps and Roberts 1978; Senior et al. 1976) and glyphosate (*Pseudomonas;* Moore et al. 1983). In the relatively few cases listed in Table 2 the first enzyme attacking the herbicide and leading to loss of phytotoxicity has been characterized. The genes located on degradative plasmids are available for gene transfer experiments, and genetically engineered bacteria are already finding application for the decontamination of polluted soil and water (Finn 1983). However, the transfer and expression of such genes in plant cells has so far not been described.

2.2 Plants

In addition to the microorganisms mentioned above, plants also offer a rich reservoir of potential herbicide resistance genes. In fact, plants appear to rival bacteria with regard to metabolic potency. Plant metabolism even of persistent environmental chemicals has been demonstrated by use of suspension-cultured cells (Sandermann et al. 1984). Plant microsomal fractions were able to oxygenate the ubiquitous plant contaminant, benzo [α] pyrene by a mechanism which differed from that of mammalian microsomal fractions (v.d. Trenck and Sandermann 1980). Wheat and soybean cells could transform the trichloromethyl group of DDT into a polar carboxyl group (Arjmand and Sandermann 1985). A unique wheat esterase for the cleavage of the mass-produced plasticizer chemical, bis-(2-ethylhexyl)-phthalate, has recently been purified to homogeneity (Krell and Sandermann 1984). With regard to the plant metabolism of herbicides, major pathways are the formation of various soluble conjugates (reviewed; Lamoureux and Frear 1979; Sandermann 1982; Shimabukuro et al. 1982) and the copolymerization into the "insoluble" plant lignin fraction (reviewed; Sandermann et al. 1983). Metabolic reactions leading to herbicide resistance of crop plants have in many cases been identified (Ashton and Crafts 1981; Hatzios and Penner 1982; Wain and Smith 1976). To cite a few examples, the resistance of wheat against 2,4-D and chlorsulfuron is attributed to hydroxylation, the resistance of corn against atrazine to conjugation with glutathione, and the resistance of rice against propanil to amidase cleavage.

Sensitive crop plants have been made herbicide-resistant by conventional plant breeding leading, for example, to simazine-resistant rape and metribuzin-resistant soybean (Faulkner 1982). The latter example has been attributed to increased deaminase activity (Fedtke and Schmidt 1983).

Herbicide resistance of weeds has been observed to arise spontaneously in the field, the best-known example being the resistance of *Amaranthus* and many additional weed species against atrazine (Bachthaler et al. 1983). This resistance has been attributed to a single DNA base change (see below).

Acquired resistance against amitrole, bentazon, 2,4-D, phenmedipham, picloram, propham and a few more herbicides has been observed in plant cell cultures or protoplasts. The biochemical basis for resistance has usually not been identified in these studies (Meredith and Carlson 1982; Negrutiu et al. 1984).

In general, physiological factors leading to plant herbicide resistance may occur in herbicide uptake and/or transport, in binding to the target site and in selective metabolism of the herbicide. Virtually nothing is known about defined plant proteins involved in herbicide uptake and/or transport. The further discussion will therefore concentrate on the roles of the target site and of metabolic enzymes.

3 Herbicide Resistance by Changes in the Target Site

Three examples for resistance by changes in the target site have so far been elucidated in biochemical detail. The case of chlorsulfuron is described in the accompanying article (Hahlbrock et al. 1984).

3.1 Atrazine

This widely used herbicide is known to act on a plastoquinone-binding membrane protein ("32K-protein") of photosystem II (Arntzen et al. 1983). The protein is coded for by the *psbA*-gene which has been cloned from atrazine-sensitive plants and from the resistant weeds, *Amaranthus hybridus* and *Solanum nigrum*. Comparison of the DNA sequences revealed that atrazine resistance was associated with a single A to G base transition, leading in turn to an amino acid change from serine to glycine (Hirschberg and McIntosh 1983). The authors suggested that genetic "transfer of resistance to now susceptible crops may have practical importance".

3.2 Glyphosate

This broad-spectrum herbicide is known to inhibit an enzyme of the shikimate pathway, 5-enolpyruvylshikimate 3-phosphate (EPSP)-synthase (Steinrücken and Amrhein 1980; Amrhein et al. 1983). The enzyme has been highly purified from *E. coli* (Duncan et al. 1984) as well as from pea seedlings (Mousdale and Coggins 1984). Both enzyme proteins had a molecular weight of about 50,000. The microbial enzyme is coded for by the *aroA*-gene which has been cloned from *E.coli* (Rogers et al. 1983) and *Salmonella* (Comai et al. 1983). Two different mechanism of glyphosate resistance have been recognized. The EPSP-synthases of resistant *Salmonella* (Comai et al. 1983) and *Aerobacter* strains (Schulz et al. 1984) were found to have greatly decreased affinity for glyphosate. The mutated *Salmonella aroA*- gene is available on a cosmid vector (Comai et al. 1983). Increased EPSP-synthase activity provides a second mechanism for apparent glyphosate resistance. This was found empirically in *Aerobacter* as well as cultured plant cells (Amrhein et al. 1983). In *E.coli*, overexpression of the enzyme was achieved in a controlled way by placing the normal *aroA*-gene on a multicopy plasmid (Rogers et al. 1983).

4 Herbicide Resistance by Selective Metabolism

Microbial enzymes which may lead to herbicide resistance have been considered in section 2.1. An even greater number of enzymes which in one step convert a herbicide to a non-phytotoxic metabolite has been isolated from plant sources. Gene transfer of such enzymes obviously provides herbicide-sensitive plant cells with a direct selection marker. The known enzymes will be described in two sections, according to two different types of selection. For herbicides inhibiting chloroplast functions "green" cultured plant cells will have to be employed for selection. This is at present an experimentally difficult task (Meredith and Carlson 1982). For herbicides acting on nonchloroplast targets (e.g., hormone analogues or biosynthetic inhibitors) the common dark-grown plant cell cultures can be employed for selection (Meredith and Carlson 1982; Negrutiu et al. 1984).

Table 3. One-step detoxification of herbicides by plant enzymes. The sites of enzymatic modification are indicated by arrows

A. Herbicides inhibiting chloroplast function

Herbicide	*Enzyme (Source)*	*Molecular weight*	*Reference*
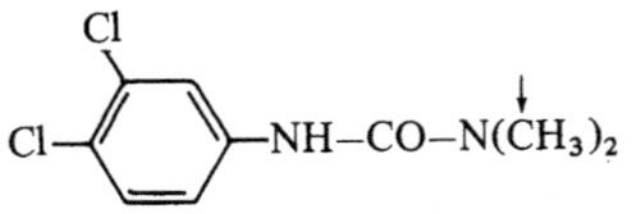 *Diuron*	N-Demethylase (cotton, soybean)	?	Frear et al. 1972
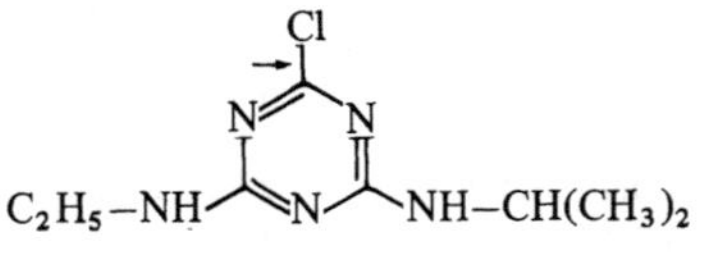 *Atrazine*	Glutathione S-transferase (corn)	40-45,000	Frear et al. 1972; Guddewar and Dauterman 1979
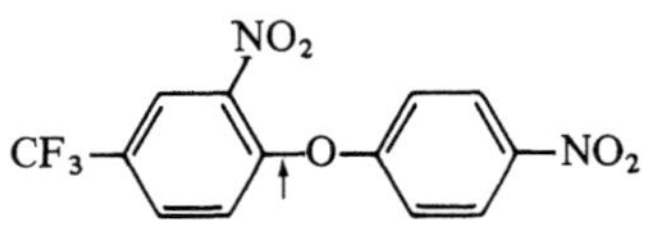 *Fluorodifen*	Glutathione S-transferase (pea)	47,000	Frear and Swanson 1973; Diesperger and Sandermann 1979
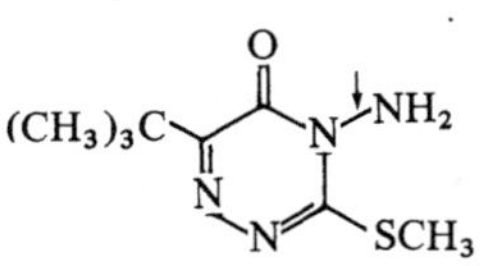 *Metribuzin*	Deaminase (soybean)	?	Fedtke and Schmidt 1983
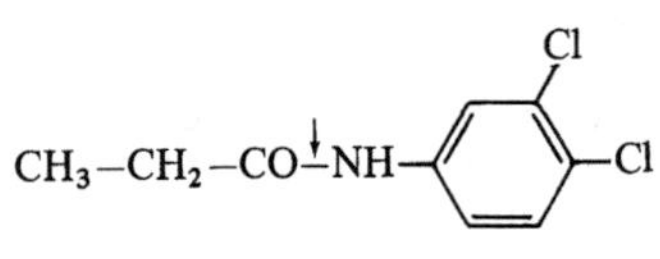 *Propanil*	Amidase (Rice, Dandelion, tulip)	?	Frear and Still, 1968; Hoagland, 1978; Gaynor and Still, 1983

B. Herbicides with nonchloroplast targets

Herbicide	*Enzyme (Source)*	*Molecular weight*	*Reference*
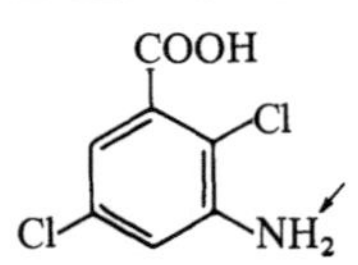 *Chloramben*	N-Glucosyltransferase (soybean)	?	Frear 1968
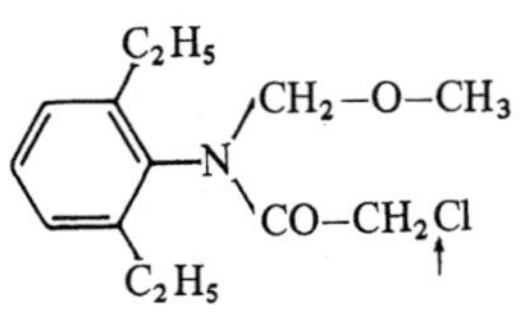 *Alachlor*	Glutathion S-transferase (corn)	50,000	Mozer et al. 1983

Table 3 (continued)

Herbicide	*Enzyme (Source)*	*Molecular weight*	*Reference*
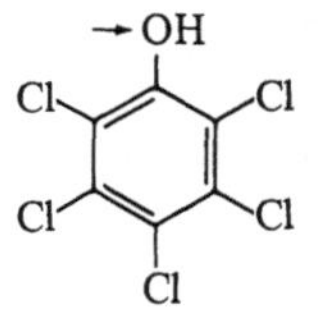 *Pentachlorophenol*	O-Glucosyltrans-ferase (soybean, wheat)	47,000	Schmitt et al. 1984

4.1 Herbicides Inhibiting Chloroplast Functions

Enzyme reactions leading in one step to nonphytotoxic metabolites are listed in Table 3A. The N-demethylation of diuron and related phenylurea herbicides, such as monuron and fluometuron, is catalyzed by a microsomal cytochrome P-450 system which has been isolated from cotton and which has so far not been solubilized (Frear et al. 1972). A glutathione S-transferase for atrazine has been purified from corn. This enzyme also accepted propazine, simazine and other 2-chlorotriazines as substrate (Frear et al. 1972; Guddewar and Dauterman 1979). The pea glutathione S-transferase catalyzing cleavage of fluorodifen (Frear and Swanson 1973) existed in molecular weight forms of 47,000 and 82,000, and could be separated from another pea glutathione S-transferase for the endogeneous substrate, cinnamic acid (Diesperger and Sandermann 1979). The soybean deaminase system for metribuzin has not been purified (Fedtke and Schmidt 1983). A similar crude deaminase system with chloridazon as substrate had previously been prepared from sugar beet (Fedtke and Schmidt 1979). The amidase for the cleavage of propanil has been isolated from rice and a few other plants (Frear and Still 1968; Hoagland 1978). The enzyme occurred in "soluble" as well as particulate cell fractions. The "soluble" portion has not been characterized in detail. The particulate fraction has recently been localized to the outer mitochondrial membrane (Gaynor and Still 1983).

4.2 Herbicides with Nonchloroplast Target Sites

Enzyme reactions leading in one step to nonphytotoxic metabolites are listed in Table 3B. An N-glucosyltransferase acting on chloramben and various chlorinated anilines has been isolated from soybean (Frear 1968). This enzyme has not been highly purified and characterized. Two glutathione S-transferases acting on alachlor have recently been highly purified from corn (Mozer et al. 1983). One of these enzymes was induced by herbicide antidotes. Both enzymes appeared to consist of two subunits of molecular weight 27-29,000. 1-Chloro-2,4-dinitrobenzene was a much more active substrate than alachlor. Atrazine was not tested so that it remained unclear whether the isolated transferases were related to the corn glutathione S-transferase for atrazine (cf. Table 3A). O-Glucosyltransferases detoxifying the strongly phytotoxic pentachlorophenol have recently been isolated and purified from cell suspension cultures of soybean and wheat (Schmitt et al. 1984). The soybean enzyme has been purified to electrophoretic homogeneity (R. Schmitt, unpublished work). In soybean as well as wheat cultures, the β-D-glucopyranoside of pentachlorophenol was further conjugated by highly active O-malonyltransferases which have also been purified from both cultures (molecular weights, approx. 43,000; Schmitt et al. 1985). Pentachlorophenol is best known as a fungicide for wood protection, but it is also a bactericide and a herbicide used mainly in rice. Its general biocidal activity allows to use pentachlorophenol as a selection agent with high activity against all organisms used in plant molecular biology, such as plants, *Escherichia coli* and *Agrobacterium*.

5 Discussion

5.1 Research Aspects

Generation of herbicide resistance through gene transfer involves the following experimental steps: identification of a protein leading to resistance (e.g., modified receptor or metabolic enzyme), isolation of the corresponding gene, vector construction, transfer of the gene into the crop plant and regulated gene expression. There is as yet no example where this entire experimental sequence has been carried out. However, the above sections show that successful examples for each individual step of the experimental sequence exist. The manipulation of chloroplast functions presents special difficulties with regard to gene transfer as well as selection procedures (LeBaron and Gressel 1982). Therefore, gene transfer may first succeed with nuclear resistance genes for herbicides with high selection power, such as pentachlorophenol or glyphosate.

Use of the transferred herbicide resistance as a direct selection marker should allow a stringent monitoring for the resistance gene itself, as well as any other gene which is placed into a suitable vector in combination with the resistance gene. Herbicide resistance genes could therefore gain the same basic importance in studies on plant gene regulation that antibiotic resistance genes have in bacterial molecular genetics. Besides numerous practical applications, one can, for example, envision studies on the tissue specificity of plant gene expression and on the induction of gene expression, e.g., by light or by chemicals such as hormones or elicitors. Placing the transferred gene under such controls may in turn lead to more sophisticated practical applications. One could also analyze the effects of promoter sequences, introns, insertion elements and other regulatory DNA sequences. Gene inactivation by insertion of a Ds-element has already been successfully studied in maize by using the herbicidal chemical, allylalcohol, as a selection agent (Döring et al. 1984; Osterman and Schwartz 1981).

5.2 Toxicological Aspects

Techniques for transfer of herbicide resistance genes into crop plants may lead to an entirely new field of molecular plant breeding. In view of this agricultural perspective some thoughts on toxicology seem appropriate.

The American Environmental Protection Agency has in the case of a genetically modified *Pseudomonas* strain taken the view that the organism should be subject to regulatory review in analogy to pesticide regulations (Sun 1983). This view appears even more adequate for crop plants genetically engineered for herbicide resistance. As shown above, one may have a choice of transferring either bacterial or plant genes to achieve herbicide resistance. One could stay within the gene transfer processes of conventional plant breeding by transferring genomic plant resistance genes rather than plant cDNA or nonplant bacterial DNA. In fact, some of the plant enzymes conferring herbicide resistance may turn out to be biosynthetic enzymes for the plant pigments used as phenotypic markers in classical plant genetic studies (Mendel 1866; McClintock 1951). It should be noted that previous discussions of herbicide resistance through gene transfer have exclusively considered use of bacterial genes (Drummond 1983; Kearney 1983; Schell 1983).

A second area deserving consideration concerns metabolic changes which may be caused by transfer of resistance genes. A drastic example for a metabolic aberration resulting from gene transfer was recently discovered when the *Pseudomonas* plasmid genes for naphthalene oxidation were transferred into *E. coli* (Ensley et al. 1983). The *E. coli* cells developed a deep blue color, since naphthalene oxygenase attacked the indole moiety derived from cellular tryptophane, leading to the production of indigo dye. Previous studies had shown that the plasmid-encoded oxygenases for naphthalene (Catterall and Williams 1971) and other substrates such as toluene (Gibson et al. 1968) had rather broad substrate specificities. Plants contain a large assortment of secondary aromatic compounds which are known to be attacked by bacterial enzymes (Barz and Hösel 1975). Genetically introduced microbial enzymes could therefore act on endogeneous plant constitutents, the metabolites formed being of uncertain nutritional and toxicological relevance. A similar problem may arise with plant enzymes for the one-step detoxification of herbicides, although such enzymes usually are rather substrate-specific. Another point of concern after genetic transfer of herbicide resistance relates to the residue situation. Making the target site resistant (cf. Sect. 3.) could invite an increased application of the herbicide, leading to elevated residues. Modifying the crop plant by introduction of a herbicide metabolizing enzyme could furthermore lead to the accumulation of high amounts of herbicide metabolites in the plant tissue (cf. Sandermann 1984). Residue regulations in West Germany (Pflanzenschutzmittel-Höchstmengenverordnung 1982) refer with few exceptions only to the unmodified parent pesticides and thus appear inadequate for the elevated metabolite residues which may occur in herbicide-resistant crop plants.

In conclusion, current legal regulations should be considerably extended in order to deal with crop plants which have been genetically engineered for herbicide resistance.

Acknowledgements. This article has greatly benefited from advice and critical comments from C. Beck, J. Ebel and G. Feix (all of Freiburg), K. Hahlbrock (Köln), M. Kröger (Gießen), F. Lingens (Hohenheim), W. Reineke (Wuppertal), D. Scheel (Köln) and R. Schmitt (Freiburg). Experimental studies from this laboratory have been supported by the Deutsche Forschungsgemeinschaft (Sa 180/15-4) and in part by the Bundesministerium für Forschung und Technologie (grant 03 7270) and the Fonds der Chemischen Industrie.

References

Amrhein N, Johänning D, Schab J, Schulz A (1983) Biochemical basis for glyphosate-tolerance in a bacterium and a plant tissue culture. FEBS Lett 157:191-196

Arjmand M, Sandermann H (1985) Metabolism of DDT and related compounds in cell suspension cultures of soybean (*Glycine max* L.) and wheat (*Triticum aestivum* L.). Pesticide Biochem. Physiol., in press.

Arntzen CJ, Steinback KE, Vermaas W, Ohad I (1983) Molecular characterization of the target site(s) for herbicides which affect photosynthetic electron transport. In: Miyamoto J, Kearney PC, Matsunaka S, Hutson DH, Murphy SD (eds) Pesticide chemistry. Human welfare and the environment. Proc 5th Int Congr Pesticide Chemistry, Kyoto, Japan, vol 3. Pergamon, Oxford, pp 51-58

Ashton FM, Crafts AS (1981) Mode of action of herbicides, 2nd edn. Wiley-Interscience, New York

Bachthaler G, Kees H, Dinzenhofer B (1983) Die Ausbildung resistenter Linien von Ackerunkrautarten nach fortgesetzter Anwendung von Herbiziden, insbesondere von Triazinen. Gegenwärtiger Kenntnisstand über Ursachen und praktische Auswirkungen. Nachrichtenbl Dtsch Pflanzenschutzdienst (Braunschw) 35:161-168

Barton KA, Brill WJ (1983) Prospects in plant genetic engineering. Science (Wash DC) 219:671-676

Barton KA, Binns AN, Matzke AJM, Chilton MD (1983) Regeneration of intact tobacco plants containing full length copies of genetically engineered T-DNA, and transmission of T-DNA to R1 progeny. Cell 32:1033-1043

Barz W, Hösel W (1975) Metabolism of flavonoids. In: Harborne JB, Mabry TJ, Mabry H (eds) The Flavonoids. Chapman and Hall, London, pp 916-969

Bevan MW, Flavell RB, Chilton MD (1983) A chimaeric antibiotic resistance gene as a selectable marker for plant cell transformation. Nature (Lond) 304:184-187

Bollag JM (1974) Microbial transformation of pesticides. Adv Appl Microbiol 18:75-130

Caplan A, Herrera-Estrella L, Inze D, Van Haute E, Van Montagu M, Schell J, Zambryski P (1983) Introduction of genetic material into plant cells. Science (Wash DC) 222:815-821

Catterall FA, Williams PA (1971) Some properties of the naphthalene oxygenase from *Pseudomonas* sp. NCIB 9816. J Gen Microbiol 67:117-124

Comai L, Sen LC, Stalker DM (1983) An altered *aroA* gene product confers resistance to the herbicide glyphosate. Science (Wash DC) 221:370-371

Cripps RE, Roberts TR (1978) Microbial degradation of herbicides. In: Hill IR, Wright SJL (eds) Pesticide Microbiology. Academic, London, pp 669-730

Diesperger H, Sandermann H (1979) Soluble and microsomal glutathione S-transferase activities in pea seedlings (*Pisum sativum* L). Planta (Berl) 146:643-648

Döring HP, Freeling M, Hake et al. (1984) A *Ds*- mutation of the *Adh1* gene in *Zea mays* L. Mol Gen Genet 193:199-204

Don RH, Pemberton JM (1981) Properties of six pesticide degradation plasmids isolated from *Alcaligenes paradoxus* and *Alcaligenes eutrophus*. J Bacteriol 145:681-686

Drummond M (1983) Launching genes across phylogenetic barriers. Nature (Lond) 303:198-199

Duncan K, Lewendon A, Coggins JR (1984) The purification of 5-enolpyruvylshikimate 3-phosphate synthase from an overproducing strain of *Escherichia coli*. FEMS Microbial degradation of xenobiotics and recalcitrant compounds. Academic, London, pp 271-285

Eberspächer J, Lingens F (1981) Microbial degradation of the herbicide chloridazon. In: Leisinger T, Hütter R, Cook AM, Nüesch J (eds) Microbial degradation of xenobiotics and recalcitrant compounds. Academic, London, pp 271-285

Ensley BD, Ratzkin BJ, Osslund TD, Simon MJ, Wackett LP, Gibson DT (1983) Expression of naphthalene oxidation genes in *Escherichia coli* results in the biosynthesis of indigo. Science (Wash DC) 222:167-169

Faulkner JS (1982) Breeding herbicide-tolerant crop cultivars by conventional methods. In: LeBaron HM, Gressel J (eds) Herbicide resistance in plants. Wiley-Interscience, New York, pp 235-256

Fedtke C, Schmidt RR (1979) Characterization of the metamitron deaminating enzyme activity from sugar beet (*Beta vulgaris I*) leaves. Z Naturforsch Sect C Biosci 34:948-950

Fedtke C, Schmidt RR (1983) Behaviour of metribuzin in tolerant and susceptible soybean varieties. In: Miyamoto J, Kearney PC, Matsunaka S, Hutson DH, Murphy SD (eds) Pesticide chemistry. Human welfare and the environment. Proc 5th Int Congr Pesticide Chemistry, Kyoto, Japan, vol 3. Pergamon, Oxford, pp 177-182

Finn RK (1983) Use of specialized microbial strains in the treatment of industrial waste and in soil decontamination. Experientia (Basel) 39:1231-1236

Fraley RT, Rogers SG, Horsch et al. (1983) Expression of bacterial genes in plant cells. Proc Natl Acad Sci USA 80:4803-4807

Frear DS (1968) Herbicide metabolism in plants I. Purification and properties of UDP-glucose: arylamine N-glucosyl-transferase from soybean. Phytochemistry (Oxf) 7:381-390

Frear DS, Still GG (1968) The metabolism of 3,4-dichloropropionanilide in plants. Partial purification and properties of an aryl acylamidase from rice. Phytochemistry (Oxf) 7:913-920

Frear DS, Swanson HR (1973) Metabolism of substituted diphenylether herbicides in plants. I. Enzymatic cleavage of fluorodifen in peas (*Pisum sativum* L.). Pestic Biochem Physiol 3:473-482

Frear DS, Swanson HR, Tanaka FS (1972) Herbicide metabolism in plants. Recent Adv Phytochem 5:225-246

Gaynor JJ, Still CC (1983) Subcellular localization of rice leaf aryl acylamidase activity. Plant Physiol (Bethesda) 72:80-85

Gibson DT, Koch JR, Kallio RE (1968) Oxidative degradation of aromatic hydrocarbons by microorganisms. I. Enzymatic formation of catechol from benzene. Biochemistry 7:2653-2662

Grisebach H, Ebel J (1983) Phytoalexine und Resistenz von Pflanzen gegenüber Schadorganismen. Biologie In Unserer Zeit 13:129-136

Guddewar MB, Dauterman WC (1979) Purification and properties of a glutathione S-transferase from corn which conjugates s-triazine herbicides. Phytochemistry (Oxf) 18:735-740

Hahlbrock K, Chappell J, Scheel D (1984) Genes involved in resistance reactions in higher plants. Possible candidates for gene transfer? In: Starlinger P, Schell J (eds) Proc 35th Mosbach Colloquium. Springer, Berlin Heidelberg New York,

Hatzios KK, Penner D (1982) Metabolism of herbicides in higher plants. Burgess, Minneapolis

Harborne JB (1982) Introduction to ecological chemistry, 2nd edn. Academic, New York

Hedin PA (1982) New concepts and trends in pesticide chemistry. J Agric Food Chem 30:201-215

Hedin PA (ed) (1983) Plant resistance to insects. ACS Symp Ser 208. American Chemical Society, Washington DC

Herrera-Estrella L, De Block M, Messens E, Hernansteens JP, Van Montagu M, Schell J (1983a) Chimeric genes as dominant selectable markers in plant cells. EMBO J 2: 987-995

Herrera-Estrella L, Depicker A, Van Montagu M, Schell J (1983b) Expression of chimaeric genes transferred into plant cells using a Ti-plasmid-derived vector. Nature (Lond) 303:209-213

Hirschberg J, McIntosh L (1983) Molecular basis of herbicide resistance in *Amaranthus hybridus*. Science (Wash DC) 222:1346-1349

Hoagland RE (1978) Isolation and some properties of an aryl acylamidase from red rice, *Oryza sativa* L, that metabolizes 3',4'-dichloropropionanilide. Plant Cell Physiol 19:1019-1027

Horsch RB, Fraley RT, Rogers SG, Sanders PR, Lloyd A, Hoffmann N (1984) Inheritance of functional foreign genes in plants. Science (Wash DC) 223:496-498

Johnson LM, Talbot HW (1983) Detoxification of pesticides by microbiol enzymes. Experientia (Basel) 39:1236-1246

Karns JS, Kilbane JJ, Duttagupta S, Chakrabarty AM (1983) Metabolism of halophenols by 2,4,5-trichlorophenoxyacetic acid-degrading *Pseudomonas cepacia*. Appl Environ Microbiol 46:1176-1181

Kaufman DD, Kearney PC (1976) Microbial transformation in the soil. In: Audus LJ (ed) Herbicides. Physiology, biochemistry, ecology, vol 2. Academic, London, pp 29-64

Kearney PC (1983) Principles of biological pesticide degradation. Abstracts ACS National Meeting No 186, Abstract PEST 0005

Kilbane JJ, Chatterjee DK, Karns JS, Kellogg ST, Chakrabarty AM (1982) Biodegradation of 2,4,5-trichlorophenoxyacetic acid by a pure culture of *Pseudomonas cepacia*. Appl Environ Microbiol 44:72-78

Kobayashi H, Rittmann BE (1982) Microbial removal of hazardous organic compounds. Environ Sci Technol 16:170A-183A

Krell HW, Sandermann H (1984) Plant biochemistry of xenobiotics. Purification and properties of a wheat esterase hydrolyzing the plasticizer chemical, bis-(2-ethylhexyl)-phthalate Eur. J. Biochem. 143, 57-62

Lamoureux GL, Frear DS (1979) Pesticide metabolism in higher plants. In vitro enzyme studies. In: Paulson GD, Frear DS, Marks EP (eds) Xenobiotic metabolism. In vitro methods. ACS Symp Ser 97. American Chemical Society, Washington DC, pp 77-128

LeBaron HM, Gressel J (eds) (1982) Herbicide resistance in plants. Wiley-Interscience, New York

McClintock B (1951) Chromosome organization and genic expression. Cold Spring Harbor Symp Quant Biol 16:13-47

Mendel G (1865/1866) Versuche über Pflanzen-Hybriden. Verh Naturforsch Ver Brünn 4: 3-47

Menn JJ (1980) Contemporary frontiers in chemical pesticide research. J Agric Food Chem 28:2-8

Meredith CP, Carlson PS (1982) Herbicide resistance in plant cell cultures. In: LeBaron HM, Gressel J (eds) Herbicide resistance in plants. Wiley-Interscience, New York, pp 275-291

Moore JK, Braymer HD, Larson AD (1983) Isolation of a *Pseudomonas* sp which utilizes the phosphonate herbicide glyphosate. Appl Environ Microbiol 46:316-320

Motosugi K, Soda K (1983) Microbial degradation of synthetic organochlorine compounds. Experientia (Basel) 39:1214-1220

Mousdale DM, Coggins JR (1984) Purification and properties of 5-enolpyruvylshikimate 3-phosphate synthase from seedlings of *Pisum sativum* L. Planta (Berl) 160:78-83

Mozer TJ, Tiemeier DC, Jaworski EG (1983) Purification and characterization of corn glutathione S-transferase. Biochemistry 22:1068-1072

Negrutiu I, Jacobs M, Caboche M (1984) Advances in somatic cell genetics of higher plants-the protoplast approach in basic studies on mutagenesis and isolation of biochemical mutants. Theor Appl Genet 67:289-304

Osterman JC, Schwartz D (1981) Analysis of a controlling element mutation at the *Adh* locus of maize. Genetics 99:267-273

Pflanzenschutzmittel-Höchstmengenverordnung (1982) Bundesgesetzblatt, part I, No 22, 29.6.1982, p 745

Rogers SG, Brand LA, Holder SB, Sharps ES, Brackin MJ (1983) Amplification of the *aroA* gene from *Escherichia coli* results in tolerance to the herbicide glyphosate. Appl Environ Microbiol 46:37-43

Sandermann H (1982) Metabolism of environmental chemicals. A comparison of plant and liver enzyme systems. In: Klekowski EJ (ed) Environmental mutagenesis, carcinogenesis and plant biology, vol I. Praeger Scientific, New York, pp 1-32

Sandermann H (1984) Umweltchemikalien in Pflanzen. Umschau 84(4):115-118

Sandermann H, Scheel D, v.d. Trenck T (1983) Metabolism of environmental chemicals by plants - Copolymerization into lignin. J Appl Polymer Sci Appl Polymer Symp 37:407-420

Sandermann H, Scheel D, v.d. Trenck T (1984) Use of plant cell cultures to study the metabolism of environmental chemicals. Ecotoxicol Environ Safety 8, 167-182

Schell JS (1983) Leben mit fremden Genen. Natürliche und künstliche Übertragung genetischer Programme zwischen nichtverwandten Pflanzen. Naturwiss Rundsch 36:254-260

Schildknecht H (1981) Reiz- und Abwehrstoffe höherer Pflanzen - ein chemisches Herbarium. Angew Chem Int Ed Engl 93:164-183

Schlösser E (1983) Allgemeine Phytopathologie. Thieme, Stuttgart

Schmitt R, Kaul J, v.d. Trenck T, Schaller E, Sandermann H (1985) β-D-Glucosyl and O-malonyl-β-D-glucosyl conjugates of pentachlorophenol in soybean and wheat. Identification and enzymatic synthesis Pesticide Biochem. Physid., in press

Schulz A, Sost D, Amrhein N (1984) Insensitivity of 5-enolpyruvylshikimic acid 3-phosphate synthase to glyphosate confers resistance to this herbicide in a strain of *Aerobacter aerogenes*. Arch Microbiol 137:121-123

Senior E, Bull AT, Slater JH (1976) Enzyme evolution in a microbial community growing on the herbicide Dalapon. Nature (Lond) 263:476-479

Shimabukuro RH, Lamoureux GL, Frear DS (1982) Pesticide metabolism in plants. Reactions and mechanisms. In: Matsumura F, Krishna Murti CR (eds) Biodegradation of pesticides. Plenum, New York, pp 21-66

Stanlake GJ, Finn RK (1982) Isolation and characterization of a pentachlorophenol-degrading bacterium. Appl Environ Microbiol 44:1421-1427

Steinrücken HC, Amrhein N (1980) The herbicide glyphosate is a potent inhibitor of 5-enolpyruvylshikimic acid 3-phosphate synthase. Biochem Biophys Res Commun 94: 1207-1212

Stumpf PK, Conn EE (eds) (1981) The Biochemistry of Plants vol 7. Secondary plant products. Academic, New York

Sun M (1983) EPA revs up to regulate biotechnology. Science (Wash DC) 222:823-824

Trenck T v.d., Sandermann H (1980) Oxygenation of benzo [α] pyrene by plant microsomal fractions. FEBS Lett 119:227-231

Wain RL, Smith MS (1976) Selectivity in relation to metabolism. In: Audus LJ (ed) Herbicides. Physiology, biochemistry, ecology, Vol 2. Academic, New York, pp 279-302

Foot and Mouth Disease Virus: Genome Organization, Antigenic Variation, and New Approaches to a Safe Vaccine

H. SCHALLER[1], S. FORSS[1], K. STREBEL[1], E. BECK[1], H. O. BÖHM[2], J. LEBAN[3], and E. PFAFF[1]

1 Introduction

Foot and Mouth disease viruses (FMDV), or aphthoviruses, are picornaviruses which are the causative agents of an aggressive and economically important disease of clover-footed farm animals. Their virion contains a single-stranded RNA genome of about 8 kb with a small protein (VPg) covalently attached to its 5´-end, an internal poly (C) tract, and a poly(A) sequence at the 3´-end (Fig. 1). This RNA is of positive polarity and can act directly as a messenger RNA. Protein synthesis involves post-translational cleavage of a 260 K polyprotein which is encoded between a major translation initiation site next to the poly(C) tract and the 3´-end of the RNA. This mode of protein synthesis is common to all picornaviruses and makes them interesting model systems for studying the mechanisms of translation initiation and of protein maturation by specific proteolytic cleavages. Another subject of intensive research concerns the molecular mechanism of the rapid variation of neutralizing antigenic determinants on the FMDV capsid proteins which causes considerable problems in vaccination programs.

2 Nucleotide Sequence and Gene Expression

The FMDV genome is divided by the poly(C) tract into two parts of different size and function (Fig. 1). The small (S) segment to the 5´-side is probably involved only in initiation of viral RNA replication, and the large (L) segment 3´of the poly(C) contains all the protein coding information (Sangar et al. 1980). We have cloned and sequenced a set of overlapping cDNA clones from FMDV O1K comprising a continous sequence of 7915 nucleotides starting with 11 C residues from the poly(C) tract and ending in the 3´-terminal poly(A) sequence (Forss et al. 1984). This sequence of 7802 nonhomopolymeric nucleotides (Fig. 2) represents the complete primary structure of the L segment. Assuming additional 150 bases for poly(C) and 400 for the S segment (Harris et al. 1980) this corresponds to 94% of an FMDV genome of 8500 nucleotides.

From the kinetics of the appearance of FMDV-specific gene products it has been predicted that FMDV-RNA contains a single long open translational reading frame for a polyprotein, which in turn is a precursor for all gene products (Fig. 1). Our sequence reveals an open reading frame of 2332 codons between positions 766 and 7800 with a first possible initiation codon for the polyprotein at position 805 (see Fig. 2).

[1]Microbiology, University of Heidelberg and ZMBH, Im Neuenheimer Feld 230, 6900 Heidelberg, FRG

[2]Federal Research Institute for Animal Virus Diseases, Paul-Ehrlich-Strasse 28, 7400 Tübingen, FRG

[3]BIOGEN S.A., 1221 Geneva, Switzerland

35. Colloquium - Mosbach 1984
The Impact of Gene Transfer Techniques in Eukaryotic Cell Biology

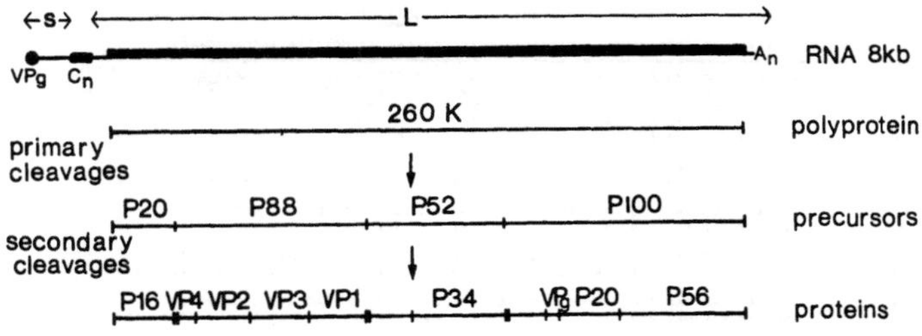

Fig. 1. Physical map of the FMDV genome and cleavage pathways of the viral proteins. *Filled circle* at the 5′-end of the genome symbolizes the viral genome-linked protein (*VPg*). *Cn* and *An* illustrate the internal poly(C) tract and the 3′poly(A) tail. The *L* (large) segment 3′ from the poly(C) encodes a polyprotein of about 2400 amino acids which is processed by successive protease cleavages into four precursors and further into the mature proteins. These are ordered in the lower line according to Harris et al. (1981)

Polyprotein synthesis also starts at a second in phase AUG in position 889 (Beck et al. 1983a). Usually the 5′-proximal AUG is used to initiate translation in an eukaryotic mRNA, which suggests that the ribosomes first recognize the capped 5′-end of the RNA and then traverse downstream until an AUG is encountered (Kozak 1983). Clearly this model is not applicable to FMDV, since positions 805 and 889 are the ninth and tenth AUG's downstream from the poly(C) (Fig. 2). However, these two start sites differ from other AUG codons in the FMDV sequence in that they are preceded at a short distance by a stretch of 11 pyrimidines, which are interrupted by no more than one purine residue. These pyrimidine runs, which are also present upstream from the presumed P10 gene (see below) and the translational start site in the poliovirus genome (Fig. 3), may be important for the recognition by ribosomes or initiation factors of an uncapped mRNA with a long untranslated leader sequence. They also show significant base complementarity to the 3′-end of the 18S ribosomal RNA from eukaryotes.

3 The Deduced Protein Map

The long open reading frame encodes a polypeptide with a maximal size of 2332 amino acids and a calculated molecular weight of 258.9 K in excellent agreement with the 260 K determined experimentally for the picornaviral polyproteins (Summers et al. 1972). As mentioned before, the nucleotide sequence suggests that two sites for intiation of translation may be utilized in FMDV, generating two closely related "leader" proteins, L and L′ (P20a and P16) which differ by 28 amino acids at their N-terminus and have molecular weights of 24.4 K and 21.2 K, respectively. Recent attempts to express the FMDV L gene in *E. coli* indicate that its encoded amino acid sequence possesses a protease activity capable to cleave the L/P1 junction from a larger L/P1 precursor (Strebel et al. 1984).

The primary product following the leader protein in the polyprotein is P1 (M.W. 79.3 K), formerly called P88, the precursor of the capsid proteins which are arranged in the order VP4-VP2-VP3-VP1 (Sangar 1979). Protein data from FMDV O1K (Strohmaier 1978; Kurz et al. 1981, amino acids underlined in Fig. 2) unambiguously locate the limits of VP1, VP2, and VP3. These capsid proteins have a large fraction of charged amino acids (about 20%). In contrast to VP1-VP3, no sequence information

```
92   (C)11 AAGTTTTACCGTCGTTCCCGACGTAAAAGGGAGGTAACCACAAGCTTGAAACCGTCCGGCCCGACGTAAAAGGGTGGTAACCACAAGCTTACTGCCGTCTTTCCCGACGTTAAA

208  GGGATGAAACCACAAGACTTACCTTCGCTCGGAAGTAAAACGACAAACACACACAGTTTTGCCCGTTTTCATGAGAAATGGGACGTCTGCGCACGAAACGCGCCGTCGCTTGAGGAGGAC
        MetLysProGlnAspLeuProSerLeuGlySerLysThrThrAsnThrHisSerPheAlaArgPheHisGluLysTrpAspValCysAlaArgAsnAlaProSerLeuGluGluAsp

326  TTGTACAAACACGATCTATGCAGGTTTCCCCAACTGACACAAACCGTGCAACTTGAAACTCCGCCTGGTCTTTCCAGGTCTAGAGGGGTAACATTTTGTACTGTGTTTGACTCCACGCTC
     LeuTyrLysHisAspLeuCysArgPheProGlnLeuThrGlnThrValGlnLeuGluThrProProGlyLeuSerArgSerArgGlyValThrPheCysThrValPheAspSerThrLeu

446  GATCCACTAGCGAGTGTTAGTAGCGGTACTGCTGTCTCGTAGCGGAGCATGTTGGCCGTGGGAACACCTCCTTGGTAACAAGGACCCACGGGGCCGAAAGCCATGTCCTAACGGACCCAA
     AspProLeuAlaSerValSerSerGlyThrAlaValSer***

566  CATGTGTGCAACCCCAGCACGGCAGCTTTACTGTGAAACCCACTTCAAGGTGACATTGATACTGGTACTCAAACACTGGTGACAGGCTAAGGATGCCCTTCAGGTACCCCGAGGTAACAA

686  GCGACACTCGGGATCTGAGAAGGGGACTGGGACTTCTTTAAAGTGCCCAGTTTAAAAAGCTTCTACGCCTGAATAGGTGACCGGAGGCCGGCACCTTTCCTTTTATAACCACTGAACAC

P20a 805  ATG AAT ACA ACT GAC TGT TTT ATC GCT TTG GTA CAG GCT ATC AGA GAG ATT AAA GCA CTT TTT CTA TCA CGC ACC ACA GGG AAA ATG GAA        P20a
(L)       Met Asn Thr Thr Asp Cys Phe Ile Ala Leu Val Gln Ala Ile Arg Glu Ile Lys Ala Leu Phe Leu Ser Arg Thr Thr Gly Lys Met Glu   30   P16

895  CTG ACA CTG TAC AAC GGT GAG AAG AAG ACC TTT TAC TCC AGG CCC AAC AAC CAC GAC AAC TGC TGG TTG AAC GCC ATC CTC CAG TTG TTC
     Leu Thr Leu Tyr Asn Gly Glu Lys Lys Thr Phe Tyr Ser Arg Pro Asn Asn His Asp Asn Cys Trp Leu Asn Ala Ile Leu Gln Leu Phe   60

985  AGG TAC GTT GAA GAA CCA TTC TTC GAC TGG GTC TAC AGT TCG CCT GAG AAC CTC ACG CTT GAA GCC ATC AAG CAG TTG GAG GAT CTC ACA
     Arg Tyr Val Glu Glu Pro Phe Phe Asp Trp Val Tyr Ser Ser Pro Glu Asn Leu Thr Leu Glu Ala Ile Lys Gln Leu Glu Asp Leu Thr   90

1075 GGA CTT GAA CTG CAT GAG GGT GGA CCA CCT GCT CTC GTG ATC TGG AAC ATC AAG CAC TTG CTC CAC ACC GGC ATC GGC ACC GCC TCG CGA
     Gly Leu Glu Leu His Glu Gly Gly Pro Pro Ala Leu Val Ile Trp Asn Ile Lys His Leu Leu His Thr Gly Ile Gly Thr Ala Ser Arg   120

1165 CCC AGC GAG GTG TGC ATG GTG GAT GGT ACG GAC ATG TGC TTG GCT GAT TTC CAT GCT GGC ATT TTC CTT AAG GGG CAA GAA CAC GCT GTG
     Pro Ser Glu Val Cys Met Val Asp Gly Thr Asp Met Cys Leu Ala Asp Phe His Ala Gly Ile Phe Leu Lys Gly Gln Glu His Ala Val   150

1255 TTT GCG TGT GTC ACC TCC AAC GGG TGG TAC GCG ATT GAC GAT GAG GAC TTC TAC CCC TGG ACG CCG GAC CCG TCC GAC GTT CTG GTG TTT
     Phe Ala Cys Val Thr Ser Asn Gly Trp Tyr Ala Ile Asp Asp Glu Asp Phe Tyr Pro Trp Thr Pro Asp Pro Ser Asp Val Leu Val Phe   180

1345 GTC CCG TAC GAT CAA GAA CCA CTC AAC GGG GAA TGG AAA GCC AAG GTT CAA CGC AAG CTC AAA GGG GCT GGA CAA TCC AGT CCA GCG ACC
     Val Pro Tyr Asp Gln Glu Pro Leu Asn Gly Glu Trp Lys Ala Lys Val Gln Arg Lys Leu Lys Gly Ala Gly Gln Ser Ser Pro Ala Thr   210

1435 GGC TCG CAG AAC CAA TCT GGC AAT ACT GGC AGC ATA ATA AAC AAC TAC TAC ATG CAG CAG TAT CAA AAC TCC ATG GAC ACA CAG CTT GGT        VP4
     Gly Ser Gln Asn Gln Ser Gly Asn Thr Gly Ser Ile Ile Asn Asn Tyr Tyr Met Gln Gln Tyr Gln Asn Ser Met Asp Thr Gln Leu Gly   240

P88 1525 GAC AAC GCA ATC AGT GGA GGC TCT AAC GAG GGC TCC ACC GAC ACA ACC TCC ACC CAC ACA ACC AAC ACC CAG AAC AAT GAC TGG TTC TCC
(P1)     Asp Asn Ala Ile Ser Gly Gly Ser Asn Glu Gly Ser Thr Asp Thr Thr Ser Thr His Thr Thr Asn Thr Gln Asn Asn Asp Trp Phe Ser   270

1615 AAA CTT GCC AGC TCT GCT TTC AGC GGT CTT TTC GGC GCT CTT CTC GCC GAC AAG AAG ACA GAG GAG ACC ACT CTC CTC GAA GAC CGC ATC        VP2
     Lys Leu Ala Ser Ser Ala Phe Ser Gly Leu Phe Gly Ala Leu Leu Ala Asp Lys Lys Thr Glu Glu Thr Thr Leu Leu Glu Asp Arg Ile   300

1705 CTC ACC ACC CGT AAC GGC CAC ACC ACG TCG ACA ACC CAG TCA AGC GTT GGA GTC ACA TAC GGG TAC GCA ACA GCT GAA GAT TTT GTG AGC
     Leu Thr Thr Arg Asn Gly His Thr Thr Ser Thr Thr Gln Ser Ser Val Gly Val Thr Tyr Gly Tyr Ala Thr Ala Glu Asp Phe Val Ser   330

1795 GGA CCG AAC ACT TCC GGT CTC GAA ACC AGA GTT GTG CAG GCA GAA CGG TTT TTC AAA ACC CAC CTC TTC GAC TGG GTC ACC AGT GAC TCA
     Gly Pro Asn Thr Ser Gly Leu Glu Thr Arg Val Val Gln Ala Glu Arg Phe Phe Lys Thr His Leu Phe Asp Trp Val Thr Ser Asp Ser   360

1885 TTC GGA CGT TGC CAC CTC CTG GAA CTC CCG ACC GAC CAC AAA GGT GTC TAC GGC AGC CTG ACT GAC TCG TAT GCA TAT ATG AGA AAC GGC
     Phe Gly Arg Cys His Leu Leu Glu Leu Pro Thr Asp His Lys Gly Val Tyr Gly Ser Leu Thr Asp Ser Tyr Ala Tyr Met Arg Asn Gly   390

1975 TGG GAT GTC GAG GTC ACC GCG GTT GGC AAC CAG TTC AAC GGA GGG TGC CTT CTG GTC GCA ATG GTA CCA GAG CTT TAT TCT ATC CAA AAG
     Trp Asp Val Glu Val Thr Ala Val Gly Asn Gln Phe Asn Gly Gly Cys Leu Leu Val Ala Met Val Pro Glu Leu Tyr Ser Ile Gln Lys   420

2065 AGG GAA CTG TAC CAG CTC ACA CTT TTC CCT CAC CAG TTC ATC AAC CCA CGC ACG AAC ATG ACT GCG CAC ATC ACA GTG CCC TTT GTT GGC
     Arg Glu Leu Tyr Gln Leu Thr Leu Phe Pro His Gln Phe Ile Asn Pro Arg Thr Asn Met Thr Ala His Ile Thr Val Pro Phe Val Gly   450

2155 GTC AAC CGC TAC GAC CAG TAC AAG GTT CAC AAG CCT TGG ACC CTT GTG GTT ATG GTT GTA GCC CCT CTG ACC GTC AAC ACT GAA GGT GCC
     Val Asn Arg Tyr Asp Gln Tyr Lys Val His Lys Pro Trp Thr Leu Val Val Met Val Val Ala Pro Leu Thr Val Asn Thr Glu Gly Ala   480

2245 CCT CAG ATC AAG GTG TAT GCC AAC ATT GCC CCA ACC AAC GTG CAC GTC GCG GGT GAG TTT CCT TCC AAG GAG GGA ATA TTC CCC GTG GCC        VP3
     Pro Gln Ile Lys Val Tyr Ala Asn Ile Ala Pro Thr Asn Val His Val Ala Gly Glu Phe Pro Ser Lys Glu Gly Ile Phe Pro Val Ala   510

2335 TGT AGC GAC GGC TAT GGT GGC CTG GTG ACC ACG GAC CCG AAG ACG GCT GAC CCC GTT TAT GGG AAA GTG TTC AAC CCC CCC CGC AAC CAG
     Cys Ser Asp Gly Tyr Gly Gly Leu Val Thr Thr Asp Pro Lys Thr Ala Asp Pro Val Tyr Gly Lys Val Phe Asn Pro Pro Arg Asn Gln   540

2425 TTG CCG GGG CGT TTT ACC AAC CTC CTT GAT GTG GCT GAG GCA TGC CCG ACG TTT CTG CGC TTC GAG GGT GGC GTA CCG TAC GTG ACC ACG
     Leu Pro Gly Arg Phe Thr Asn Leu Leu Asp Val Ala Glu Ala Cys Pro Thr Phe Leu Arg Phe Glu Gly Gly Val Pro Tyr Val Thr Thr   570

2515 AAA ACG GAC TCG GAC AGG GTG CTT GCT CAG TTT GAC ATG TCT TTG GCA GCA AAA CAA ATG TCA AAC ACC TTC CTC GCA GGT CTT GCG CAG
     Lys Thr Asp Ser Asp Arg Val Leu Ala Gln Phe Asp Met Ser Leu Ala Ala Lys Gln Met Ser Asn Thr Phe Leu Ala Gly Leu Ala Gln   600

2605 TAC TAC ACA CAG TAC AGT GGC ACC ATC AAC CTG CAC TTC ATG TTC ACA GGA CCC ACT GAC GCG AAG GCG CGT TAC ATG GTT GCC TAC GCC
     Tyr Tyr Thr Gln Tyr Ser Gly Thr Ile Asn Leu His Phe Met Phe Thr Gly Pro Thr Asp Ala Lys Ala Arg Tyr Met Val Ala Tyr Ala   630

2695 CCA CCA GGC ATG GAG CCG CCC AAG ACA CCT GAG GCG GCC GCG CAC TGC ATT CAT GCT GAA TGG GAC ACT GGG TTA AAC TCA AAG TTT ACT
     Pro Pro Gly Met Glu Pro Pro Lys Thr Pro Glu Ala Ala Ala His Cys Ile His Ala Glu Trp Asp Thr Gly Leu Asn Ser Lys Phe Thr   660

2785 TTT TCC ATC CCC TAC CTC TCG GCC GCC GAT TAC GCG TAC ACC GCG TCT GGC GTG GCC GAG ACC ACA AAT GTG CAG GGA TGG GTC TGC TTG
     Phe Ser Ile Pro Tyr Leu Ser Ala Ala Asp Tyr Ala Tyr Thr Ala Ser Gly Val Ala Glu Thr Thr Asn Val Gln Gly Trp Val Cys Leu   690

2875 TTT CAA ATT ACA CAT GGC AAG GCC GAC GGC GAC GCT CTG GTC GTA CTG GCT AGT GCT GGT AAA GAC TTT GAG CTA AGG CTG CCG GTG GAC
     Phe Gln Ile Thr His Gly Lys Ala Asp Gly Asp Ala Leu Val Val Leu Ala Ser Ala Gly Lys Asp Phe Glu Leu Arg Leu Pro Val Asp   720

2965 GCC CGT GCG GAA ACC ACT TCT GCG GGC GAG TCA GCG GAT CCT GTC ACC ACC ACC GTT GAA AAC TAC GGT GGC GAA ACA CAG ATC CAG AGG        VP1
     Ala Arg Ala Glu Thr Thr Ser Ala Gly Glu Ser Ala Asp Pro Val Thr Thr Thr Val Glu Asn Tyr Gly Gly Glu Thr Gln Ile Gln Arg   750

3055 CGC CAA CAC ACG GAC GTC TCG TTC ATC ATG GAC AGA TTT GTG AAG GTG ACA CCG CAA AAC CAA ATT AAC ATT TTG GAC CTC ATG CAG ATT
     Arg Gln His Thr Asp Val Ser Phe Ile Met Asp Arg Phe Val Lys Val Thr Pro Gln Asn Gln Ile Asn Ile Leu Asp Leu Met Gln Ile   780

3145 CCA TCA CAC ACT TTG GTG GGA GCA CTC CTA CGC GCG TCC ACT TAC TAC TTC TCT GAC TTG GAG ATA GCA GTA AAA CAC GAG GGA GAC CTC
     Pro Ser His Thr Leu Val Gly Ala Leu Leu Arg Ala Ser Thr Tyr Tyr Phe Ser Asp Leu Glu Ile Ala Val Lys His Glu Gly Asp Leu   810

3235 ACC TGG GTT CCA AAT GGA GCG CCC GAA AAG GCG TTG GAC AAC ACC ACC AAC CCA ACT GCT TAC CAC AAG GCA CCA CTC ACC CGG CTT GCC
     Thr Trp Val Pro Asn Gly Ala Pro Glu Lys Ala Leu Asp Asn Thr Thr Asn Pro Thr Ala Tyr His Lys Ala Pro Leu Thr Arg Leu Ala   840

3325 CTG CCC TAC ACT GCG CCC CAC CGC GTG TTG GCA ACC GTG TAC AAC GGT GAG TGC AGG TAC AAC AGA AAT GCT GTG CCC AAC TTG AGA GGT
     Leu Pro Tyr Thr Ala Pro His Arg Val Leu Ala Thr Val Tyr Asn Gly Glu Cys Arg Tyr Asn Arg Asn Ala Val Pro Asn Leu Arg Gly   870

3415 GAC CTT CAG GTG TTG GCT CAA AAG GTG GCA CGG ACG CTG CCT ACC TCC TTC AAC TAC GGT GCC ATC AAA GCG ACC CGG GTC ACC GAG TTG
     Asp Leu Gln Val Leu Ala Gln Lys Val Ala Arg Thr Leu Pro Thr Ser Phe Asn Tyr Gly Ala Ile Lys Ala Thr Arg Val Thr Glu Leu   900

3505 CTT TAC CGG ATG AAG AGG GCC GAA ACA TAC TGT CCA AGG CCC TTG CTG GCA ATC CAC CCA ACT GAA GCC AGA CAC AAA CAG AAA ATT GTG
     Leu Tyr Arg Met Lys Arg Ala Glu Thr Tyr Cys Pro Arg Pro Leu Leu Ala Ile His Pro Thr Glu Ala Arg His Lys Gln Lys Ile Val   930

3595 GCA CCG GTG AAA CAG ACT TTG AAT TTT GAC CTT CTC AAG TTG GCG GGA GAC GTC GAG TCC AAC CCT GGG CCC TTC TTT TTC TCC GAC GTT        P12
     Ala Pro Val Lys Gln Thr Leu Asn Phe Asp Leu Leu Lys Leu Ala Gly Asp Val Glu Ser Asn Pro Gly Pro Phe Phe Phe Ser Asp Val   960

P52 3685 AGG TCG AAC TTC TCC AAA CTG GTG GAA ACC ATC AAC CAG ATG CAG GAG GAC ATG TCA ACA AAA CAC GGG CCT GAC TTT AAC CGG TTA GTG
(P2)     Arg Ser Asn Phe Ser Lys Leu Val Glu Thr Ile Asn Gln Met Gln Glu Asp Met Ser Thr Lys His Gly Pro Asp Phe Asn Arg Leu Val   990

3775 TCC GCA TTT GAG GAG TTG GCC ATT GGA GTG AAA GCC ATC AGA ACC GGT CTC GAC GAA GCC AAA CCC TGG TAC AAG CTT ATC AAG CTC CTA
     Ser Ala Phe Glu Glu Leu Ala Ile Gly Val Lys Ala Ile Arg Thr Gly Leu Asp Glu Ala Lys Pro Trp Tyr Lys Leu Ile Lys Leu Leu   1020

3865 AGC CGC CTG TCG TGC ATG GCC GCT GTG GCA GCA CGG TCC AAG GAC CCA GTC CTT GTG GCC ATC ATG CTG GCC GAC ACC GGT CTC GAG ATT
     Ser Arg Leu Ser Cys Met Ala Ala Val Ala Ala Arg Ser Lys Asp Pro Val Leu Val Ala Ile Met Leu Ala Asp Thr Gly Leu Glu Ile   1050

3955 CTG GAC AGC ACC TTC GTC GTG AAG AAG ATC TCC GAC TCG CTC TCC AGT CTC TTT CAC GTG CCG GCC CCC GTC TTC AGT TTC GGA GCA CCG
     Leu Asp Ser Thr Phe Val Val Lys Lys Ile Ser Asp Ser Leu Ser Ser Leu Phe His Val Pro Ala Pro Val Phe Ser Phe Gly Ala Pro   1080

4045 GTC CTG TTG GCC GGG TTG GTC AAA GTT GCC TCG AGT TTC TTC CGG TCC ACA CCC GAA GAC CTT GAG AGA GCA GAG AAA CAG CTC AAA GCA        P34
     Val Leu Leu Ala Gly Leu Val Lys Val Ala Ser Ser Phe Phe Arg Ser Thr Pro Glu Asp Leu Glu Arg Ala Glu Lys Gln Leu Lys Ala   1110

4135 CGT GAC ATC AAC GAC ATC TTC GCC ATT CTC AAG AAC GGC GAG TGG CTG GTC AAA CTG ATC CTT GCC ATC CGC GAC TGG ATT AAG GCT TGG
     Arg Asp Ile Asn Asp Ile Phe Ala Ile Leu Lys Asn Gly Glu Trp Leu Val Lys Leu Ile Leu Ala Ile Arg Asp Trp Ile Lys Ala Trp   1140

4225 ATC GCC TCA GAA GAG AAG TTT GTC ACC ATG ACA GAC TTG GTG CCT GGC ATC CTT GAA AAG CAG CGG GAC CTG AAC GAC CCG AGC AAG TAC
     Ile Ala Ser Glu Glu Lys Phe Val Thr Met Thr Asp Leu Val Pro Gly Ile Leu Glu Lys Gln Arg Asp Leu Asn Asp Pro Ser Lys Tyr   1170

4315 AAG GAA GCC AAG GAG TGG CTC GAC AAC GCG CGC CAA GCG TGT TTG AAG AGC GGG AAC GTC CAC ATT GCC AAC CTG TGC AAA GTG GTC GCA
     Lys Glu Ala Lys Glu Trp Leu Asp Asn Ala Arg Gln Ala Cys Leu Lys Ser Gly Asn Val His Ile Ala Asn Leu Cys Lys Val Val Ala   1200

4405 CCA GCA CCC AGC AAG TCG AGG CCC GAA CCC GTG GTT GTT TGC CTC CGC GGC AAA TCT GGC CAG GGC AAG AGC TTC CTT GCA AAC GTG CTT
     Pro Ala Pro Ser Lys Ser Arg Pro Glu Pro Val Val Val Cys Leu Arg Gly Lys Ser Gly Gln Gly Lys Ser Phe Leu Ala Asn Val Leu   1230

4495 GCA CAG GCA ATT TCC ACC CAC TTC ACC GGC AGA ATC GAC TCA GTG TGG TAC TGC CCA CCT GAC CCT GAC CAC TTC GAC GGT TAC AAC CAG
     Ala Gln Ala Ile Ser Thr His Phe Thr Gly Arg Ile Asp Ser Val Trp Tyr Cys Pro Pro Asp Pro Asp His Phe Asp Gly Tyr Asn Gln   1260

4585 CAA ACC GTT GTT GTG ATG GAT GAT TTG GGC CAG AAC CCT GAC GGC AAG GAC TTC AAA TAC TTT GCC CAA ATG GTC TCG ACC ACA GGG TTT
     Gln Thr Val Val Val Met Asp Asp Leu Gly Gln Asn Pro Asp Gly Lys Asp Phe Lys Tyr Phe Ala Gln Met Val Ser Thr Thr Gly Phe   1290

4675 ATC CCG CCC ATG GCA TCA CTC GAG GAC AAA GGT AAA CCT TTC AAC AGC AAA GTC ATC ATC GCG ACC ACC AAC TTG TAC TCG GGC TTC ACC
     Ile Pro Pro Met Ala Ser Leu Glu Asp Lys Gly Lys Pro Phe Asn Ser Lys Val Ile Ile Ala Thr Thr Asn Leu Tyr Ser Gly Phe Thr   1320

4765 CCG AGG ACC ATG GTA TGT CCC GAC GCA CTG AAC CGG AGG TTT CAC TTT GAC ATC GAT GTG AGT GCT AAG GAT GGG TAC AAA ATT AAC AGC
     Pro Arg Thr Met Val Cys Pro Asp Ala Leu Asn Arg Arg Phe His Phe Asp Ile Asp Val Ser Ala Lys Asp Gly Tyr Lys Ile Asn Ser   1350

4855 AAA TTG GAC ATT ATC AAA GCA CTC GAA GAC ACC CAC GCC AAC CCA GTG GCA ATG TTT CAA TAC GAC TGT GCC CTT CTC AAC GGC ATG GCC
     Lys Leu Asp Ile Ile Lys Ala Leu Glu Asp Thr His Ala Asn Pro Val Ala Met Phe Gln Tyr Asp Cys Ala Leu Leu Asn Gly Met Ala   1380

4945 GTT GAA ATG AAG AGA ATG CAA CAA GAC ATG TTC AAG CCT CAA CCA CCC CTC CAG AAT GTG TAC CAG CTT GTT CAG GAG GTG ATC GAT CGG
     Val Glu Met Lys Arg Met Gln Gln Asp Met Phe Lys Pro Gln Pro Pro Leu Gln Asn Val Tyr Gln Leu Val Gln Glu Val Ile Asp Arg   1410
```

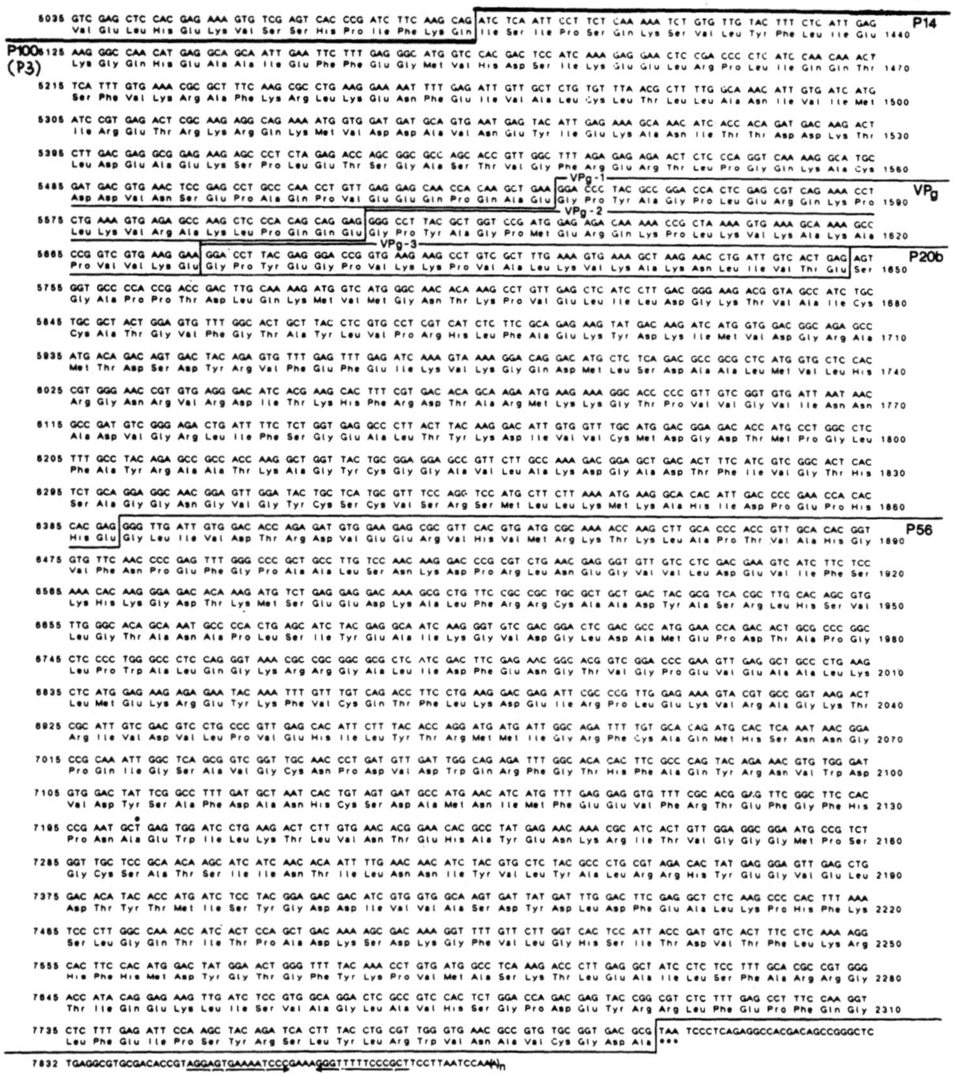

Fig. 2a,b. The nucleotide sequence and encoded information of the L segment of FMDV RNA strain O1K (U residues are shown as T). The amino acid sequence corresponds to the total polyprotein of 2332 amino acids. A translated amino acid sequence is also shown for a hypothetical small polypeptide encoded upstream from the polyprotein. The numbers to the right represent amino acid positions in the polyprotein. Predicted limits of the primary precursors are indicatd to the *left*, and those of the stable viral polypeptides to the *right*. *Dashed lines* represent borders where the supporting data are weak. Underlined amino acids have been determined experimentally for this virus strain (Strohmaier 1978; Kurz et al. 1981). The nucleotide positions to the left use the BamH1 site at position 3000 as a reference (Küpper et al. 1981) AUG codons in the 5′-region are *underlined*, a palindromic sequence at the 3′-end is *underlined by arrows*

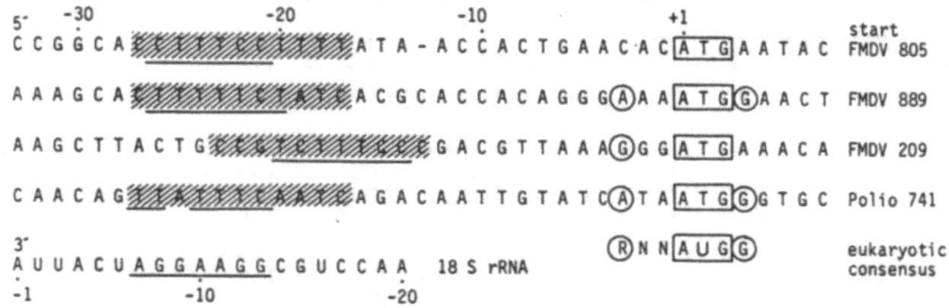

Fig. 3. Comparison of the close surroundings of the potential initiation sites for translation in the genomes of FMDV and poliovirus. The map positions of the start sites, given to the right, refer to position 1 in the boxed ATG-codons. *Circles* indicate nucleotides common to the eukaryotic consensus sequence (Kozak 1983). The pyrimidine blocks preceding the ATG are marked by the *hatched areas*, the bases complementary to the 3'-end of eukaryotic 18S rRNA are underlined. The poliovirus sequence is from Kitamura et al. (1981)

exists for the fourth small capsid protein VP4, whose N-terminus is chemically blocked. Its position could therefore only be predicted from the apparent size of VP4 and from similarity in the amino acid sequence to the VP1/P52 junction, since both sites are presumed to be cleaved by the same protease. These data suggest that VP4 is cleaved from P20a at amino acids 217/218 (Beck et al. 1983a) or at positions 213/214.

Much less information was available for the nonstructural proteins encoded in the P2 (P52) and P3 (P100) precursors (Fig. 1). Except for the exactly mapped VPg genes (Forss and Schaller 1982) only the approximate positions of P34 and P56a had been allocated from the kinetics of their biosynthesis and peptide maps (Sangar 1979), whereas precursor/product relationships had indicated that proteins P12 and P20b mapped next to P34 and P52, respectively (A. King, personal communication). Therefore, initially only a rough protein map could be established, using size estimations from polyacrylamide gel electrophoresis and homologies to poliovirus polypeptides for which the order was known (Kitamura et al. 1981). The exact coding limits of the individual functional proteins were predicted from the cleavage specificity for Glu (Gln)/Gly (Ser, Thr) linkages of the FMDV protease (Forss and Schaller 1982; Forss et al. 1984). In some cases (P52, P12, P34, and P14) minor corrections were made using very recent data from Grubman and collaborators, who determined the amino acid sequence at the N-terminus of the polypeptides synthesized in vitro and corresponding to P52, P12, P34, P14, and P20b (personal communication, and P56 (Robertson et al. 1983) from FMDV A12. The map thus obtained for FMDV (Figs. 2 and 4) was recently confirmed by immunoprecipitation of polypeptides of the predicted size from FMDV O1K infected BHK cells, using antisera against bacterially synthesized polypeptides that correspond to the respective segments of the polyprotein (Strebel et al. 1984).

The biochemical properties of the FMDV proteins predicted from the amino acid sequence in Fig. 2 are listed in Table 1, which also includes a new generalized nomenclature for picornavirus genes. In accordance with the data from Strebel et al., mentioned above, these data indicate that the nonstructural genes are organized as follows: P2 (formerly P52), the precursor from the middle part of the polyprotein has a calculated molecular weight of 54.5 K. It contains two stable proteins, 2b (P12) and 2c (P34), with unknown functions. The N-terminal limit of P2 was originally set next to the carboxy-terminus of VP1 (Fig. 2) which had been accurately identified for FMDV O1K by Kurz et

Table 1. Predicted map positions and biochemical properties of FMDV polypeptides as deduced from the nucleotide sequence

Poly-peptide	Map co-ordinate [a]	Map position	No. of amino acids	Molecular weight	Net charge	Function
Poly-protein	-	805-7800	2332	258946.8	+37	Precursor
P20a	L	805-1455	217	24367.4	-5	Protease (?)
P16	L'	889-1455	189	21243.1	-7	
P88	P1	1456-3615	720	79305.6	+13	Precursor
VP4	1a	1456-1662	69	7362.3	-3	Capsid protein
VP2	1b	1663-2316	218	24410.4	+4	" "
VP3	1c	2317-2976	220	23746.4	-2	" "
VP1	1d	2977-3615	213	23840.5	+14	" "
P52	P2	3616-5079	488	54501.1	+8	Precursor
(P12)	2b	3664-4125	154	16255.1	+2	?
P34	2c	4126-5079	318	35892.9	+8	?
P100	P3	5080-7800	907	100844.7	+21	Precursor
(P14)	3a	5080-5538	153	17355.2	-5	?
VPg-1	3b-1	5539-5607	23	2604.5	+3	Genome-linked
VPg-2	3b-2	5608-5679	24	2622.4	+4	protein
VPg-3	3b-3	5680-5751	24	2579.3	+3	"
P20b	3c	5752-6390	213	23012.4	+9	Protease
P56	3d	6391-7800	470	52760.9	+7	RNA polymerase

[a] Nomenclature suggested for the picornaviral polypeptides at the 3rd Meeting of the European Study Group of Picornaviruses, Urbino Italy, Sept. 5-10 1983

al. (1981). However, as shown for FMDV A12, the proteins P2 and P12 start 16 amino acids downstream from the C-terminus of VP1 (Fig. 2). The spacer element missing may be degraded during the processing events, or it may represent a very short, as yet undetected, functional polypeptide. In poliovirus this region appears to have expanded to a gene of substantial size, probably gene 2a, coding for a protein of 149 amino acids (Fig. 4). The carboxy-terminal precursor P3 (formerly P100, calculated molecular weight 100.8 K) comprises the sequence between amino acids 1426 and 2332. It is processed into six proteins: 3a (P14) of 17.4 K, three VPg's (in tandem) of 2.3 K each, a protease 3c (P20b), of 23.0 K, and an RNA polymerase, 3d (P56a), of 52.7 K.

4 Protease Cleavage Sites

The present map of the FMDV genome (Figs. 2 and 4) predicts that at least 12 sites in the protein sequence need to be cleaved to give rise to mature viral gene products. As can be taken from Fig. 2, seven of these sites (VP2/VP3/VP1, P14/VPg2/VPg3/P20b/P56) show similar amino acid sequences, suggesting that they are cleaved by a single (viral) protease, recognizing the consensus sequence Glu/Gly (Ser, Thr) and probably being encoded in gene 3c (P20b). This specificity is similar to that displayed by the poliovirus which cleaves between Gln and Gly residues at 8 out of 11 processing sites in the polio polyprotein

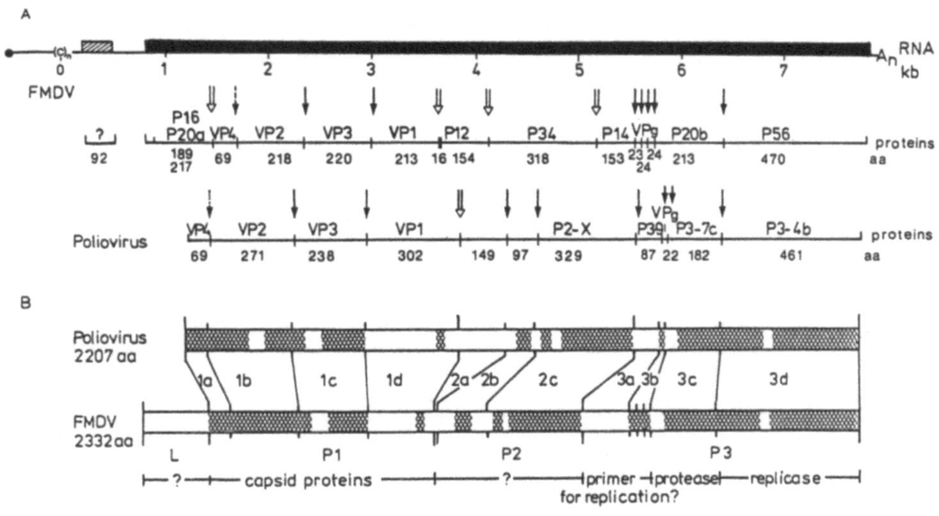

Fig. 4A,B. Comparison of the FMDV and poliovirus genomes. A Schematic map of gene organization and protein processing for FMDV (Forss 1983) and poliovirus (Kitamura et al. 1981). *Open arrows* indicate cleavages possibly executed by cellular proteases, while *filled arrows* represent processing by the Glx/Gly viral proteases. *Dashed arrows* illustrate morphogenetic cleavages occuring during virus particle maturation. The FMDV polypeptides are termed as in Fig. 2; the number of amino acids is given for each protein below the lines; B Sequence homology between corresponding parts of the two polyproteins. Shaded segments indicate more than 25% homology. Related gene products are connected by vertical lines and identified using the new general nomenclature for picornaviruses.(3rd Meeting of the European Study Group of Picornaviruses, Urbino, Italy, September 1983)

(Kitamura et al. 1981, Fig. 4a). The sequences around the remaining five cleavage sites in FMDV (P20a/VP4/VP2, P1/P2/P3) show little sequence homology to each other, and are thought to be recognized by cellular proteases. However, recent experiments suggest that cleavage at these sites may also involve protease P20b (Klump et al., Chap. V, 19, this Vol.) and/or a second protease encoded in the L segment of the FMDV genome (Strebel et al. 1984).

5 A Minor Unlinked FMDV Gene?

The sequence coding for the polyprotein is preceded by 713 nucleotides of apparently untranslated RNA of known sequence, and by some additional 550 nucleotides of yet uncloned RNA comprising the poly(C) tract and the S fragment (Figs. 1 and 2). This seems to be exceptionally long for a leader segment of a small, otherwise compact, viral genome, and suggests that additional short coding sequences may exist on this segment of the viral genome. At present the most likely candidate for such an unlinked gene is a sequence of 92 translatable codons that follows the first AUG after the poly(C) tract (pos. 209 in Fig. 2). A polypeptide of 10 K (tentatively named P10) has recently been identified among the products from in vitro translation of FMDV RNA by immunoprecipitation with antisera directed against the P10 amino acid sequence (Strebel et al. 1984). If confirmed in vivo, these result would classify FMDV RNA as a polycistronic eukaryotic mRNA.

6 Homology to Poliovirus

Gene organization and processing mechanisms predicted for FMDV from the nucleotide sequence in Fig. 2 are summarized in Fig. 4 and compared to those from poliovirus, a member of the enterovirus family, and the only other picornavirus for which the complete coding sequence is presently available. As outlined in Fig. 4A, the overall organization is well conserved between the two genomes, but FMDV differs from poliovirus by an increase in size of its genome by about 1000 nucleotides most of which are added in the 5`-proximal part of the genome. Within the polyprotein region the two genomes differ drastically only by the addition of the L gene and two extra VPg genes in FMDV and by the addition of a third extended gene (2a or 2b) in the central part of the polio genome. Other corresponding genes often differ in size. Thus, the capsid proteins are larger in poliovirus, while FMDV has expanded its nonstructural proteins in the P3 segment.

The alignment of the two genomes was refined using sequence homologies between functionally corresponding segments. Using a dot matrix program, significant homology was detected on the amino acid level throughout most of the coding region, and to a lower extent also on the nucleotide level, indicating common structural features of general functional importance (Forss 1983). As shown in Fig. 5, these sequence homologies are very high in certain parts of the two nonstructural proteins, the polymerase (3d), and protein 2c (x/P34). Less, but still high homology was detected between the protease genes and the capsid proteins VP2 and VP3, indicating that the functional specificity of these proteins was less stringently conserved during picornaviral evolution. The evolutionary divergence is most pronounced in proteins 1d, 2a/2b, and 3a. Protein 1d (VP1) is the capsid protein most exposed at the surface of the virion and, as a consequence of the pressure of the host's immune system, its sequence is highly variable also between different aphthoviruses giving rise to seven serotypes and many more subtypes. In contrast proteins 2a/2b and 3a, although variable between FMDV and poliovirus, are highly conserved between two FMDV serotypes as exemplified by a comparison of FMDV O1K and C1O (unpublished results). Therefore, we conclude that these latter proteins play an important role in the FMDV lifecycle but co-evolved with their targets which are most likely FMDV encoded. Following the same argument, we predict that the more generally conserved picornaviral proteins (2c and 3d) depend in their functions on the interaction with host factors of conserved structure such as the ribosome. In this context, it is of note that sequences common to FMDV and poliovirus in proteins 2c and 3d are also present in Cow Pea Mosaic Virus, a plant virus which has thus been correlated with picornaviruses (Franssen et al. 1984).

The comparison of the two picornaviral genomes in Fig. 4b also indicates that these differ in size predominantly in regions with low sequence homology. Sometimes extended blocks of nucleotides are also found inserted into well conserved genes like in gene 1b (VP2). Together with the addition of complete genes, like the L gene or the extra VPg genes in FMDV, these data indicate that evolution of picornaviral genomes involved the insertion or deletion of RNA segments of several hundred nucleotides.

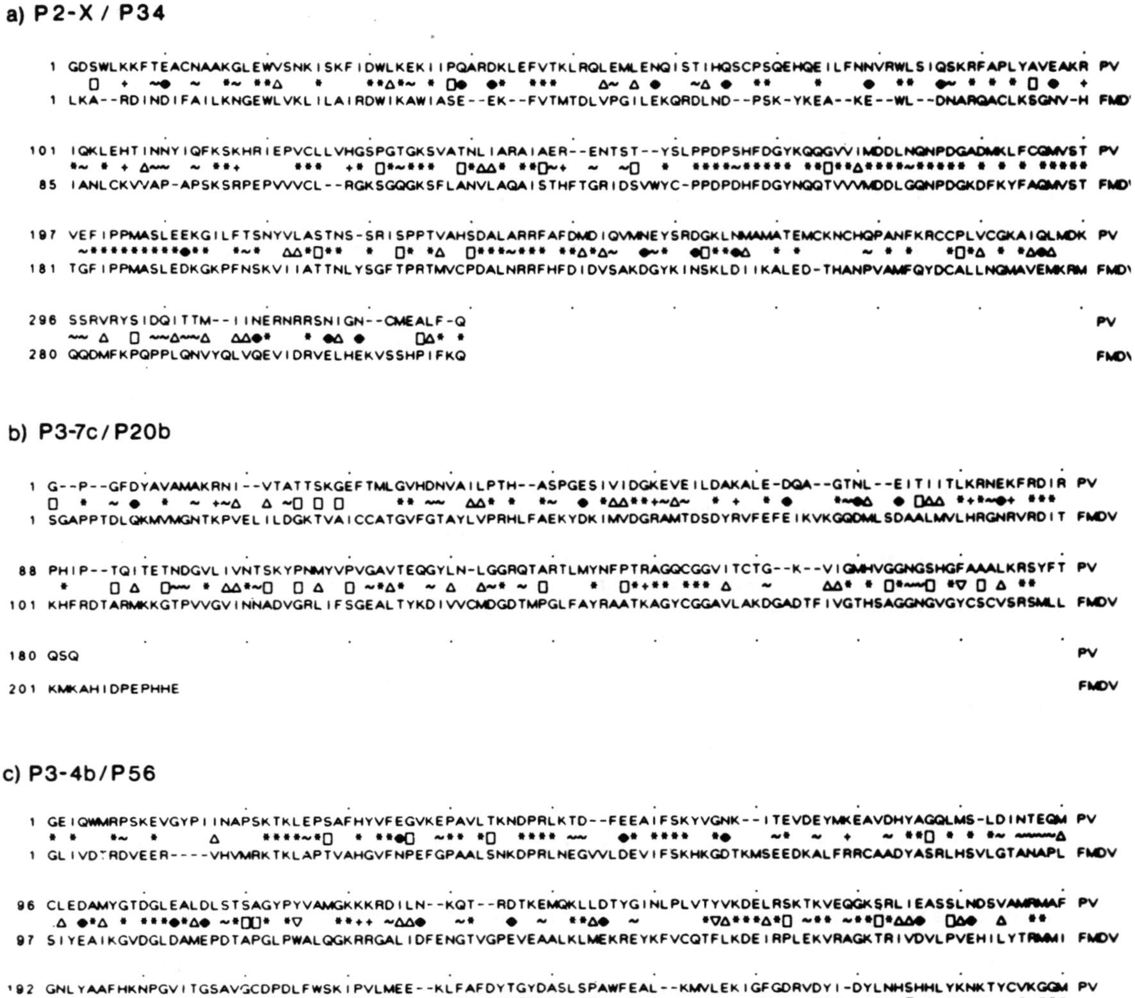

Fig. 5a-c. Amino acid sequence homologies in a P2c (P34/-"x"). b protease, and c polymerase regions of FMDV and poliovirus (Kitamura et al. 1981). For the C-terminus of the polymerase, data for EMCV (Drake et al. 1982) has been included. A limited number of empty spacings (*dashes*) have been introduced into the sequences in order to improve alignment. Identical amino acids are indicated by (✱) and similar amino acids are specified: (□), STPAG; (θ), NDEQ; (∼), (□, θ); (+), HRK; (Δ), MILV; (∇), FYW

7 Antigenic Variation of FMDV and a Potential Subunit Vaccine

FMDV is able to escape the host's immune system by a high variability of its surface antigens giving rise to seven distinct serotypes and 60 subtypes of the virus. This antigenic variation correlates with an extensive variation of the amino acid sequence of the three major structural proteins VP1, VP2, and VP3 which are caused by viable point mutants known to occur with high probability in FMDV RNA (Domingo et al. 1980), as well as in other RNA genomes, because of the intrinsically imprecise mechanism of RNA replication (Holland et al. 1982). In FMDV, the variation of the encoded amino acid sequence is highest in VP1 (up to 30% between two serotypes, Beck et al. 1983b). This protein is therefore probably the most exposed subunit on the surface of the virion, a notion that is also supported by the finding that it carries epitopes capable of inducing serotype-specific neutralizing antibodies in test animals.

Use of the isolated VP1 as safe subunit vaccine was suggested several years ago, but large amounts of the pure VP1 protein have only become available through the recent advances in recombinant DNA technology. To express the protein in *E. coli* we fused the VP1 gene to a highly active bacterial expression unit which gave rise to a hybrid protein containing also the first 99 amino acids of the replicase gene of bacteriophage MS2 (Küpper et al. 1981; Küpper et al. 1982). This fusion protein was synthesized in *E. coli* to levels of up to 20% of total cellular protein (about 10^6 molecules per bacterial cell) and subsequently shown to be at least as effective in vaccination experiments as the subunit isolated from the virus. However, its efficacy was low if compared to that of the same sequence presented as part of the virion and further improvement in antigen presentation will be required to reach the ultimate goal of a multivalent highly protective subunit vaccine.

8 Antigenic Determinants and a Potential Synthetic Vaccine

Antigenic determinants in a protein can be mimicked by short carrier-linked peptides, and antibodies against such peptides can recognize the corresponding sequential run of amino acids within the native or denatured protein (Sela et al. 1967). More recently this approach has been successfully applied to correlate nucleic acid sequences with conjectured amino acid sequences of known or hypothetical gene products (Walter et al. 1980; Lerner et al. 1981). In addition, peptides with antiviral or antimicrobial immunogenicity have been identified for a number of pathogenic microorganisms and are being explored as potential synthetic vaccines (for a review see Sutcliffe et al. 1983).

In FMDV a hexadecapeptide segment from structural protein VP1 (peptide A, Fig. 6) has been identified as the major neutralizing antigenic determinant on the virion (Pfaff et al. 1982; Bittle et al. 1982). As indicated in Fig. 6, this sequence is part of a VP1 segment that is highly variable between FMDV serotypes and subtypes (Beck et al. 1983; Cheung et al. 1983), and is exposed on the surface of the intact FMDV virion to enzymatic attack. It has been suggested that the neutralizing epitopes on FMDV are based on a particular α-helical structure of the hexadecapeptide sequence and essentially all of the neutralizing antibodies present in the hyperimmune serum from FMDV-infected, reconvalescent cattle were found to absorb to a peptide A sepharose column (Pfaff et al. 1982). Polyvalent or monoclonal anti-

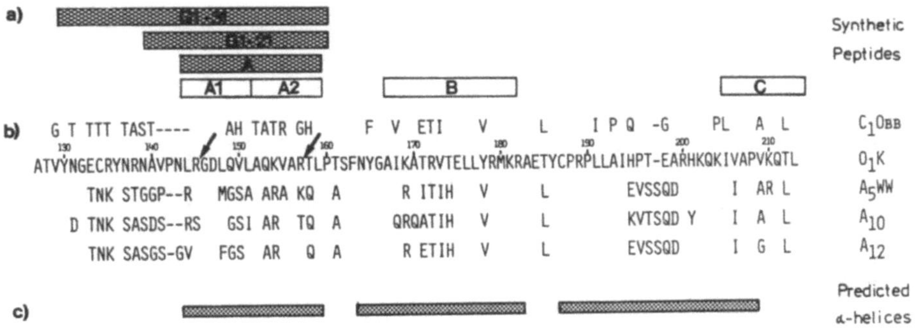

Fig. 6a-c. Structural and serological analysis of the carboxy-terminal part of capsid protein VP1 (Beck et al. 1983b; Pfaff et al. 1982). a Synthetic peptides tested for serological cross-reaction with FMDV. Only the peptides (*hatched*) were able to induce neutralizing antibodies to FMDV O1K; b Only differences to the O1K sequence are specified. *Dashes* represent deletions inserted to optimize alignment of the sequences. *Arrows* indicate tryptic cleavage sites exposed in the virion (Strohmaier et al. 1982); c Sequences predicted to attain a α-helical conformation

peptide A antibodies recognize and neutralize FMDV serotype-specifically and recently a differentiation of FMDV strains even from within a subtype has been achieved using monoclonal antibodies recognizing peptide A (Thiel, Strohmaier, Pfaff, unpublished results).

Taken together, these results indicate that peptides covering the peptide A sequence are capable of mimicking the structure of a neutralizing epitope in the virus, thus eliciting immune responses that should protect test animals against FMDV infection. These peptides may thus serve as the basis for a safe, chemically defined synthetic FMDV vaccine. Preliminary experiments with Guinea Pigs have indeed demonstrated a serotype-specific protective immuno response (Table 2), and depending on the formulation of the antigen, protection against a challenge with live virus was obtained even after a single inoculation (Bittle et al. 1982; Leban et al., unpublished results). However, similar vaccination experiments with cattle required at least two immunizations which make the synthetic vaccine at present commercially less attractive. Therefore, as probably also true for other potential synthetic vaccines (Sutcliffe et al. 1983), further research efforts must be directed towards the development of improved carrier molecules and new adjuvants, or alternatively to analyse the secondary structure of the neutralizing epitope and to stabilize the peptide in this particular conformation.

Acknowledgments. This research was supported by grants from BIOGEN S.A., and more recently by the Deutsche Forschungsgemeinschaft (Forschergruppe Genexpression), and the Fond der Chemischen Industrie.

Table 2. Protection of immunized guinea pigs against challenge with life virus

Antigen	Dose (µg)	Neutralization index (log 10) (a)	(b)	Animals protected / challenged
Peptide A	20	<0.5	1.3(0.7-1.7)	8/8
Peptide A	200	<0.5	1.7(0.7-2.0)	8/8
Peptide 61-21	200	<0.5	1.7	4/4
Peptide 61-31	200	<0.5	1.3	3/3
FMDV O1K	10	1.3	>2.5	4/4
KLH	-		<0.5	0/4

Peptides were coupled to KLH and used for two successive inoculations as described (Pfaff et al. 1982). FMDV was inactivated by ethylene imine treatment. The neutralizing activity is given for the pooled sera from all test animals 3 weeks after the first (a) and the second (b) inoculation. Values in parenthesis give the range of activity in the individual animals

References

Beck E, Forss S, Strebel K, Cattaneo R, Feil G (1983a) Structure of the FMDV translation start site and of the structural proteins. Nucleic Acids Res 11:7873-7885

Beck E, Feil G, Strohmaier K (1983b) The molecular basis of the antigenic variation of foot and mouth disease virus. EMBO J 2:555-559

Bittle JL, Houghten RA, Alexander H, Shinnick TM, Sutcliffe JG, Lerner RA (1982) Protection against foot and mouth disease by immunization with a chemically synthesized peptide predicted from the viral nucleotide sequence. Nature (Lond) 298:30-33

Black DN, Stephenson P, Rowlands DJ, Brown F (1979) Sequence and location of the poly C tract in aphtho-and cardiovirus RNA. Nucleic Acids Res 6:2381-2390

Cheung A, Delamarter J, Weiss S, Küpper H (1983) Comparison of the major antigenic determinants of different serotypes of foot and mouth disease virus. J Virol 48: 451-459

Drake NL, Palmenberg AC, Gosh A, Omilianowski DR, Kaesberg P (1982) Identification of the polyprotein termination site on encephalomyelocarditis viral RNA. J Virol 41:726-729

Domingo E, Davila M, Ortin J (1980) Nucleotide sequence heterogeneity of the RNA from a natural population of foot and mouth disease virus. Gene (Amst) 11:333-346

Forss S, Schaller H (1982) A tandem repeat gene in a picornavirus. Nucleic Acids Res 10:6441-6450

Forss S (1983) Structure and organization of the foot and mouth disease virus genome. Thesis, University of Heidelberg

Forss S, Strebel K, Beck E, Schaller H (1984) Nucleotide sequence and genome organization of foot and mouth disease virus. Nucleic Acid Res 12:6587-6601

Franssen H, Leunissen J, Goldbach R, Lomonosoff G, Zimmern D (1984) Homologous sequence in non-structural proteins from cowpea mosaic virus and picornaviruses. EMBO J 3:855-861

Harris TJR (1980) Comparison of the nucleotide sequence of the 5`-end of RNAs from nine aphthoviruses, including representatives of the seven serotypes. J Virol 36:659-664

Harris TJR, Brown F, Sangar DV (1981) Differential precipitation of foot and mouth disease virus protein made in vivo and in vitro by hyperimmune and virus particle Guinea Pig antisera. Virology 112:91-98

Holland J, Spindler K, Horodyski F, Grabau E, Nichol S, VandePol S (1982) Rapid evolution of RNA genomes. Science (Wash DC) 215:1577-1585
Kitamura N, Semler BL, Rothenberg PG et al (1981) Primary structure, gene organization and polypeptide expression of poliovirus RNA. Nature (Lond) 219:547-553
Kozak M (1983) Comparison of initiation of protein synthesis in procaryotes, eucaryotes and organelles. Microbiol Rev 47:1-45
Kurz C, Forss S, Strohmaier K, Schaller H (1981) Nucleotide sequence and corresponding amino acid sequence of the gene for the major antigen of foot and mouth disease virus. Nucleic Acids Res 9:1919-1930
Küpper H, Keller W, Kurz C et al. (1981) Cloning of cDNA of major antigen of foot and mouth disease virus and expression in *E. coli*. Nature (Lond) 289:555-559
Küpper H, Delamarter J, Otto B, Schaller H (1982) Expression of major foot and mouth disease antigen in *E. coli*. Proc 4th Int Symp Genetics of Industrial Microorganisms, Japan
Lerner RA, Sutcliffe JG, Shinnick TM (1981) Antibodies to chemically synthesized peptides predicted from DNA sequences as probes of gene expression. Cell 23:309-310
Pfaff E, Mussgay M, Böhm HO, Schulz GE, Schaller H (1982) Antibodies against a preselected peptide recognize and neutralize foot and mouth disease virus. EMBO J 1:869-874
Robertson BH, Morgan DO, Moore DM et al. (1983) Identification of amino acid and nucleotide sequence of the foot and mouth disease virus RNA polymerase. Virology 126:614-623
Sangar DV (1979) The replication of picornaviruses. J Gen Virol 45:1-13
Sangar DV, Black DN, Rowlands DJ, Harris TJR, Brown F (1980) Location of the initiation site for protein synthesis on foot and mouth disease virus RNA by in vitro translation of defined fragments of the RNA. J Virol 33:59-68
Sela M, Schechter B, Schachter I, Borek A (1967) Cold Spring Harbor Symp Quant Biol 32:537-545
Strebel K, Beck E, Strohmaier K, Schaller H (to be published 1974) Characterization of FMDV gene products with antisera against bacterially, synthesized fusion proteins
Strohmaier K (1978) The N-terminal sequence of three coat proteins of foot and mouth disease virus. Biochem Biophys Res Commun 85:1640-1645
Strohmaier K, Franze R, Adam KH (1982) Location and characterization of the endogenic portion of the FMDV immunizing protein. J Gen Biol 59:295-306
Summers DF, Shaw EN, Stewart ML, Maizel JV (1972) Inhibition of cleavage of large poliovirus specific precursor proteins in infected Hela cells by inhibitors of proteolytic enzymes. J Virol 1:880-884
Sutcliffe JG, Shinnick TM, Green N, Lerner RA (1983) Antibodies that React with predetermined sites on proteins. Science (Wash DC) 219:660-666
Walter G, Scheidtmann KH, Carbone A, Laudano AP, Doolittle RF (1980) Antibodies specific for the carboxy- and amino-terminal regions of simian virus 40 large tumor antigen. Proc Natl Acad Sci USA 77:5197-5200

Structure, Proteolytic Processing, and Neutralization Antigenic Sites of Poliovirus

D. C. DIAMOND[1], R. HANECAK[1], B. L. SEMLER[2], B. A. JAMESON[2], J. BONIN[2], and E. WIMMER[2]

1 Introduction

Poliovirus is a small naked icosahedral virion, composed of 60 copies each of 4 capsid proteins enclosing one 7.5 kb single-stranded, messenger-sense RNA. The RNA has a tract of poly(A) at the 3' end (Yogo and Wimmer 1972) and is covalently linked to a small protein, VPg, which appears to be involved in priming RNA replication (Takagami et al. 1983; Wimmer 1982). The complete sequence of the RNA of all three serotype and the fine structure of the genetic map have been determined (Kitamura et al. 1981; Semler et al. 1981a, 1981b; Emini et al. 1982a; Racaniello and Baltimore 1981; Nomoto et al. 1982; Cann et al. 1983; Stanway et al. 1983, 1984; Toyoda et al. 1984).

The viral genome encodes a single "polyprotein" which is cleaved by at least two proteinases, one of which is virus-encoded (Hanecak et al. 1982). Proteolytic processing yields the structural and nonstructural proteins of the virus (Fig. 1). In addition to VPg and the virus encoded proteinase, $3C^{pro}$, there are two other major nonstructural proteins; $3D^{pol}$, a primer and template-dependent RNA polymerase, and 2C, a protein of unknown function found associated with membrane-bound replication complexes. All four of these proteins and their precursors may be directly involved in replication.

The virus is susceptible to neutralization by the binding of antibodies to its capsid. Although the dominant neutralization antigenic sites (N-Ags; for definitions, see Wimmer et al. 1984) against which neutralizing monoclonal antibodies (N-McAbs) have been obtained, appear to reside on the capsid protein VP1 (Emini et al 1982b; Evans et al. 1983; Minor et al. 1983) other N-Ags have been identified on all three exposed capsid proteins (Fig. 2). There exist at least six N-Ags on VP1 and one each on VP2 and VP3 (Emini et al. 1983a, 1983b; Emini et al., submitted; Jameson et al., in preparation).

The recent advances in biotechnology have opened new vistas for the study of poliovirus. These are discussed in light of our recent results.

2 The Viral Proteinase

The polyprotein of poliovirus is known to be cleaved at 12 sites. $3C^{pro}$ has been shown to be responsible for nine of these cleavages, i.e., that of glutamine-glycine (Q-G) amino acid pairs (Hanecak et al.

[1] Department of Molecular Biology and Biochemistry, University of California, Irvine, California 92717, USA

[2] Department of Microbiology and Molecular Genetics, College of Medicine, University of California, Irvine, California 92717, USA

[3] Department of Microbiology. School of Medicine, State University of New York at Stony Brook, Stony Brook, New York 11794, USA

35. Colloquium - Mosbach 1984
The Impact of Gene Transfer Techniques in Eukaryotic Cell Biology

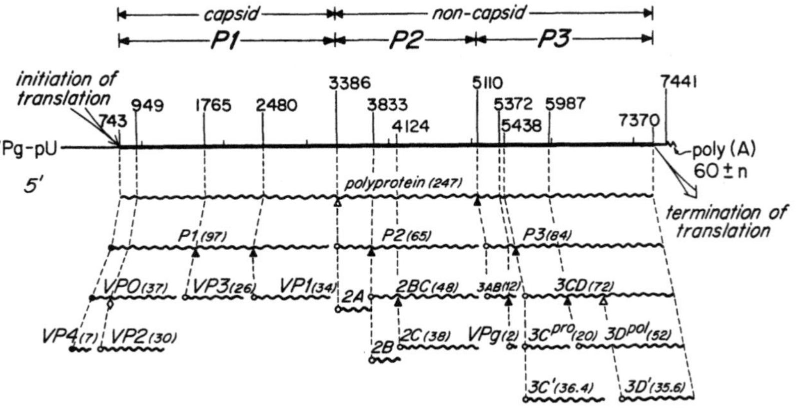

Fig. 1. Gene organization and protein processing of poliovirus. Virion RNA, terminated at the 5' end with the protein-linked protein VPg and at the 3' end with poly(A), is shown as a *solid line*, the translated region being more pronounced than the non-coding regions. *N* indicates the sites at which initiation (N743) and termination (N7367) of translation occur. *Numbers above the virion RNA line* refer to the first nucleotide of the codon specifying the N-terminal amino acid for the viral specific proteins. The coding region has been divided into three segments (*P1, P2, P3*), corresponding to the primary cleavage products of the polyprotein. Polypeptides are presented as *waved lines*. *Numbers in parentheses* are molecular weights calculated from the amino acid sequences. Carboxy-terminal "trimming" does not occur. The N-terminus is glycine in all cases except VP2 where it is serine. The carboxy-terminal amino acid of 3CD is phenylalanine. *Closed circles* indicate the N-termini are blocked. *Closed triangles* Q-G; *open triangles* Y-G; *open diamond* N-S cleavage sites. Polypeptides 3C' and 3D' are products of an alternate cleavage mode of 3D. The nomenclature used corresponds to that recommended by Rueckert and Wimmer (1984) (modified after Kitamura et al. 1981; Hanecak et al. 1982, 1984; Semler et al. 1981b)

1982). 3C^{pro} is itself generated by this type of cleavage but because the virion does not contain 3C^{pro} it has been speculated that the viral proteinase is generated by intramolecular cleavage. That the occurrence of a Q-G pair alone is not sufficient for cleavage by 3C^{pro} is demonstrated by several observations: 3C^{pro} acts at only nine of the 13 Q-G pairs in the polyprotein; it does not give rise to general proteolysis of host cell proteins; cleavage occurs in statu nascendi and completed polyprotein, synthesized under conditions which promote misfolding at 43°C, is not a substrate for processing (Baltimore 1971). No common amino acid sequences have been recognized which distinguish between those Q-G pairs which are cleaved and those which are not. It thus appears that the three-dimensional structure of a potential processing site is critical in determining if it will be cleaved. This sensitivity of proteolytic processing to conformation has often been advanced as a rationale for the absence of complementation between mutants which has confounded attempts to analyze the functions of poliovirus proteins by classical genetic techniques.

VP4
1 M G A Q V S S Q K V G A H E N S N R A Y G G S T I N Y T T I N Y Y R D S A S N A
41 A S K Q D F S Q D P S K F T E P I K D V L I K T A P M L N S P N I E A C G Y S D
VP2
81 R V L Q L T L G N S T I T T Q E A A N S V V A Y G R W P E Y L R D S E A N P V D
121 Q P T E P D V A A C R F Y T L D T V S W T K E S R G W W W K L P D A L R D M G L
161 F G Q N M Y Y H Y L G R S G Y T V H V Q C N A S K F H Q G A L G V F A V P E M C
201 L A G D S N T T T M H T S V Q N A N P G E K G G T F T G T F T P D N N Q T S P A
9
241 R R F C P V D Y L L G N G T L L G N A F V F P H Q I I N L R T N N C A T L V L P
281 Y V N S L S I D S M V K H N N W G I A I L P L A P L N F A S E S S P E I P I T L
VP3
321 T I A P M C C E F N G L R N I T L P R L Q G L P V M N T P G S N Q Y L T A D N F
361 Q S P C A L P E F D V T P P I D I P G E V K N M M E L A E I D T M I P F D L S A
401 T K K N T M E M Y R V R L S D K P H T D D P I L C L S L S P A S D P R L S H T M
8 12
441 L G E I L N Y Y T H W A G S L K F T F L F C G S M M A T G K L L V S Y A P P G A
481 D P P K K R K E A M L G T H V I W D I G L Q S S C T M V V P W I S N T T Y R Q T
521 I D D S F T E G G Y I S V F Y Q T R I V V P L S T P R E M D I L G F V S A C N D
VP1
561 F S V R L L R D T T H I E Q K A L A Q G L G Q M L E S M I D N T V R E T V G A A
5
601 T S R D A L P N T E A S G P T H S K E I P A L T A V E T G A T N P L V P S D T V
641 Q T R H V V Q H R S R S E S S I E S F F A R G A C V T I M T V D N P A S T T N K
1 4 3 2
681 D K L F A V W K I T Y K D T V Q L R R K L E F F T Y S R F D M E L T F V V T A N
721 F T E T N N G H A L N Q V Y Q I M Y V P P G A P V P E K W D D Y T W Q T S S N P
14
761 S I F Y T Y G T A P A R I S V P Y V G I S N A Y S H F Y D G F S K V P L K D Q -
801 S A A L G D S L Y G A A S L N D F G I L A V R V V N D H N P T K V T S K I R V Y
841 L K P K H I R V W C P R P P R A V A Y Y G P G V D Y K D G T L T P L S T K D L T
881 T Y

Fig. 2. Amino acid sequence of the P1 (capsid) region of poliovirus, type 1 (Mahoney). *Underlined regions* have been incorporated into synthetic peptides and represent N-Ags. Bonin et al., in preparation

3 Expression of $3C^{pro}$

The expression of individual poliovirus proteins from plasmid vectors offers a new strategy for their study. Since poliovirus is translated monocistronically the individual proteins lack promoters and start and stop codons. All control sequences must be provided by the expression vector. These considerations make it difficult to express anything but a fusion protein with some segment of the polyprotein embedded in it. Authentic termini can be obtained only by proteolytic processing. Thus it was of interest to see whether or not $3C^{pro}$ could excise itself from such a fusion protein.

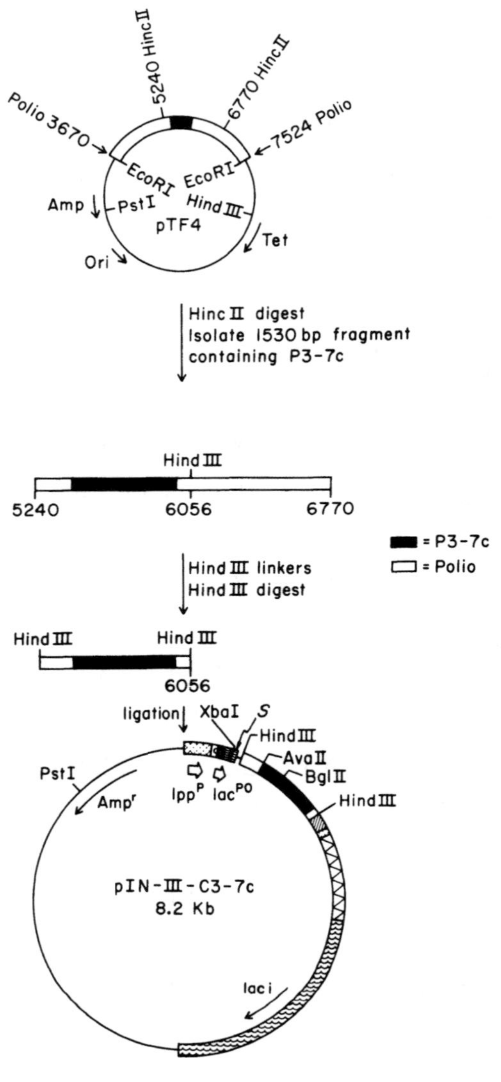

Fig. 3. Schematic representation of pIN-III-C3-7c plasmid construction. (P3-7c are names for $3C^{pro}$ from older nomenclatures, see Rueckert and Wimmer 1984). The approximate location of several restriction endonuclease cleavage sites are shown for plasmid pTF4 (Semler et al., in press). *Open bars* represent poliovirus genomic sequences flanking the region of the viral genome encoding $3C^{pro}$ (P3-7c) designated by a *solid bar*. The position of the origin of replication (ori), the -lactamase gene of pBR322 (Amp) and the gene for tetracycline resistance are indicated by *arrows*. The construction of the pIN-III-C3-7c recombinant vector is described in detail in Hanecak et al. (1984). The location of the lipoprotein promoter sequence and the lactose promoter-operator region are indicated by *blocked arrows*. *Box with diagonal lines* represents the 3' untranslated regions of the bacterial lipoprotein and contains the stop codon for protein synthesis. *Dotted box* designates the location of transcription termination signals. *Box with wavy lines* represents the gene for the lactose repressor (lac i). *S* denotes the location of the region encoding the signal peptide for lipoprotein (Hanecak et al. 1984)

The expression vector was selected from M. Inouye's pIN series. These bacterial expression vectors utilize the promoter for lipoprotein, the most abundant protein of *E. coli* (Masui et al. 1983). Specifically we used pIN-III-C3 which also contains the lac UV5 promoter-operator sequences and the lac repressor gene (Fig. 3) thus enabling regulated expression of an inserted gene. The insertion site in this construction is located a short way downstream from the lipoprotein signal peptide (as opposed to within or adjacent to it). Thus the fusion product should be directed to the outer membrane where it would be protected from cytoplasmic proteinases.

A HincII-HindIII fragment spanning the region encoding $3C^{pro}$ (Fig. 3) was excised from a poliovirus cDNA-containing plasmid and inserted into the vector with HindIII linkers. The expected product consisted of 273 amino acids from poliovirus with 3 amino acids on the C-terminus and 35 amino acids (including 20 in the signal peptide) on the N-terminus contributed by the vector. It was anticipated that these few amino acids would not influence protein folding so much as to inhibit processing.

Upon induction with IPTG three major anti-3C^{pro} precipitable polypeptides are produced (Fig. 4). Polypeptide 1 is the putative primary translation product after cleavage of the signal peptide and has a relatively short half-life. Polypeptides 2 and 3 are both stable and polypeptide 3 comigrates with 3C^{pro}. To determine whether authentic poliovirus-like processing was occurring, N-terminal radiochemical microsequence analysis was carried out. This analysis showed that both polypeptides 2 and 3 have the same N-terminus as 3C^{pro}. Carboxypeptidase A digestion of polypeptide 3 confirmed that its C-terminus is the same as 3C^{pro}. Because there is no apparent precursor-product relationship between polypeptides 2 and 3, the production of polypeptide 3 may depend on which Q-G pair in polypeptide 1 is cleaved first.

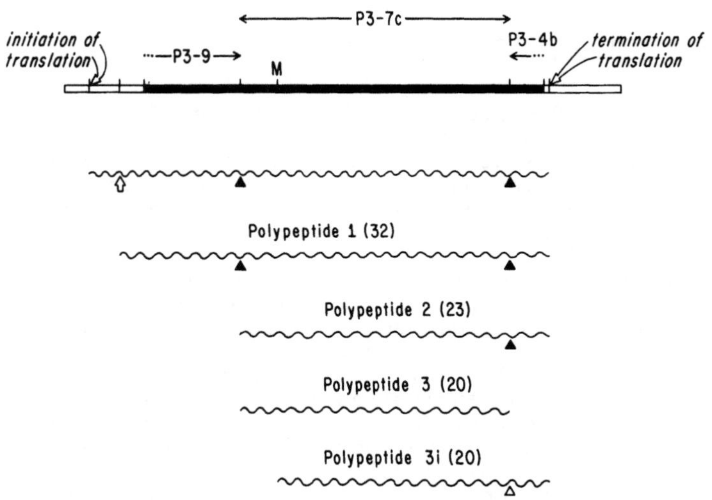

Fig. 4. Processing scheme for poliovirus-specific polypeptides synthesized in bacteria. The message is shown as a *box*. *Open boxes* represent pIN-III-C3 specific regions present within the message. The poliovirus genomic RNA sequence is shown as a *closed box*. The methionine residue present within the protein 3C^{pro} at nucleotide 5516 is indicated by *M*. Polypeptides are denoted by *wavy lines*. *Arrow* indicates the site of cleavage of the signal peptide. Cleavages of Q-G pairs of amino acids are indicated by *closed triangles*. The unutilized Q-G cleavage site in polypeptide 3i is shown as an *open triangle*. *Numbers in parentheses* are the molecular weights (in kilodaltons) calculated from the amino acid sequences assuming an average molecular weight of 110 for each amino acid. *P3-9*, *P3-7c*, and *P3-4b* correspond to 3AB, 3C^{pro}, and 3D^{pol} respectively. (Hanecak et al. 1984)

Demonstration that processing was due to autocatalytic cleavage and not to the fortuitous action of some bacterial enzyme was accomplished by linker insertion mutation. A SacI linker was inserted at the unique BglII site, thereby causing the insertion of 4 amino acids, 2 neutral, 2 basic, within 3C^{pro} and abolishing the production of polypeptides 2 and 3. In addition to polypeptide 1, a minor band, polypeptide 3i, was also still observed. Polypeptide 3i was first observed as a contaminant in the microsequencing of polypeptide 3. It is produced by internal initiation promoted by a fortuitous Shine-Dalgarno sequence adjacent to a methionine codon within the coding region of $3C^{pro}$.

Similar results have been obtained with a larger construction containing most of the P2 region (Fig. 1) in addition to those sequences in pINIII-C3-7c. Processing is observed at the expected sites and is abolished by the insertion of 4 neutral amino acids effected by the insertion of a BglII linker at the BglII site (Ariga et al., unpublished).

This expression system is now being used to study the enzymology of $3C^{pro}$. Processing is inhibited by Zn^{2+}, as it is in HeLa cell infections, as well as by cysteine-proteinase inhibitors (Nicklin and Wimmer, unpublished). Sequence alignment of $3C^{pro}$ with proteinases of other picornaviuses reveals a conserved cysteine and histidine also supporting the idea that $3C^{pro}$ is a cysteine proteinase (Argos et al., submitted).

4 Antigenicity and Expression of Capsid Proteins

Co-expression of $3C^{pro}$ and capsid proteins has been suggested as a means of obtaining material for a subunit vaccine. Although two very effective polio vaccines already exist they are neither problem free nor easily applicable world wide. The main advantages of a subunit vaccine would be the elimination of the possibility of contamination with live virulent virus or reversion from an attenuated to virulent phenotype, and the potentially lower costs of production associated with recombinant DNA techniques. However, attempts to express VP1 containing sequences as fusion proteins in *E. coli* have not given promising results. Aside from problems of poor yield and enzymatic degradation, the neutralizing antigenic potential of the fusion proteins, as tested directly (Enger-Valk et al. 1984) or by reactivity with hyperimmune sera and neutralizing monoclonal antibodies (N-mcAb) (van der Werf et al. 1983), was severely limited. This is not surprising since purified capsid proteins of poliovirus give poor neutralizing titers (Chow and Baltimore 1982; Dernick et al. 1983).

Another approach toward the development of a virus-free polio vaccine which we find promising is the use of synthetic peptides derived from the sequence of the exposed capsid proteins. The primary consideration in choosing a sequence to synthesize was that it be hydrophilic and bounded by structure breaking amino acids (Chou and Fasman 1974; Hopp and Woods 1981). Interserotypic variation and variation in response to human passage (peptide 14) have also been used as criteria for selecting potential N-Ags (Fig. 2). When peptides corresponding to such sites were synthesized it was found that most of them were capable of eliciting a weak neutralizing response (Table 1). When peptides 3 and 9 were coupled to the same carrier a synergistic high titer neutralizing response was obtained (Jameson and Wimmer, unpublished results). All of these peptides (Table 1) were capable of priming the immune system for a high titer neutralizing response upon a single subsequent injection of a subimmunogenic dose of virus (Emini et al. 1983).

N-mcAbs are known to react with only two of these N-Ags on VP1, corresponding to amino acids 70-80 and 93-103 (Emini et al. 1983c). Each group of N-mcAbs recognizes a functionally nonoverlapping neutralization epitope (N-Ep). This raises the possibility that the other sites rarely stimulate an immune response in their natural context. It is conceivable that the virus has evolved to mask its other surface structures from the immune system with mutable immunodominant sites.

Table 1. Synthetic peptides

Peptide	Capsid Protein	Amino acids	Elicits neutralizing response	Primes neutralizing response
1	VP1	70-75	-	+
2	VP1	97-103	-	+
3	VP1	93-103	+	+
4	VP1	70-80	-	+
5	VP1	11-17	-	+
9	VP2	192-203	+	+
12	VP3	75-81	+	+
14	VP1	141-147	+	+
15	VP1	113-121	+	+
16	VP1	165-172	+	+

It is known that the apparently immunodominant sites of poliovirus vary during the course of an infection (Crainic et al. 1983). Since any population of virus contains some level of variants, a vaccine which elicited a response to those structures which were less able to vary would appear to be advantageous (Wimmer et al. 1984). It may even be possible to obtain a response against all serotypes. Peptide 15 represents a site which is conserved among the three poliovirus serotypes and is capable of eliciting a neutralizing response equally effective against all of them (Jameson et al., unpublished). A single reagent effective against multiple serotypes would greatly facilitate the production of vaccines against many viruses for which vaccines are currently impracticable or ineffective.

The dependence of neutralizing antigenicity on tertiary and quaternary structure implied by the weak response elicited by isolated capsid proteins, noted above, is further supported by sequence analysis of N-mcAb resistant viral variants. We have found that the mutations associated with the resistance phenotype do not map to the epitope, nor do they necessarily abolish antibody binding. For a group of Mahoney strain specific N-mcAbs resistance has been confered by two mutations in VP3; amino acid 60 (Thr>Lys) or amino acid 73 (Ser >Cys) (Diamond and Wimmer, unpublished). The former mutation abolishes binding by some of the antibodies (the latter does not), but the variants with this amino acid change are now neutralized by Sabin strain specific N-mcAbs (Diamond and Wimmer, unpublished). It is unlikely, however, that this single mutation causes the attenuated phenotype of the Sabin strain. These data indicate an unexpected complexity of the relations between capsid structure, attenuation, antibody binding, and neutralization. In the light of the tight capsid packing it is not unreasonable that a single mutation could elicit conformational changes that result in loss of neutralizability at a distant N-Ag (Moudallal et al. 1982).

5 Conclusion

Expression of $3C^{pro}$ in *E. coli* gives rise to an authentic proteinase activity. Co-expression of $3C^{pro}$ and other poliovirus proteins in bacteria may provide a source of individual, mature viral proteins, free of both natural host and other viral activities. The activities of the proteins can thus be analyzed in the bacteria directly following induction, as described for 3C^{pro}. Expression of various subsets of the viral genome may facilitate the dissection of the processes of viral replication. Alternatively, large amounts of these proteins can be produced for purification and subsequent use in in vitro systems or for antibody production. While capsid proteins produced in this manner may not be useful as a vaccine they could be useful in studies of viral morphogenesis. Clearly the techniques of gene transfer will lead to a greater understanding of poliovirus.

References

Argos P, Kamer G, Nicklin MJH, Wimmer E (1984) Homology between proteins of the animal picornaviruses and a plant comovirus. Identification of a conserved His-Cys pair in the sequences of the viral proteinase (submitted for publication)

Baltimore D (1971) Polio is not dead. In: Pollard M (ed) Perspectives in virology, vol 3. Academic, New York, pp 1-14

Bonin J, Jameson BA, Emini EA, Diamond DC, Wimmer E (to be published) Synthetic peptides as potential vaccines against picornaviral disease.

Cann AJ, Stanway G, Hauptmann R, Minor PD, Schild GC, Almond JW (1983) Poliovirus type 3: molecular cloning of the genome and nucleotide sequence of the region encoding the protease and polymerase proteins. Nucleic Acids Res 11:1267-1287

Chou PY, Fasman GD (1974) Prediction of protein conformation. Biochemistry 13:222-245

Chow M, Baltimore D (1982) Isolated poliovirus capsid protein VPl induces a neutralizing response in rats. Proc Natl Acad Sci USA 79:7518-7521

Crainic R, Couillin P, Blondel B, Cabu N, Boue A, Horodniceanu F (1983) Natural variation of poliovirus epitopes. Infect Immunol 41:1217

Dernick R, Heukeshoven J, Hilbrig M (1983) Induction of neutralizing antibodies by all three structural poliovirus polypeptides. Virology 130:243-246

Emini EA, Jameson BA, Wimmer E (1984) Peptide induction of poliovirus neutralizing antibodies: identification of a new antigenic site on coat protein VP2 (subitted for publication)

Emini EA, Elzinga M, Wimmer E (1982a) Carboxy-terminal analysis of poliovirus proteins: termination of poliovirus RNA translation and location of unique poliovirus polyprotein cleavage sites. J Virol 42:194-199

Emini EA, Jameson BA, Lewis AJ, Larsen GR, Wimmer E (1982b) Poliovirus neutralization epitopes: analysis and localization with neutralizing monoclonal antibodies. J Virol 43:997-1005

Emini EA, Dorner AJ, Dorner LF, Jameson BA, Wimmer E (1983a) Identification of a poliovirus neutralization epitope through use of neutralizing antiserum raised against a purified viral structural protein. Virology 124:144-151

Emini EA, Jameson BA, Wimmer E (1983b) Priming for an induction of anti-poliovirus neutralizing antibodies by synthetic peptides. Nature (Lond) 304:699-703

Emini EA, Kao SY, Lewis AJ, Wimmer E, Crainic R (1983c) The functional basis of poliovirus neutralization determined with monospecific neutralizing antibodies. J Virol 46:466-474

Enger-Valk BE, Jore J, Pouwels PH, van der Marel P, van Wezel TL (1984) Expression in *Escherichia coli* of capsid protein VP1 of poliovirus type 1. In: Chanock RM, Lerner RA (eds) Modern approaches to vaccines. Cold Spring Harbor, pp 173-178

Evans DMA, Minor PD, Schild GC, Almond JW (1983) Critical role of an eight-amino acid sequence of VP1 in neutralization of poliovirus type 3. Nature (Lond) 304:459-462

Hanecak R, Semler BL, Anderson CW, Wimmer E (1982) Proteolytic processing of poliovirus polypeptides: antibodies to polypeptide P3-7c inhibit cleavage at glutamine-glycine pairs. Proc Natl Acad Sci USA 79:3973-3977

Hanecak R, Semler BL, Ariga H, Anderson CW, Wimmer E (to be published 1984) Expression of a cloned gene segment of poliovirus in *E. coli:* evidence for autocatalytic production of the viral proteinase. Cell

Hopp TP, Woods KR (1981) Prediction of protein antigenic determinants from amino acid sequences. Proc Natl Acad Sci USA 78:3824-3828

Jameson BA, Kew O, Bonin J, Wimmer E (to be published) Identification of a new neutralization site on poliovirus type 1 using viral isolates from infant vaccinees

Kitamura N, Semler BL, Rothberg PG et al. (1981) Primary structure, gene organization and polypeptide expression of poliovirus RNA. Nature (Lond) 291:547-553

Masui Y, Coleman J, Inouye M (1983) Multipurpose expression cloning vehicles in *E. coli*. In: Inouye M (ed) Experimental manipulation of gene expression. Academic New York, pp 15-32

Minor PD, Schild GC, Bootman J et al. (1983) Location and primary structure of the antigenic site for poliovirus neutralization. Nature (Lond) 301:674-679

Mondallal ZA, Briand JB, van Regenmortel MAV (1982) Monoclonal antibodies as probes of the antigenic structure of tobacco mosaic virus. EMBO J 1:1005-1010

Nomoto A, Omata T, Toyoda H et al. (1982) Complete nucleotide sequence of the attenuated poliovirus Sabin 1 strain genome. Proc Natl Acad Sci USA 79:5793-5797

Racaniello VR, Baltimore D (1981) Molecular cloning of poliovirus cDNA and determination of the complete nucleotide sequence of the viral genome. Proc Natl Acad Sci USA 78:4887-4891

Rueckert RR, Wimmer E (1984) Systematic nomenclature of picornavirus proteins. J Virol 50:957-959

Semler BL, Anderson CW, Kitamura N, Rothberg PG, Wishart WL, Wimmer E (1981a) Poliovirus replication proteins: RNA sequence encoding P3-1b and the sites of proteolytic processing. Proc Natl Acad Sci USA 78:3464-3468

Semler BL, Hanecak R, Anderson CW, Wimmer E (1981b) Cleavage sites in the polypeptide precursors of poliovirus protein P2-X. Virology 114:589-594

Semler BL, Dorner AJ, Wimmer E (to be published 1984) Production of infectious poliovirus from cloned cDNA is dramatically increased by SV40 transcription and replication signals. Nucleic Acids Res

Stanway G, Hughes PJ, Mountford RC, Reeve P, Minor PD, Schild GC, Almond JW (1984) Comparison of the complete nucleotide sequences of the genomes of the neurovirulent poliovirus P3/Leon/37 and its attenuated Sabin vaccine derivative P3/Leon $12a_1b$. Proc Natl Acad Sci USA 81:1539-1543

Takegami T, Semler BL, Anderson CW, Wimmer E (1983) Membrane fractions active in poliovirus RNA replication contain VPg precursor polypeptides. Virology 128:33-47

Toyoda H, Kohara M, Kataoka Y, Suganuma T, Omata T, Imura N, Nomoto A (1984) The complete nucleotide sequence of all three poliovirus serotype genomes (Sabin): implication for the genetic relationship, gene function and antigenic determinants. J Mol Biol 174:561-585

van der Werf S, Wychowski C, Bruneau P, Blondel B, Crainic R, Horodniceanu F, Girard M (1983) Localization of a poliovirus type 1 neutralization epitope in viral capsid polypeptide VP1. Proc Natl Acad Sci USA 80:5080-5084

Wimmer E (1982) Genome-linked proteins of viruses. Cell 28:199-201

Wimmer E, Jameson BA, Emini EA (1984) Poliovirus antigenic sites and vaccines. Nature (Lond) 308:19

Yogo Y, Wimmer E (1972) Polyadenylic acid at the 3'-terminus of poliovirus RNA. Proc Natl Acad Sci USA 69:1877-1882

Expression of the Foot and Mouth Disease Virus Protease in *E. coli*

W. Klump, J. Soppa, O. Marquardt, and P. H. Hofschneider[1]

1 Introduction

The coding sequences of the genome of foot and mouth disease virus (FMDV) form a continuous open reading frame of about 7 kb, which is translated into a single polyprotein of about 250,000 M.W. It is processed post-translationally by the action of a virus-encoded protease and possibly also host-specific proteases (for a review see Sangar 1979).

Understanding of this process requires elucidation of the role the virus protease plays in maturation of FMDV proteins. We therefore established a system in *E. coli* cells (Klump et al. 1984) that (1) separates the virus protease activity from interfering host-specific protease activities and (2) is suitable for protein processing.

By this means proteins were produced which range in size from 45,000 to 180,000 M.W. and have identical NH_2-terminal amino acid sequences specific for the capsid protein VP1. Their COOH-termini differ according to the different length of FMDV cDNA which is expressed. The polyprotein fragments produced in *E. coli* exhibit some cleavage sites which are identical to those present in the polyprotein produced in virus-infected BHK-cells (Sangar 1979). Analogous to the processing of the viral precursors in BHK-cells, the virus-specific proteins expressed in *E. coli* should be cleaved by the virus-encoded protease if it is expressed in *E. coli* in an active form.

2 Plasmid Constructions

2.1 The Expression System, Construction of pPLVP1

E. coli clones have been described which carry FMDV O_1K-specific cDNA (Küpper et al. 1981; Fig. 1A). In the same report subcloning of the gene encoding the capsid protein VP1 (BamHI-HindIII cDNA fragment of clone 1034, see Fig. 1A) into the vector pPLc24 (Remaut et al. 1981), which enables the expression of VP1 in *E. coli*, has been described. In this construction designated pPLVP1 the FMDV reading frame is in phase with that of the 5'-adjacent gene fragment of phage MS2 replicase, which has all signals necessary for initiation of translation. Transcription depends on the leftward promoter p_L of bacteriophage λ, which is controlled by the λ repressor cI857, a temperature-sensitive mutant repressing at 28° but not at $42^{\circ}C$. Hence vector-dependent expression is induced by shifting the temperature of the bacterial cultures from 28° to $42^{\circ}C$.

[1]Max-Planck-Institut für Biochemie, 8033 Martinsried, FRG

35. Colloquium - Mosbach 1984
The Impact of Gene Transfer Techniques in Eukaryotic Cell Biology

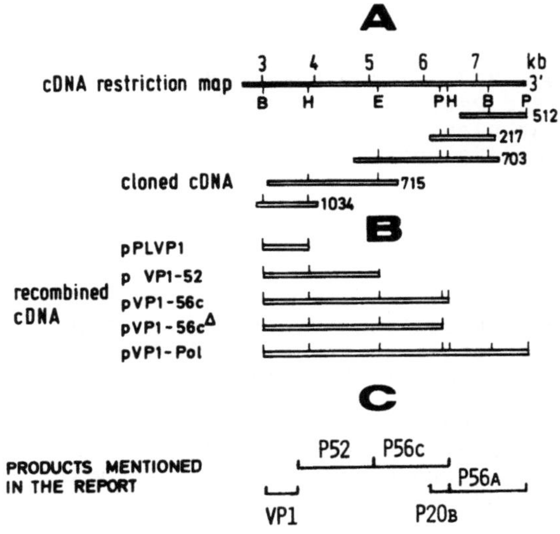

Fig. 1A-C. Correlation of FMDV cDNA fragments to the physical map of FMDV. A Localization of cDNA fragments present in *E. coli* clones according to the restriction map of FMDV cDNA essential as described by Küpper et al. (1981); B Recombined cDNA fragments mentioned in this report; C Localization of relevant FMDV proteins according to the physical map of FMDV. Length of cDNA is given in kilobases (kb). The relevant cleavage sites for restriction enzymes (B, BamHI; E, EcoRI; H, HindIII; P, PstI) are indicated. Applying the picornavirus nomenclature proposed at the third Meeting of the European Study Group on Molecular Biology of Picornaviruses (1983, Urbino, Italy). The new notation is given in parentheses following the current name of the protein: VP1 (1d); P52 (P2); P56c (3abc); P20b (3c); P56a (3d)

2.2 Construction of pVP1-56c

The HindIII-EcoRI cDNA fragment of clone 715 and the EcoRI-HindIII cDNA fragment of clone 703 (Fig. 1A) were inserted into pPLVP1 at the HindIII site adjacent to the BamHI-HindIII cDNA fragment. The recombinant plasmid contains the FMDV genes encoding VP1 and the precursors P52 and P56c (Fig. 1B,C).

2.3 Construction of pVP1-52

Plasmid pVP1-52 is derived from pVP1-56c by deletion of the 1300 bp EcoRI-HindIII cDNA fragment represented in clone 703 (Fig. 1A). The plasmid pVP1-52 contains the FMDV genes encoding VP1 and the precursor P52 (Fig. 1B,C).

2.4 Construction of pVP1-56c$^{\Delta}$

Plasmid pVP1-56c$^{\Delta}$ is derived from pVP1-56c by deletion of the 150 bp PstI-HindIII cDNA fragment (Fig. 1B).

2.5 Construction of pVP1-Pol

The PstI-BamHI cDNA fragment of clone 217 and the BamHI-PstI cDNA fragment of clone 512 were inserted at the PstI site present in the cDNA fragment of pVP1-56c. The recombinant plasmid contains the FMDV genes encoding VP1, the precursors P52 and P56c and the RNA dependent RNA polymerase.

2.6 Construction of 1.11

1.11 contains two expression vectors (pPLc24) and two different FMDV cDNA inserts (see Fig. 5A). One insert consists of the BamHI-HindIII cDNA fragment of clone 1034, the HindIII-EcoRI cDNA fragment of clone 715 and the EcoRI-BamHI cDNA fragment of clone 703. It contains the genes for VP1, the precursors P52 and P56c and part of the RNA dependent RNA polymerase. The other insert contains the BamHI-HindIII cDNA fragment of 1034 in phase to the reading frame and the HindIII-EcoRI cDNA fragments of the clones 715 and 703 which are both oriented opposite to the viral reading frame. Thus no FMDV cDNA sequence additional to the VP1 gene can be expressed from this insert.

3 Expression of Virus-Specific Proteins

Virus-specific proteins were detected by immunoblot analysis using antisera to FMDV O_1K. These sera, obtained from goats immunized with inactivated virus particles, contain mainly FMDV-specific antibodies directed against the virus capsid protein VP1. Since all the virus-specific proteins expressed in *E. coli* contain the VP1 domain, the expression of these proteins can be detected and their size determined.

Immunoblot analysis demonstrated that the size of the virus-specific proteins encoded by pPLVP1 and pVP1-52 is proportional to the viral reading frame (Fig. 2, track 1 and 2; track 4: background control). However, from pVP1-56c with a reading frame for a 145,000 M.W. protein, a virus-specific protein of 34,000 was expressed (Fig. 2, tracks 3). A 34,000 M.W. VP1-specific protein was also detected when expression from the open reading frame of pVP1-Pol (180,000 M.W., see Fig. 1B) was analyzed by immunoblot (not shown). The occurrence of the 34,000 VP1-specific protein may indicate the action of the virus-encoded protease which cleaves the virus-specific polyproteins of 145,000 or 180,000 M.W. respectively. The defined size of the 34,000 protein suggests that the protease recognizes specific cleavage sites. The other protein fragments resulting from cleavage of the virus-specific proteins cannot be detected since the anti-FMDV-serum reacts with protein VP1 only.

4 Mapping of the FMDV Protease Gene

Since the VP1-specific 34,000 M.W. cleavage product suggests the expression of an active viral protease, this system facilitates the mapping of the protease gene. This was achieved when 150 bp were deleted from the 3'-end of the FMDV-specific sequence of pVP1-56c. Gene expression with the mutant pVP1-56c$^{\Delta}$ revealed a 140,000 VP1-specific protein by immunoblot (Fig. 2, track 5). The experimentally observed

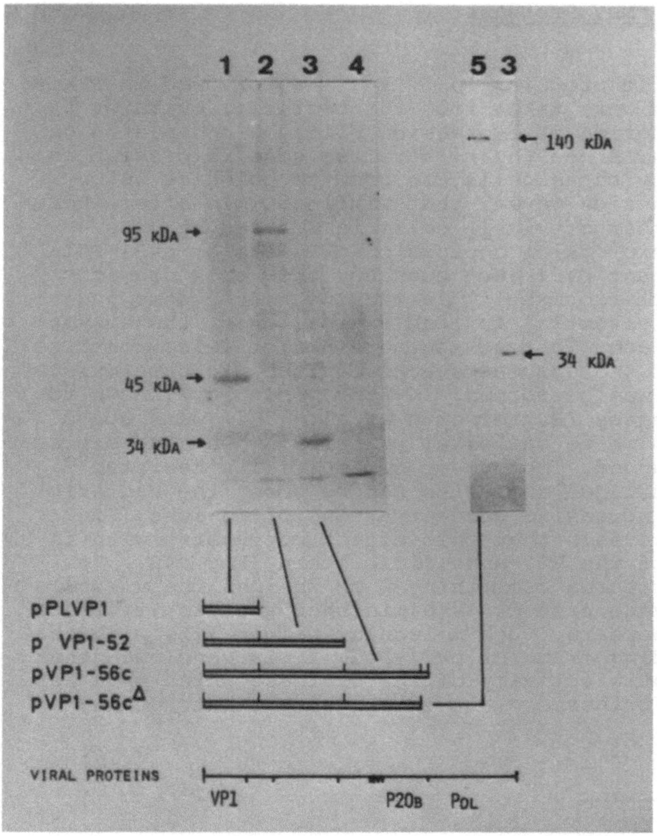

Fig. 2. Immunoblot analysis of FMDV-specific proteins in *E. coli* extracts. *E. coli* C600/pcI857 cells containing the expression plasmids pPLVP1, pVP1-52, pVP1-56c, pVP1-56c$^{\Delta}$ or pPLc24 (control), respectively, were cultured overnight at 28°C. Expression was induced by 1:5 dilution with medium (1% Bacto-tryptone/0.7% yeast extract/ 0.5% NaCl) prewarmed to 42°C and was stopped 30 min later by chilling the cultures in ice. Total *E. coli* proteins were subjected to SDS/PAGE (11% gel), and subsequently blotted to nitrocellulose filters. The immunoblot was obtained using goat anti-O_1K as probe. *Track 1* pPLVP1; *track 2* pVP1-52; *tracks 3* pVP1-56c; *track 4* pPLc24; *track 5* pVP1-56c$^{\Delta}$. Molecular masses of detected virus-specific proteins are given in kilodalton (kDa)

molecular weight fits with that calculated from the open reading frame. It is concluded that the deletion affects the protease gene so that only inactive virus protease is expressed. The deletion correlates with the position of protein P20b in the physical map of FMDV (Fig. 1C). It is suggested that this protein is the protease. The same location of the virus protease gene, between the VPg gene and the polymerase gene, has been demonstrated for two other picornaviruses, poliovirus (Hanecak et al. 1982) and encephalomyocarditis virus (Palmenberg et al. 1979).

5 Bacteriotoxicity of the FMDV Protease

Expression of virus-specific proteins in *E. coli* was followed in a kinetic experiment. Aliquots were taken from the bacterial cultures at different times after induction of expression, diluted and plated on agar containing ampicillin. Since the β-lactamase gene is present in the expression vector, only those cells can grow to colonies which harbor the vector. The kinetics reveal that within 30 min after induction of expression the number of viable cells is reduced by 3 to 4 orders of magnitude when expression depends on pVP1-56c or pVP1-Pol (Fig. 3). The deletion mutant pVP1-56c$^{\Delta}$ does not show this dramatic toxic effect. Thus the bacteriotoxicity is induced by the same nucleotide sequence which is essential for the expression of the enzymatically active virus protease. To test whether in fact selection pressure against expression of the enzymatic active virus protease exists, the following experiment was performed. Zn-ions, which are known to inhibit picornaviral proteases (Butterworth et al. 1974) were added in concentrations from 1-3 mM to bacterial cultures, when gene expression from pVP1-Pol was induced. The number of ampicillin-resistant cells was determined as outlined above. As can be seen, the bacteriotoxicity is reduced with increasing concentrations of Zn-ions (Fig. 4A). Immunoblot analysis reveals that this effect correlates with inhibition of the cleavage of the VP1-specific protein (Fig. 4B). In extracts from bacterial cultures containing 2 mM Zn-ions the virus-specific 34,000 M.W. cleavage product is diminished whereas virus-specific protein of 180,000 appears. Such molecular weight fits with that deduced from the open reading frame of pVP1-Pol. It is concluded that Zn-ions inhibit the enzymatic activity of the protease which then is no longer bacteriotoxic. Furthermore, it is suggested that the bac-

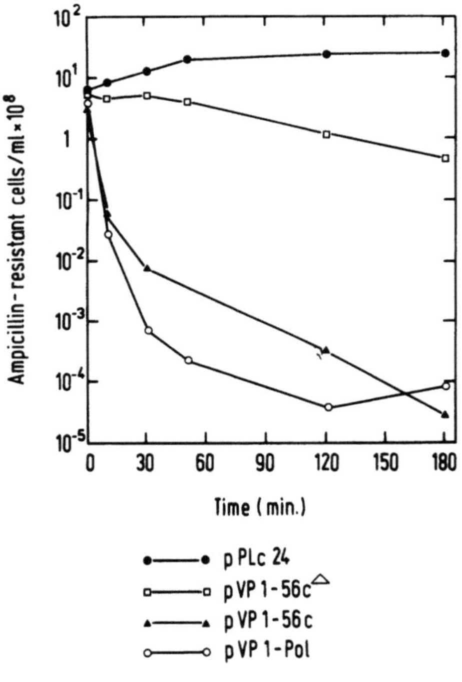

Fig. 3. Effect of FMDV gene expression on the growth of *E. coli*. *E. coli* C600/pcI857 cells containing the expression plasmids pVP1-56c, pVP1-56c$^{\Delta}$, pVP1-Pol or pPLc24 (control), respectively, were cultured as described in Fig. 2. The number of viable ampicillin-resistant cells was determined as outlined in the text. The diagram shows the extrapolated actual number of viable ampicillin-resistant cells in the culture. Expression was induced at 0 min on the time scale

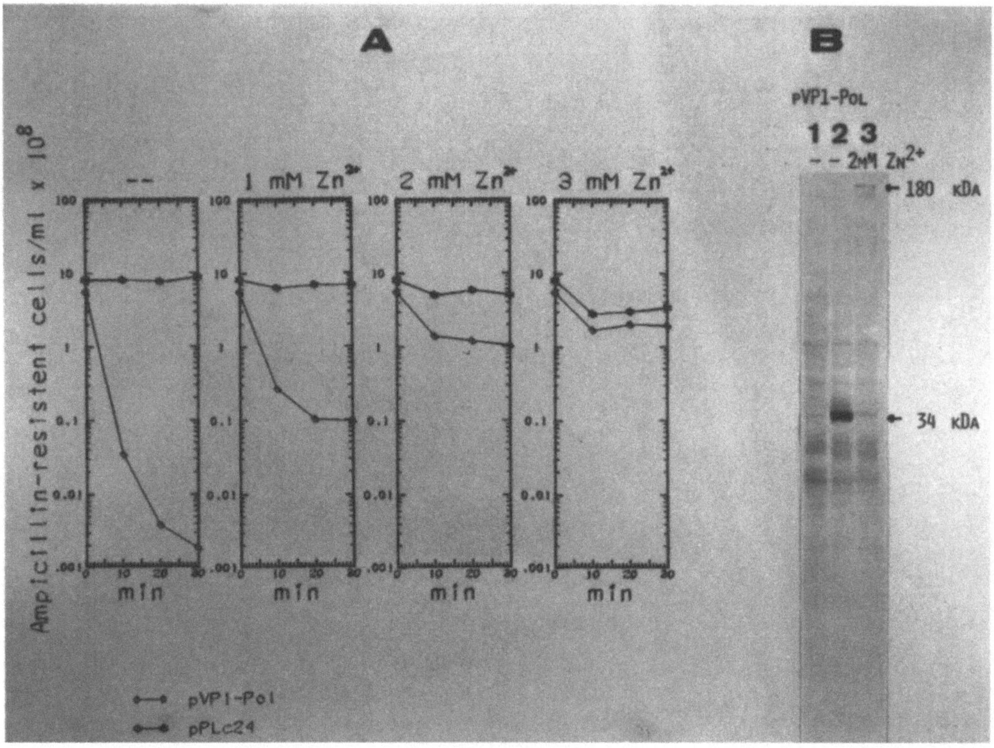

Fig. 4A,B. Effect of Zn-ions on the bacteriotoxicity and enzymatic activity of the FMDV protease. A Growth kinetic of *E. coli* cells. Zn-ions were added at different concentrations to bacterial cultures of C600/pcI857/pVP1-Pol resp. /pPLc24 before induction of expression. At 10-min intervals after induction of expression the number of ampicillin-resistant cells was determined as described in the text. The four diagrams show the extrapolated actual number of viable ampicillin-resistant cells containing pVP1-Pol and pPLc24 (control) respectively. Above the diagrams the concentration of Zn-ions is indicated. Expression was induced at 0 min on the time scale; B Immunoblot of *E. coli* extracts. Aliquots of extracts derived from C600/pcI857/pVP1-Pol were subjected to PAGE and blottet. *Track 1* before induction of expression; *track 2* 30 min after induction of expression without Zn-ions; *track 3* 30 min after induction of expression using medium containing 2 mM Zn-ions. Molecular masses of detected virus-specific proteins are given in kDa

teriotoxicity observed in Fig. 3 is caused by action of the viral protease on *E. coli* proteins.

This is of interest, since it may indicate that enzyme and substrate need not be fused as is the case in the FMDV polyprotein. Whether in fact the virus protease converts a substrate which originates from a separate expression unit was investigated experimentally. Use was made of an expression plasmid which contains two expression units with FMDV inserts of different size (Fig. 5A). The viral protease gene is included in only one of these inserts (Fig. 5B). Virus-specific expression products were determined by immunoblot analysis. Without Zn-ions only the VP1-specific 34,000 M.W. cleavage product is detectable (Fig. 5C, track 1). By inhibiting the viral protease with Zn-ions the synthesis of both primary virus-specific products of 45,000 and 180,000 M.W. respectively, is demonstrated (Fig. 5C, tracks 2-4).

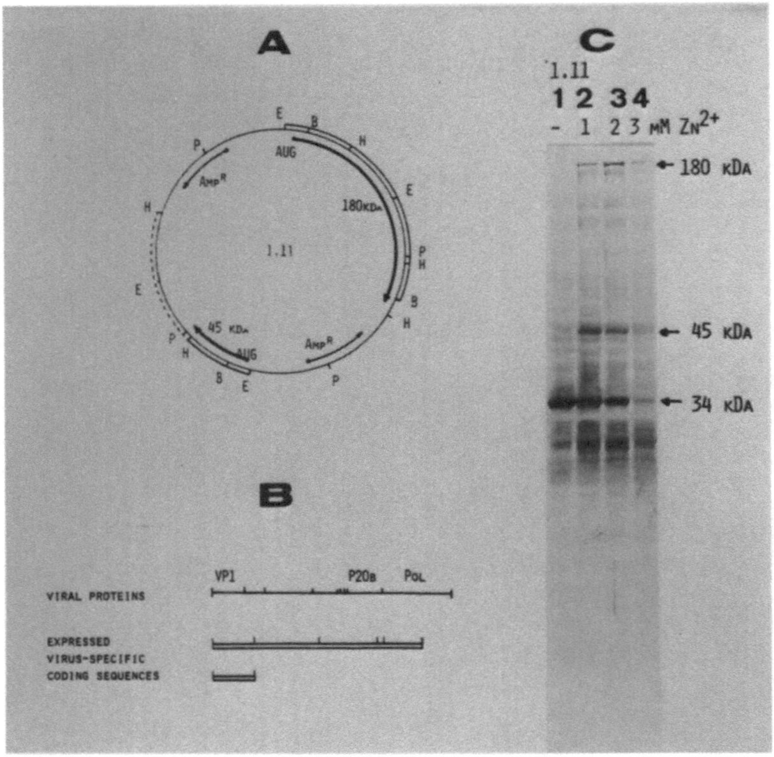

Fig. 5A-C. Construction map of 1.11 and expression of vector encoded virus-specific proteins. A Vector 1.11 is depicted, *arrows* indicating open reading frames, the ones orientated *counterclockwise* supplying resistance against ampicillin (Amp^r), *clockwise* encoding virus-specific proteins with indication given for their molecular masses. *Double lined segments* represent the FMDV cDNA fragments; *interrupted double line* is a FMDV cDNA fragment inserted in counterclockwise orientation, thus not being expressed. Cleavage sites for restriction enzymes are indicated as described in Fig. 1; B Correlation of the FMDV cDNA fragments inserted in 1.11 to a corresponding segment of the physical map of FMDV; C Immunoblot showing virus-specific proteins of *E. coli* C600/pcI857/1.11 extracts isolated 30 min after induction of expression from bacterial cultures containing different concentrations of Zn-ions (0-3 mM). Molecular masses of detected virus-specific proteins are given in kDa

6 The Results Led to the Following Conclusions:

1. The FMDV protease expressed in bacteria is capable of cleaving virus-specific proteins. These are cleaved regardless of whether they are fused to the protease domain or expressed from a separate gene. Thus the system allows for investigations on the processing of virus-specific polyproteins in *E. coli*.

2. The bacteriotoxicity caused by expression of the FMDV protease suggests that bacterial proteins have substrate properties. Likewise, nonviral proteins may also be cleaved in picornavirus infected cells and this may be the molecular basis of some of the cytopathic effects observed after infection.

3. In infected cells the picornavirus protease cleaves specific amino acid sequences. It is shown here that specific cleavage occurs also in an experimental system. Efforts to develop an in vitro FMDV protease system thus are stimulated. This may become useful in protein analysis.

Acknowledgments. We are grateful to Dr. K. Strohmaier, Tübingen, for supplying us with antisera. We wish to thank Miss M. Wechsler for excellent technical assistance. The cloning and growth of recombinant bacteria was carried out under L3B1 safety conditions as advised by the German Zentrale für biologische Sicherheit.

References

Butterworth BE, Korant BD (1974) Characterization of the large picornaviral polypeptides produced in the presence of zinc ion. J Virol 14:282-291

Hanecak R, Semler BL, Anderson CW, Wimmer E (1982) Proteolytic processing of poliovirus polypeptides: antibodies to polypeptide P3-7c inhibit cleavage at glutamine-glycine pairs. Proc Natl Acad Sci USA 79:3973-3977

Klump W, Marquardt O, Hofschneider PH (1984) Biologically active protease of foot and mouth disease virus is expressed from cloned viral cDNA in *E. coli*. Proc Natl Acad Sci USA 81:3351-3355

Küpper H, Keller W, Kurz C et al. (1981) Cloning of cDNA of major antigen of foot and mouth disease virus and expression in *E. coli*. Nature (Lond) 289:555-559

Palmenberg AC, Pallansch MA, Rueckert RR (1979) Protease required for processing picornaviral coat protein resides in the viral replicase gene. J Virol 32:770-778

Remaut E, Stanssens P, Fiers W (1981) Plasmid vectors for high-efficiency expression controlled by the p_L promoter of coliphage lambda. Gene (Amst) 15:81-93

Sangar DV (1979) The replication of picornaviruses. J Gen Virol 45:1-13

Y. Y. Gleba, K. M. Sytnik

Protoplast Fusion

Genetic Engineering in Higher Plants

Editor: **R. Shoeman**

1984. 62 figures. X, 220 pages.
(Monographs on Theoretical and Applied Genetics, Volume 8)
ISBN 3-540-13284-8

Contents: Introduction. - Techniques of Parasexual Hybridization. - Protoplast Fusion and Parasexual Hybridization of Higher Plants. - Transmission Genetics of Parasexual Hybridization in Closely Related Crosses. - Protoplast Fusion and Hybridization of Distantly Related Plant Species. - Use of Somatic Hybridization. - Conclusion. - References. - Subject Index.

Protoplast Fusion examines the potential of higher plant cell fusion as a technique for the genetic modification of higher plant cells and higher plants. The first book entirely devoted to somatic hybridization in higher plants, it summarizes available information in much more detail than previously published reviews. The most important feature of the book is the consistent genetic approach for analysis of the results of somatic cell fusion. This is particularly important for describing and evaluating data in the area of research, where most results are still produced by plant physiologists rather than plant geneticists.
The book contains many results not previously available in English. Originally published in Russian 1982, it was revised considerably and up-dated for this English edition.

MIX
Papier aus verantwortungsvollen Quellen
Paper from responsible sources
FSC® C105338

If you have any concerns about our products,
you can contact us on
ProductSafety@springernature.com

In case Publisher is established outside the EU,
the EU authorized representative is:
Springer Nature Customer Service Center GmbH
Europaplatz 3, 69115 Heidelberg, Germany

Printed by Libri Plureos GmbH
in Hamburg, Germany